Quantum Theory of the Optical and Electronic Properties of Semiconductors

H. Haug
Institut für Theoretische Physik
Universität Frankfurt
Robert-Mayer-Str. 8
6000 Frankfurt/Main
Fed. Rep. Germany

Stephan W. Koch
Optical Sciences Center
& Department of Physics
University of Arizona
Tucson, AZ 85721
USA

Published by

World Scientific Publishing Co. Pte. Ltd.

P O Box 128, Farrer Road, Singapore 9128

USA office: 687 Hartwell Street, Teaneck, NJ 07666

UK office: 73 Lynton Mead, Totteridge, London N20 8DH

Library of Congress Cataloging-in-Publication data is available.

QUANTUM THEORY OF THE OPTICAL AND ELECTRONIC PROPERTIES OF SEMICONDUCTORS

ISBN 981-02-0024-2
981-02-0249-0 (pbk)

Printed in Singapore by JBW Printers & Binders Pte. Ltd.

To

Barbara and Rita

PREFACE

The electronic properties of semiconductors form the basis of the latest, and current technological revolution, namely the development of ever smaller and more powerful computing devices, which affect not only the potential of modern science but practically all aspects of our daily life. This dramatic development is based on the ability to engineer the electronic properties of semiconductors and to miniaturize devices down to the limits set by quantum mechanics, thereby allowing a large scale integration of many devices on a single semiconductor chip.

Parallel with the development of electronic semiconductor devices, and no less spectacular, has been the technological use of the optical properties of semiconductors. The fluorescent screens of television tubes are based on the optical properties of semiconductor powders, the red light of GaAs light emitting diodes is known to all of us from the displays of domestic appliances, and semiconductor lasers are used to read optical discs and to write in laser printers. Furthermore fiber-optic communications whose light sources, amplifiers and detectors are again semiconductor electro-optical devices, are expanding the capacity of the communication networks dramatically.

These few examples suffice to illustrate the need for a textbook on the quantum theory of the electronic and optical properties of semiconductors and semiconductor devices. There is a growing demand for solid-state physicists, material and optical engineers who understand enough of the basic microscopic theory of semiconductors to be able to use effectively the possibilities to engineer, design and optimize electro-optical devices with certain desired characteristics.

Semiconductor properties are very sensitive to the addition of free carriers, which can be introduced into the system by doping the crystal with atoms from another group in the periodic system, electronic injection or optical excitation. In addition, modern crystal growth techniques make it possible to grow layers of semiconductor material which are narrow enough to confine the electron motion in one dimension. In such *quantum-well* structures the electron wavefunctions are quantized like the standing waves of a particle in a square well potential. The task to calculate these eigenfunctions is usually one of the first problems in an introductory quantum mechanical course. If one is only interested in energies close to the ground state, one can neglect in sufficiently narrow quantum wells all energetically higher standing waves. Then the electron motion perpendicular to the quantum-well layer is suppressed, the semiconductor is quasi-two-dimensional. In this sense it is possible to make low-dimensional systems such as quantum wells, quantum wires and quantum dots which are effectively two, one and zero dimensional. This new aspect of semiconductor physics is taken into account in this book by evaluating various electronic and optical properties in different dimensions.

Students and researchers who have had a general background in quantum mechanics should be able to follow this book. The necessary many-particle physics techniques, such as field quantization and Green's function techniques are developed explicitly. Wherever possible we emphasize the motivation of a certain derivation and the physical meaning of an obtained result, avoiding as much as possible the discussion of formal mathematical aspects of the theory.

The electronic properties of a semiconductor are primarily determined by transitions within one energy band, i.e. by intraband transitions, which describe the transport of carriers in real space. Optical properties, on the other hand, are connected with transitions between the valence and conduction bands, i.e. with interband transitions. However a strict separation is impossible. Electronic devices such as a *p-n* diode can only be understood if one considers also interband transitions, and many optical devices cannot be understood if one does not take into account the effects of intraband scattering, of transport and diffusion. Therefore we speak generally about electro-optical properties, with somewhat more emphasize on the optical properties than on conventional electronic properties. In view of our intention to describe the basic underlying physical phenomena, we will mainly be concerned with simple models of intrinsic semiconductors and leave the rich spectrum of phenomena introduced by complicated, realistic band structures, impurities, surfaces etc. to other more specialised textbooks.

This book is based on lectures given by the authors at the J. W. Goethe Universität, Frankfurt (FRG), and at the University of Arizona, Tucson (USA). It is written for researchers or graduate-level students as an introduction to the quantum theory of semiconductors. The book (or parts of it) can serve as graduate-level textbook for use in solid state physics courses, or for more specialized courses on electronic and optical properties of semiconductors and semiconductor devices. Especially the later chapters establish a direct link to current research in semiconductor physics.

Many of our collegues and students have helped in different ways to complete this book. We especially wish to thank our collegues L. Banyai and M. Lindberg for many scientific discussions and help in several calculations. We gratefully acknowledge I. Galbraith for a first draft of Chap. 19 and a critical reading of part of the manuscript. Furthermore, we thank our students C. Ell, Y. Hu, J. Müller, M. Pereira Jr., and D. Richardson for helpful suggestions and assistance in completing some of the figures. Last, but not least, we thank M. Sargent III for scientific collaboration and support using his Scroll System PS^{TM} Technical Word Processor to prepare the camera ready manuscript.

We acknowledge financial support from the Deutsche Forschungsgemeinschaft (DFG), the National Science Foundation (NSF), and the Optical Circuitry Cooperative (OCC) at the University of Arizona. Furthermore, we appreciate a NATO travel grant, which supported several mutual visits in Tucson and Frankfurt.

Tucson and Frankfurt
February 1990

Hartmut Haug
Stephan W. Koch

CONTENTS

Quantum Theory of the Optical and Electronic Properties of Semiconductors

Chapter 1
OSCILLATOR MODEL

Semiconducting materials have physical properties which are somewhat intermediate between those of insulators and metals. This situation makes semiconductors extremely sensitive to imperfections and impurities in the atomic lattice of the crystals. Before techniques were developed allowing well controlled crystal growth, research in semiconductors was considered by many physicists a highly suspect enterprise.

Starting with the research on Ge and Si in the 1940's, physicists learned to exploit the sensitivity of the semiconductors to the content of foreign atoms in the host lattice. They learned to dope materials with specific impurities which act as donors or acceptors of electrons and thus opened the field for developing semiconductor diodes and transistors. In addition pure semiconductors were also found to have interesting properties.

Electrons in the semiconductor ground state are bound to the ions and cannot move freely, in contrast to electrons in excited states. The ground state and the lowest excited state are separated by an energy gap and by absorbing a light quantum, an electron can be excited across the energy gap. In the spectral range around the energy gap pure semiconductors have been found to exhibit interesting linear and nonlinear optical properties. Before we discuss the quantum theory of these optical properties, we first present a classical description of a dielectric medium in which the electrons are assumed to be bound by harmonic forces to the positively charged ions. If we excite such a medium with the periodic transverse electric field of a light beam, we induce an electrical polarization due to microscopic displacement of bound charges. This so-called oscillator model for the electric polarization was introduced in the pioneering work of Lorentz, Planck and Einstein. We expect the model to yield reasonably realistic results as long as the light frequency does not exceed the frequency corresponding to the energy gap, so that the electron stays in its bound state.

We show in this chapter that the analysis of this simple model already yields a qualitative understanding of many basic aspects of light-matter interaction. Furthermore it is useful to introduce such general concepts as optical susceptibility, dielectric function, absorption and refraction, as well as Green's functions.

1-1. Optical Susceptibility

The electric field, which is assumed to be polarized in the x direction, causes a displacement x of an electron with a charge e from its equilibrium position. The resulting polarization, defined as dipole moment per unit volume, is

$$\mathscr{P} = \frac{P}{L^3} = -n_0 ex = -n_0 d, \tag{1.1}$$

where $L^3 = V$ is the volume, $d = ex$ is the electric dipole moment, and n_0 is the mean electron density per unit volume. Describing the electron under the influence of the electric field $\mathscr{E}(t)$ (parallel to x) as a damped driven oscillator, we can write Newton's equation as

$$m\frac{d^2x}{dt^2} = -2m\gamma\frac{dx}{dt} - m\omega_0{}^2 x - e\mathscr{E}(t)\ , \tag{1.2}$$

where γ is the damping constant, and m and ω_0 are the mass and resonance frequency of the oscillator, respectively. The electric field is assumed to be monochromatic with a frequency ω, i.e., $\mathscr{E}(t) = \mathscr{E}_0\cos(\omega t)$. Often it is convenient to consider a complex field[1]

$$\mathscr{E}(t) = \mathscr{E}(\omega)e^{-i\omega t} \tag{1.3}$$

and take the real part of it whenever a final physical result is calculated. With the ansatz

---

[1] It is important to note, that in physics the convention is used that periodic complex fields always have the form (1.3) with a minus sign in the exponent. In electrical engineering the opposite convention is used writing complex periodic fields as $e^{j\omega t}$, where j is like i the imaginary unit $\sqrt{-1}$.

$$x(t) = x(\omega)e^{-i\omega t} \tag{1.4}$$

we get from Eq. (1.2)

$$m\ (\omega^2+i2\gamma\omega-\omega_0^2)\ x(\omega) = e\mathcal{E}(\omega) \tag{1.5}$$

and from Eq. (1.1)

$$\mathcal{P}(\omega) = -\frac{n_0 e^2}{m}\frac{1}{\omega^2+i2\gamma\omega-\omega_0^2}\mathcal{E}(\omega)\ . \tag{1.6}$$

The complex coefficient between $\mathcal{P}(\omega)$ and $\mathcal{E}(\omega)$ is defined as the optical susceptibility $\chi(\omega)$. We obtain

$$\boxed{\chi(\omega) = -\frac{n_0 e^2}{2m\omega_0}\left[\frac{1}{\omega+i\gamma-\omega_0'} - \frac{1}{\omega+i\gamma+\omega_0'}\right]}$$

optical susceptibility (1.7)

where

$$\omega_0' = \sqrt{\omega_0^2-\gamma^2} \tag{1.8}$$

is the renormalized (shifted) resonance frequency of the damped harmonic oscillator. In general, the optical susceptibility is a tensor relating different vector components of the polarization $\mathcal{P}_i$ and the electric field $\mathcal{E}_i$. An important feature of $\chi(\omega)$ is that it becomes singular at

$$\omega = -i\gamma \pm \omega_0'\ . \tag{1.9}$$

This relation can only be satisfied if we formally consider complex frequencies[2] $\omega = \omega' + i\omega''$. We see from Eq. (1.7) that $\chi(\omega)$ has poles in the lower half of the complex frequency plane, i.e. for $\omega'' < 0$, but it is an analytic function on the real frequency axis and in the whole upper half plane. This property of the susceptibility can be related to causality, i.e., to the fact that the polarization

---

[2] Note, that throughout this book we adopt the notation that the real part of a quantity is indicated by a prime and the imaginary part by a double prime, respectively.

$\mathscr{P}(t)$ at time t can only be influenced by fields $\mathcal{E}(t-\tau)$ acting at earlier times, i.e., $\tau \geq 0$. Let us consider the most general linear relation between the field and the polarization

$$\mathscr{P}(t) = \int_{-\infty}^{t} dt' \; \chi(t,t')\mathcal{E}(t') \; . \tag{1.10}$$

The response function $\chi(t,t')$ describes the memory of the system for the action of fields at earlier times. Causality requires that fields $\mathcal{E}(t')$ which act in the future, $t'>t$, cannot influence the polarization of the system at time t. We now make a transformation to new time arguments T and τ defined as

$$T = \frac{t+t'}{2} \quad \text{and} \quad \tau = t - t' \; . \tag{1.11}$$

If the systems is in equilibrium, the memory function χ depends only on the time difference τ and not on T, which leads to

$$\mathscr{P}(t) = \int_{-\infty}^{t} dt' \; \chi(t-t')\mathcal{E}(t')$$

$$= \int_{0}^{\infty} d\tau \; \chi(\tau)\mathcal{E}(t-\tau), \tag{1.12}$$

Next we use a Fourier transformation to transform Eq. (1.12) into frequency space. Generally, a Fourier transformation $f(\omega)$ of a function $f(t)$ is defined through the relations

$$f(\omega) = \int_{-\infty}^{\infty} dt \; f(t) \; e^{i\omega t}$$

$$f(t) = \int_{-\infty}^{\infty} \frac{d\omega}{2\pi} \; f(\omega) \; e^{-i\omega t} \; . \tag{1.13}$$

Multiplying Eq. (1.12) by $e^{i\omega t}$ and integrating over t, we get

$$\mathscr{P}(\omega) = \int_0^{\infty} d\tau\ \chi(\tau)e^{i\omega\tau} \int_{-\infty}^{+\infty} dt\ \mathscr{E}(t-\tau)e^{i\omega(t-\tau)} = \chi(\omega)\ \mathscr{E}(\omega) \tag{1.14}$$

where

$$\chi(\omega) = \int_0^{\infty} d\tau\ \chi(\tau)e^{i\omega\tau}\ . \tag{1.15}$$

The convolution integral in time, Eq. (1.12) becomes a product in Fourier space, Eq. (1.14). The time-dependent response function $\chi(t)$ relates two real quantities, $\mathscr{E}(t)$ and $\mathscr{P}(t)$, and therefore has to be a real function itself. Hence, (1.15) implies directly that $\chi^*(\omega) = \chi(-\omega)$ or $\chi'(\omega) = \chi'(-\omega)$. and $\chi''(\omega) = -\chi''(-\omega)$. Moreover, it also follows that $\chi(\omega)$ is analytic for $\omega''\geq 0$, because the factor $e^{-\omega''\tau}$ forces the integrand to zero at the upper boundary, where $\tau\rightarrow\infty$.

Since $\chi(\omega)$ is an analytic function for real frequencies we can use the Cauchy relation to write

$$\chi(\omega) = \int_{-\infty}^{+\infty} \frac{d\nu}{2\pi i}\ \frac{\chi(\nu)}{\nu-\omega-i\delta} \tag{1.16}$$

where δ is a positive infinitesimal number. The integral can be evaluated using the Dirac identity (see problem (1.1))

$$\lim_{\delta\rightarrow 0}\ \frac{1}{\omega-i\delta} = P\ \frac{1}{\omega}\ + i\pi\delta(\omega) \tag{1.17}$$

where P denotes the principal value of an integral under which this relation is used. We find

$$\chi(\omega) = P\int_{-\infty}^{+\infty} \frac{d\nu}{2\pi i}\ \frac{\chi(\nu)}{\nu-\omega} + \frac{1}{2}\int_{-\infty}^{+\infty} d\nu\ \chi(\nu)\delta(\nu-\omega). \tag{1.18}$$

For the real and imaginary parts of the susceptibility, we obtain separately

$$\chi'(\omega) = P\int_{-\infty}^{+\infty} \frac{d\nu}{\pi}\,\frac{\chi''(\nu)}{\nu-\omega} \tag{1.19}$$

$$\chi''(\omega) = -\,P\int_{-\infty}^{+\infty} \frac{d\nu}{\pi}\,\frac{\chi'(\nu)}{\nu-\omega}\,. \tag{1.20}$$

Splitting the integral into two parts

$$\chi'(\omega) = P\int_{-\infty}^{0} \frac{d\nu}{\pi}\,\frac{\chi''(\nu)}{\nu-\omega} + P\int_{0}^{+\infty} \frac{d\nu}{\pi}\,\frac{\chi''(\nu)}{\nu-\omega} \tag{1.21}$$

and using the relation $\chi''(\omega) = -\,\chi''(-\omega)$, we find

$$\chi'(\omega) = P\int_{0}^{+\infty} \frac{d\nu}{\pi}\,\chi''(\nu)\left[\frac{1}{\nu+\omega} + \frac{1}{\nu-\omega}\right]. \tag{1.22}$$

Combining the two terms yields

$$\boxed{\chi'(\omega) = P\int_{0}^{+\infty} \frac{d\nu}{\pi}\,\chi''(\nu)\,\frac{2\nu}{\nu^2-\omega^2}}$$

Kramers-Kronig relation (1.23)

This is the Kramers-Kronig relation, which allows to calculate the real part of $\chi(\omega)$ if the imaginary part is known for all positive frequencies. In realistic situations one has to be careful with the use of Eq. (1.23), because the absorption spectrum is often known only in a finite frequency range. A relation similar to Eq. (1.23) can be derived for χ'' using (1.20) and $\chi'(\omega) = \chi'(-\omega)$, see problem (1.3).

1-2. Absorption and Refraction

Before we give any physical interpretation of the susceptibility obtained with the oscillator model we will establish some relations to other important optical coefficients. The displacement field $D(\omega)$ can be expressed in terms of the polarization and electric field[3]

$$D(\omega) = \mathcal{E}(\omega) + 4\pi\mathcal{P}(\omega) = [\ 1+4\pi\chi(\omega)\ ]\ \mathcal{E}(\omega) = \epsilon(\omega)\mathcal{E}(\omega) \qquad (1.24)$$

where the optical (or transverse) dielectric function $\epsilon(\omega)$ is obtained from the optical susceptibility (1.7) as

$$\boxed{\epsilon(\omega) = 1 + 4\pi\chi(\omega) = 1 - \frac{\omega_{pl}^2}{2\omega_0}\left[\frac{1}{\omega+i\gamma-\omega_0'} - \frac{1}{\omega+i\gamma+\omega_0'}\right]}$$

optical dielectric function (1.25)

Here ω_{pl} denotes the plasma frequency of an electron plasma with mean density n_0

$$\boxed{\omega_{pl} = \sqrt{\frac{4\pi n_0 e^2}{m}}}$$

plasma frequency (1.26)

The plasma frequency is the eigenfrequency of the electron plasma density oscillations around the position of the ions. To illustrate this fact let us consider an electron plasma of density $n(\mathbf{r},t)$ close to equilibrium. The equation of continuity is

$$e\,\frac{\partial n}{\partial t} + \mathrm{div}\ \mathbf{j} = 0 \qquad (1.27)$$

with the current density

$$\mathbf{j}(\mathbf{r},t) = e\ n(\mathbf{r},t)\ \mathbf{v}(\mathbf{r},t)\ . \qquad (1.28)$$

[3] We use cgs units throughout this book.

The source equation for the electric field is

$$\mathrm{div}\ \mathcal{E} = 4\pi e\ (n - n_0) \tag{1.29}$$

and Newton's equation[4] for free carriers can be written as

$$m \frac{\partial \mathbf{v}}{\partial t} = e\ \mathcal{E}. \tag{1.30}$$

We now linearize Eqs. (1.27) - (1.29) around the equilibrium state where the velocity is zero and no fields exist. Inserting

$$n = n_0 + \delta\ n_1 + O(\delta^2)$$

$$\mathbf{v} = \delta\ \mathbf{v}_1 + O(\delta^2)$$

$$\mathcal{E} = \delta\ \mathcal{E}_1 + O(\delta^2) \tag{1.31}$$

into Eqs. (1.27) - (1.30) and keeping only terms linear in δ we obtain

$$\frac{\partial n_1}{\partial t} + n_0\ \mathrm{div}\ \mathbf{v}_1 = 0\ , \tag{1.32}$$

$$\mathrm{div}\ \mathcal{E}_1 = 4\pi e n_1\ , \tag{1.33}$$

and

$$m\ \frac{\partial \mathbf{v}_1}{\partial t} = e\ \mathcal{E}_1. \tag{1.34}$$

The equation of motion for n_1 can be derived by taking the time derivative of Eq. (1.32) and using Eqs. (1.33) and (1.34) to get

$$\frac{\partial^2 n_1}{\partial t^2} = -\ n_0\ \mathrm{div}\ \frac{\partial \mathbf{v}_1}{\partial t} = -\ \frac{n_0 e}{m}\ \mathrm{div}\ \mathcal{E}_1 = -\omega_{pl}{}^2\ n_1. \tag{1.35}$$

---

[4] Note, that we ignore the effects of a finite electron pressure, which would lead to additional terms in Eq. (1.30) and to the inclusion of spatial plasma variations. A more detailed discussion of the plasma hydrodynamics can be found in Jackson (1975), Sect. 10.8.

This simple harmonic oscillator equation is the classical equation for charge density oscillations with the eigenfrequency ω_{pl} around the equilibrium density n_0.

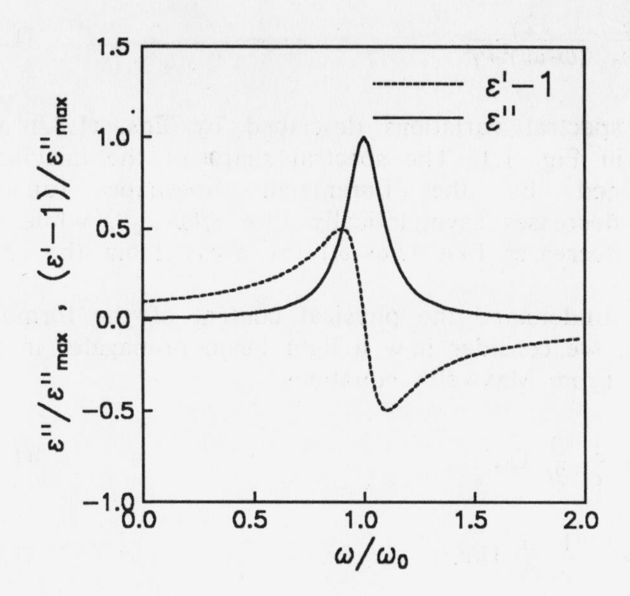

Fig. 1.1: Dispersion of the real and imaginary part of the dielectric function, Eq. (1.37) and (1.38), respectively. The broadening is taken as γ/ω_0=0.1 and $\epsilon''_{max}=\omega_{pl}^2/2\gamma\omega_0$.

Returning to the discussion of the optical dielectric function (1.25), we note that $\epsilon(\omega)$ has poles at $\omega = \pm\omega_0'-i\gamma$, describing the resonant and the nonresonant part, respectively. If we are interested in the optical response in the spectral region around ω_0 and if ω_0 is sufficiently large, the nonresonant part gives only a small contribution and it is often a good approximation to neglect it completely.

In order to simplify the resulting expressions we now consider only the resonant part of the dielectric function and assume $\omega_0 >> \gamma$, so that $\omega_0 \cong \omega_0'$ and

$$\epsilon(\omega) = 1 - \frac{\omega_{pl}^2}{2\omega_0} \frac{1}{\omega+i\gamma-\omega_0} . \tag{1.36}$$

For the real part of the dielectric function we thus get the relation

$$\epsilon'(\omega) - 1 = -\frac{\omega_{pl}^2}{2\omega_0} \frac{\omega-\omega_0}{(\omega-\omega_0)^2+\gamma^2} , \tag{1.37}$$

while the imaginary part has the following resonance structure

$$\epsilon''(\omega) = \frac{\omega_{pl}^2}{4\omega_0} \frac{2\gamma}{(\omega-\omega_0)^2+\gamma^2} \tag{1.38}$$

Examples of the spectral variations described by Eqs. (1.37) and (1.38) are shown in Fig. 1.1. The spectral shape of the imaginary part is determined by the Lorentzian lineshape function $2\gamma/((\omega-\omega_0)^2+\gamma^2)$. It decreases asymptotically like $1/(\omega-\omega_0)^2$, while the real part of $\epsilon(\omega)$ decreases like $1/(\omega-\omega_0)$ far away from the resonance.

In order to understand the physical content of the formulae for $\epsilon'(\omega)$ and $\epsilon''(\omega)$, we consider how a light beam propagates in the dielectric medium. From Maxwell's equations

$$\text{curl } \mathbf{B}(\mathbf{r},t) = \frac{1}{c} \frac{\partial}{\partial t} \mathbf{D}(\mathbf{r},t) \tag{1.39}$$

$$\text{curl } \mathcal{E}(\mathbf{r},t) = -\frac{1}{c} \frac{\partial}{\partial t} \mathbf{H}(\mathbf{r},t) \tag{1.40}$$

we find with $\mathbf{B}(\mathbf{r},t) = \mathbf{H}(\mathbf{r},t)$, which holds at optical frequencies,

$$\text{curl curl } \mathcal{E}(\mathbf{r},t) = -\frac{\partial}{\partial t} \text{curl } \mathbf{H}(\mathbf{r},t) = -\frac{1}{c^2} \frac{\partial^2}{\partial t^2} \mathbf{D}(\mathbf{r},t) . \tag{1.41}$$

Using curl curl = grad div - Δ, we get for a transverse electric field with div $\mathcal{E}(\mathbf{r},t) = 0$, the wave equation

$$\Delta\mathcal{E}(\mathbf{r},t) - \frac{1}{c^2} \frac{\partial^2}{\partial t^2}\mathbf{D}(\mathbf{r},t) = 0 . \tag{1.42}$$

Here $\Delta \equiv \nabla^2$ is the Laplace operator. A Fourier transformation of Eq. (1.42) with respect to time yields

$$\Delta\mathcal{E}(\mathbf{r},\omega) + \frac{\omega^2}{c^2}\epsilon'(\omega)\mathcal{E}(\mathbf{r},\omega) + i\frac{\omega^2}{c^2}\epsilon''(\omega)\mathcal{E}(\mathbf{r},\omega) = 0 . \tag{1.43}$$

For a plane wave propagating with wavenumber $k(\omega)$ and extinction coefficient $\kappa(\omega)$ in the z direction,

$$\mathcal{E}(\mathbf{r},\omega) = \mathcal{E}_0(\omega)\, e^{i[k(\omega)+i\kappa(\omega)]z} , \tag{1.44}$$

we get from Eq. (1.43)

$$[k(\omega) + i\kappa(\omega)]^2 = \frac{\omega^2}{c^2} [\epsilon'(\omega) + i\epsilon''(\omega)] . \tag{1.45}$$

Separating real and imaginary part of this equation yields

$$k^2(\omega) - \kappa^2(\omega) = \frac{\omega^2}{c^2} \epsilon'(\omega) , \tag{1.46}$$

$$2\kappa(\omega)k(\omega) = \frac{\omega^2}{c^2} \epsilon''(\omega) . \tag{1.47}$$

Next, we introduce the index of refraction $n(\omega)$ as the ratio between the wavenumber $k(\omega)$ in the medium and the vacuum wavenumber $k_0=\omega/c$

$$k(\omega) = n(\omega) \frac{\omega}{c} \tag{1.48}$$

and the absorption coefficient $\alpha(\omega)$ as

$$\alpha(\omega) = 2\kappa(\omega) . \tag{1.49}$$

The absorption coefficient determines the decay of the intensity $I \propto |\mathcal{E}|^2$ in real space. $1/\alpha$ is the length, over which the intensity decreases by a factor $1/e$. From Eqs. (1.46) - (1.49) we obtain the relations

$$\boxed{n(\omega) = \sqrt{\frac{1}{2}\left[\epsilon'(\omega) + \sqrt{\epsilon'^2(\omega) + \epsilon''^2(\omega)}\right]}}$$

index of refraction (1.50)

and

$$\boxed{\alpha(\omega) = \frac{\omega}{n(\omega)c}\,\epsilon''(\omega)}$$

absorption coefficient (1.51)

Hence, Eqs. (1.38) and (1.51) yield a Lorentzian absorption line, and Eqs. (1.37) and (1.50) describe the corresponding frequency-dependent index of refraction. Note that for $\epsilon''(\omega) << \epsilon'(\omega)$, which is usually true in semiconductors, Eq. (1.50) simplifies to

$$n(\omega) \cong \sqrt{\epsilon'(\omega)}. \tag{1.52}$$

Furthermore, if the refractive index $n(\omega)$ is only weakly frequency dependent for the ω-values of interest, one may approximate Eq. (1.51) as

$$\alpha(\omega) \cong \frac{\omega}{n_b c}\,\epsilon''(\omega) = \frac{4\pi\omega}{n_b c}\,\chi''(\omega)\ , \tag{1.53}$$

where n_b is the background refractive index.

For the case $\gamma \to 0$, i.e., vanishing absorption linewidth, the lineshape function approaches a delta function (see problem 1.3)

$$\lim_{\gamma \to 0} \frac{2\gamma}{(\omega-\omega_0)^2+\gamma^2} = 2\pi\delta(\omega-\omega_0)\ . \tag{1.54}$$

In this case we get

$$\epsilon''(\omega) = \pi\,\frac{\omega_{pl}^2}{2\omega_0}\,\delta(\omega-\omega_0) \tag{1.55}$$

and the real part becomes

$$\epsilon'(\omega) = 1 - \frac{\omega_{pl}^2}{2\omega_0}\,\frac{1}{\omega-\omega_0}\,sgn(\omega-\omega_0)\ , \tag{1.56}$$

where $sgn(\omega-\omega_0)$ is the sign of $(\omega-\omega_0)$.

1-3. Retarded Green's Function

A slightly more general way of solving the initial value problem

$$m\left[\frac{\partial^2}{\partial t^2} + 2\gamma\frac{\partial}{\partial t} + {\omega_0}^2\right]x(t) = -e\mathcal{E}(t) \tag{1.57}$$

is obtained by using the Green's function of Eq. (1.57). The so-called *retarded Green's function* $G(t-t')$ is defined as the solution of Eq. (1.57), where $e\mathcal{E}(t)$ is replaced by a delta function

$$m\left[\frac{\partial^2}{\partial t^2} + 2\gamma\frac{\partial}{\partial t} + {\omega_0}^2\right]G(t-t') = \delta(t-t') \ . \tag{1.58}$$

Fourier transformation yields

$$\boxed{G(\omega) = -\frac{1}{m}\frac{1}{\omega^2+i2\gamma\omega-{\omega_0}^2} \cong -\frac{1}{2m\omega_0}\frac{1}{\omega+i\gamma-\omega_0}}$$

retarded Green's function (1.59)

where we restricted ω to $\omega \cong \omega_0$ so that $\omega^2 - \omega_0^2 \cong 2\omega_0(\omega-\omega_0)$ and $\gamma\omega \cong \gamma\omega_0$. In terms of $G(t-t')$, the solution of Eq. (1.57) is then

$$x(t) = -\int_{-\infty}^{+\infty} dt' G(t-t')e\mathcal{E}(t') \tag{1.60}$$

as can be verified by inserting (1.60) into (1.57). Note, that the general solution of an inhomogeneous linear differential equation is the sum of the solution (1.60) of the inhomogeneous equation plus the solution of the homogeneous equation. However, since we are only interested in the induced polarization, we just keep the solution (1.60).

In general, the retarded Green's function $G(t-t')$ has the properties

$$G(t-t') = \begin{bmatrix} 0 \\ finite \end{bmatrix} \text{ for } \begin{bmatrix} t'>t \\ t'\leq t \end{bmatrix} \tag{1.61}$$

or

$$G(\tau) \propto \theta(\tau)$$

where $\theta(\tau)$ is the unit-step or Heavyside function

$$\theta(\tau) = \begin{bmatrix} 0 \\ 1 \end{bmatrix} \text{ for } \begin{bmatrix} \tau<0 \\ \tau\geq 0 \end{bmatrix} . \tag{1.62}$$

For $\tau<0$ we can close in (1.60) the integral by a circle with an infinite radius in the upper half of the complex frequency plane since

$$\lim_{|\omega|\to\infty} e^{i(\omega'+i\omega'')|\tau|} = \lim_{|\omega|\to\infty} e^{i\omega'\tau} e^{-\omega''|\tau|} = 0 \tag{1.63}$$

As can be seen from (1.59), $G(\omega)$ has no poles in the upper half plane making the integral zero for $\tau<0$. For $\tau\geq 0$ we have to close the contour integral in the lower half plane and get the finite result[5]

$$G(\tau) = -\frac{1}{2m\omega_0}\theta(\tau)\int_{C_\ell}\frac{d\omega}{2\pi}\frac{e^{-i\omega\tau}}{\omega+i\gamma-\omega_0} = i\theta(\tau)\frac{1}{2m\omega_0}e^{-(i\omega_0+\gamma)\tau} \tag{1.64}$$

The property that $G(\tau) = 0$ for $\tau<0$ is the reason for the name *retarded Green's function* which is often indicated by a superscript r, i.e.,

$$G^r(\tau) = 0 \text{ for } \tau<0 \quad \leftrightarrow \quad G^r(\omega) = \text{analytic for } \omega''\geq 0 . \tag{1.65}$$

The Fourier transform of Eq. (1.60) is

$$x(\omega) = -\int_{-\infty}^{+\infty} dt \int_{-\infty}^{+\infty} dt'\ e^{i\omega(t-t')}\ G(t-t')\ e^{i\omega t'}\ e\mathcal{E}(t')$$

---

[5] Contour integrals along the real axis closed by an infinite circle in the upper or lower half plane are denoted by integrals with an index C_u or C_ℓ , respectively.

$$= - e\, G(\omega)\, \mathcal{E}(\omega) \; . \tag{1.66}$$

With $\mathcal{P}(\omega) = -en_0x(\omega) = \chi(\omega)\mathcal{E}(\omega)$ we obtain

$$\chi(\omega) = n_0e^2G(\omega) \tag{1.67}$$

or

$$\chi(\omega) = - \frac{n_0e^2}{2m\omega_0} \frac{1}{\omega+i\gamma-\omega_0} \tag{1.68}$$

in agreement with the resonant part of (1.7). This agreement also justifies our treatment of the complex field in the first section of this chapter.

This concludes the introductory chapter. In summary, we have discussed the most important optical coefficients, their interrelations, analytic properties, and explicit forms in the oscillator model. It turns out that this model is often sufficient for a qualitatively correct description of isolated optical resonances. However, as we progress to describe the optical properties of semiconductors, we will see the necessity to modify and extend this simple model in many respects.

REFERENCES

For further reading we recommend:

J.D. Jackson, *Classical Electrodynamics*, 2nd *ed.*, Wiley, New York, (1975)

L.D. Landau and E.M. Lifshitz, *The Classical Theory of Fields*, 3rd *ed.*, Addison-Wesley, Reading, Mass. (1971)

L.D. Landau and E.M. Lifshitz, *Electrodynamics of Continuous Media*, Addison-Wesley, Reading, Mass. (1960)

PROBLEMS

Problem 1.1: Prove the Dirac identity

$$\frac{1}{r \mp i\epsilon} = P\, \frac{1}{r} \pm i\pi\delta(r) \; , \tag{1.69}$$

where $\epsilon \to 0$ and use of the formula under an integral is implied. *Hint:* Write Eq. (1.69) under the integral from $-\infty$ to $+\infty$ and integrate in pieces from $-\infty$ to $-\epsilon$, from $-\epsilon$ to $+\epsilon$ and from $+\epsilon$ to $+\infty$.

Problem 1.2: Derive the Kramers-Kronig relation relating $\chi''(\omega)$ to the integral over $\chi'(\omega)$.

Problem 1.3: Show that the Lorentzian

$$f(\omega) = \frac{1}{\pi} \frac{\gamma}{(\omega-\omega_0)^2 + \gamma^2} \tag{1.70}$$

approaches the delta function $\delta(\omega-\omega_0)$ for $\gamma \to 0$.

Problem 1.4: Verify Eq. (1.56) by evaluating the Kramers-Kronig transformation of Eq. (1.55). Note that only the resonant part of Eq. (1.22) should be used in order to be consistent with the resonant term approximation in Eq. (1.36).

Problem 1.5: Use Eq. (1.13) to show that

$$\int_{-\infty}^{\infty} d\omega \; e^{i\omega t} = 2\pi \, \delta(t).$$

Chapter 2
ATOM IN A CLASSICAL LIGHT FIELD

Semiconductors like all crystals are periodic arrays of one or more types of atoms. A prototype of a semiconductor is a lattice of group IV atoms, e.g. Si or Ge, which have four electrons in the outer electronic shell. These electrons participate in the covalent binding of a given atom to its four nearest neighbors which sit in the corners of a tetrahedron around the given atom. The bonding states form the valence bands which are separated by an energy gap from the energetically next higher states forming the conduction band.

In order to understand the similarities and the differences between optical transitions in a semiconductor and in an atom, we will first give an elementary treatment of the optical transitions in an atom. This chapter also serves to illustrate the difference between a quantum mechanical derivation of the polarization and the classical theory of Chap. 1.

2-1. Atomic Optical Susceptibility

The stationary Schrödinger equation of a single electron in an atom is

$$\mathcal{H}_0 \, \psi_n(\mathbf{r}) = \hbar\epsilon_n \, \psi_n(\mathbf{r}) \, , \tag{2.1}$$

where $\hbar\epsilon_n$ and ψ_n are the energy eigenvalues and the corresponding eigenfunctions, respectively. For simplicity we discuss the example of the hydrogen atom which has only a single electron. The Hamiltonian $\mathcal{H}_0$ is then given by the sum of the kinetic energy operator and the Coulomb potential in the form

$$\mathcal{H}_0 = -\frac{\hbar^2\nabla^2}{2m} - \frac{e^2}{r} \, . \tag{2.2}$$

An optical field couples to the dipole moment of the atom and introduces time-dependent changes of the wavefunction

$$i\hbar \frac{\partial}{\partial t}\psi(\mathbf{r},t) = [\ \mathcal{H}_0 + \mathcal{H}_I(t)\ ]\ \psi(\mathbf{r},t) \tag{2.3}$$

with

$$\mathcal{H}_I(t) = -\ ex\mathcal{E}(t) = -\ d\mathcal{E}(t)\ . \tag{2.4}$$

Here d is the operator for the electric dipole moment and we assumed that the homogeneous electromagnetic field is polarized in x direction. Expanding the time-dependent wavefunctions into the stationary eigenfunctions of Eq. (2.1)

$$\psi(\mathbf{r},t) = \sum_m a_m(t)\ e^{-i\epsilon_m t}\ \psi_m(\mathbf{r})\ , \tag{2.5}$$

inserting into Eq. (2.3), multiplying from the left by $\psi_n^*(\mathbf{r})$ and integrating over space, we find for the coefficients a_n the equation

$$i\hbar\ \frac{da_n}{dt} = -\ \mathcal{E}(t) \sum_m e^{i\epsilon_{nm}t}\ \langle n|d|m\rangle\ a_m\ , \tag{2.6}$$

where

$$\epsilon_{nm} = \epsilon_n - \epsilon_m \tag{2.7}$$

is the energy difference and

$$\langle n|d|m\rangle = \int d^3r\ \psi_n^*(\mathbf{r})\ d\ \psi_m(\mathbf{r}) \equiv d_{nm} \tag{2.8}$$

is the electric dipole matrix element. Assuming that the electron was initially at $t \to -\infty$ in the state $|\ell\rangle$, i.e.,

$$a_n(t \to -\infty) = \delta_{n,\ell} \tag{2.9}$$

and solving Eq. (2.6) iteratively, we obtain for $n \neq \ell$ in first order

$$i\hbar \frac{da_n}{dt} = - \mathcal{E}(t)\, d_{n\ell}\, e^{i\epsilon_{n\ell} t} \ . \tag{2.10}$$

Our quantum mechanical system defined by Eqs. (2.2) - (2.3) does not contain any damping. We therefore have to switch-on the field adiabatically to avoid unphysical switch-on perturbations

$$\mathcal{E}(t) = \frac{1}{2} \mathcal{E}_\omega \left[e^{-i(\omega+i\eta)t} + \text{c.c.} \right] , \tag{2.11}$$

where c.c. stands for the complex conjugate of the previous term. The asymptotically small η assures that the field vanishes as $t \to -\infty$. Inserting Eq. (2.11) into Eq. (2.10) we obtain

$$a_n(t) = - \frac{d_{n\ell}}{2\hbar} \mathcal{E}_\omega \left[\frac{e^{-i(\omega-\epsilon_{n\ell})t}}{\omega+i\eta-\epsilon_{n\ell}} - \frac{e^{i(\omega+\epsilon_{n\ell})t}}{\omega-i\eta+\epsilon_{n\ell}} \right] , \tag{2.12}$$

where we let $\eta \to 0$ in the exponents after the integration. As it turns out, the crucial role of the infinitesimally small η is to shift the poles into the proper part of the complex plane.

If we want to generate results in higher-order perturbation theory, we have to continue the iteration by inserting the first-order result into the RHS of (2.6) and calculate this way the changes of the wavefunction up to quadratic and higher order in the electric field. Here, we will limit ourselves to the terms linear in the field. This approximation is called linear response.

The field-induced polarization is given as the expectation value of the dipole operator

$$\mathcal{P}(t) = - N_0 \int d^3r\ \psi^*(\mathbf{r},t)\ d\ \psi(\mathbf{r},t) \ , \tag{2.13}$$

where N_0 is the density of noninteracting atoms. Linear in the field, we obtain

$$\mathscr{P}(t) = -\frac{N_0}{2\hbar}\sum_n \left[\frac{|d_{n\ell}|^2}{\omega+i\eta+\epsilon_{n\ell}} - \frac{|d_{n\ell}|^2}{\omega+i\eta-\epsilon_{n\ell}}\right] \mathscr{E}_\omega e^{-i\omega t} + \text{c.c.}\,, \tag{2.14}$$

and the resulting optical susceptibility is

$$\chi(\omega) = -\frac{N_0}{\hbar}\sum_n |d_{n\ell}|^2 \left[\frac{1}{\omega+i\eta+\epsilon_{n\ell}} - \frac{1}{\omega+i\eta-\epsilon_{n\ell}}\right]$$

atomic optical susceptibility (2.15)

where the complex conjugate term gives rise to a factor of 2.

2-2. Oscillator Strength

If we compare the first term of Eq. (2.15) with the result of the oscillator model, Eq. (1.7), we see that both expressions have similar structures. However, in comparison with the oscillator model the atom is represented not by one but by many oscillators with different eigenfrequencies $\epsilon_{\ell n}$. To see this, we rewrite the expression (2.15), pulling out the same factors which appear in the oscillator result, Eq. (1.7),

$$\chi(\omega) = -\frac{N_0 e^2}{2m}\sum_n \frac{f_{n\ell}}{\epsilon_{n\ell}} \left[\frac{1}{\omega+i\eta-\epsilon_{\ell n}} - \frac{1}{\omega+i\eta+\epsilon_{\ell n}}\right]. \tag{2.16}$$

Hence, each partial oscillator has the strength of

$$f_{n\ell} = \frac{2m}{\hbar}|x_{n\ell}|^2 \epsilon_{n\ell}$$

oscillator strength (2.17)

Adding the strengths of all oscillators by summing over all the final states n, we find

$$\sum_n f_{n\ell} = \frac{2m}{\hbar} \sum_n \langle n|x|\ell\rangle\langle\ell|\, x|n\rangle(\epsilon_n - \epsilon_\ell)$$

$$= \frac{2m}{\hbar^2} \sum_n \langle n|x|\ell\rangle\langle\ell|\, [x, \mathscr{H}_0]|n\rangle \ , \tag{2.18}$$

where

$$[x, \mathscr{H}_0] = x\mathscr{H}_0 - \mathscr{H}_0 x = -\frac{\hbar^2}{2m}\left[x\frac{d^2}{dx^2} - \frac{d^2}{dx^2}x\right] = \frac{\hbar^2}{m}\frac{d}{dx} \tag{2.19}$$

is the commutator of x and $\mathscr{H}_0$. Using the completeness relation $\Sigma_n\ |n\rangle\langle n| = 1$ yields

$$\sum_n f_{n\ell} = 2\ \langle\ell|\ \frac{d}{dx}x|\ell\rangle \ . \tag{2.20}$$

Applying the chain rule and integrating by parts, one can show that

$$\langle\ell|\ \frac{d}{dx}x\ |\ell\rangle = \langle\ell|\ell\rangle + \langle\ell|\, x\frac{d}{dx}|\ell\rangle$$

$$= 1 - \langle\frac{d}{dx}x\ell|\ell\rangle \ . \tag{2.21}$$

Since $f_{n\ell}$ is real, $\langle\frac{d}{dx}x\ell|\ell\rangle$ has to be equal to $\langle\ell|\ \frac{d}{dx}x|\ell\rangle$. Therefore, Eq. (2.21) yields

$$2\langle\ell|\ \frac{d}{dx}x|\ell\rangle = 1 \ . \tag{2.22}$$

Hence, the oscillator strength fulfills the sum rule

$$\boxed{\sum_n f_{n\ell} = 1}$$

oscillator strength sum rule (2.23)

In the picture of the oscillator model one may say that in an atom one oscillator is distributed over many partial oscillators, each having the strength $f_{n\ell}$.

From Eq. (2.15) we see that the imaginary part of the dielectric function for the atom is

$$\epsilon''(\omega) = \omega_{pl}^2 \sum_n \frac{f_{n\ell}}{\epsilon_{n\ell}} 2\pi \left[\delta(\omega-\epsilon_{n\ell}) - \delta(\omega-\epsilon_{\ell n}) \right] \tag{2.24}$$

Since $|\ell\rangle$ is the occupied initial state and $|n\rangle$ are the final states, we see that the first term in Eq. (2.24) describes light absorption. Energy conservation requires

$$\hbar\epsilon_n = \hbar\omega + \hbar\epsilon_\ell \ , \tag{2.25}$$

i.e., an optical transition from the lower state $|\ell\rangle$ to the energetically higher state $|n\rangle$ takes place if the energy difference $\hbar\epsilon_{n\ell}$ is equal to the energy $\hbar\omega$ of a light quantum, called a photon. In other words, a photon is absorbed and the atom is excited from the initial state $|\ell\rangle$ to the final state $|n\rangle$. This interpretation of our result is the correct one, but to be fully appreciated it actually requires also the quantum mechanical treatment of the light field.

The second term on the RHS of Eq. (2.24) describes negative absorption causing amplification of the light field, i.e., optical gain. This is the basis of laser action, as we will see in Chap. 15. In order to produce optical gain, the system has to be prepared in a state $|\ell\rangle$ which has a higher energy than the final state $|n\rangle$, because the energy conservation expressed by the delta function in the second term on the RHS of (2.24) requires

$$\hbar\epsilon_\ell = \hbar\omega + \hbar\epsilon_n \ . \tag{2.26}$$

If the energy of a light quantum equals the energy difference $\hbar\epsilon_{\ell n}$, stimulated emission occurs. In order to obtain stimulated emission in a real system, one has to invert the system so that it is initially in an excited state rather than in the ground state.

2-3. Optical Stark Shift

Until now we have only calculated and discussed the linear response of an atom to a weak light field. For the case of two atomic levels interacting with the light field, we will now determine the response at arbitrary field intensities. Calling these two levels $n = 1,2$ with $\epsilon_2 > \epsilon_1$, we get from Eq. (2.6) using Eq. (2.11)

$$i\hbar \frac{da_1}{dt} = - e^{i(\omega-\epsilon_{21})t} \frac{d_{12}\mathcal{E}_\omega}{2} a_2 , \tag{2.27}$$

$$i\hbar \frac{da_2}{dt} = - e^{-i(\omega-\epsilon_{21})t} \frac{d_{21}\mathcal{E}_\omega}{2} a_1 , \tag{2.28}$$

where we took only the resonant terms into account. This approximation is called the *rotating-wave approximation* (RWA). For simplicity we first treat the case of exact resonance, $\omega = \epsilon_{21}$. Then we have

$$\frac{d^2a_2}{dt^2} = i \frac{d_{21}}{2\hbar} \mathcal{E}_\omega \frac{da_1}{dt} = - \left| \frac{d_{12}\mathcal{E}_\omega}{2\hbar} \right|^2 a_2 = - \frac{\omega_R^2 a_2}{4} \tag{2.29}$$

where

$$\boxed{\omega_R = \frac{|d_{21}\mathcal{E}|}{\hbar}}$$

Rabi frequency (2.30)

is the Rabi frequency. The solution of (2.29) is of the form

$$a_2(t) = a_2(0)\, e^{\pm i\omega_R t/2} . \tag{2.31}$$

For $a_1(t)$ we get the equivalent result. Inserting the solutions for a_1 and a_2 back into Eq. (2.5) we see that the original frequencies ϵ_1 and ϵ_2 become $\epsilon_1 \pm \omega_R/2$ and $\epsilon_2 \pm \omega_R/2$, respectively. Hence, as indicated in Fig. 2.1 one has not just one but three optical transitions with the frequencies ϵ_{21}, and $\epsilon_{21} \pm \omega_R$, respectively. In other words, under the influence of the light field the single transition possible in a two-level atom splits into a triplet. The splitting is proportional to the field strength and the electric dipole moment and can therefore be observed only for reasonably strong fields.

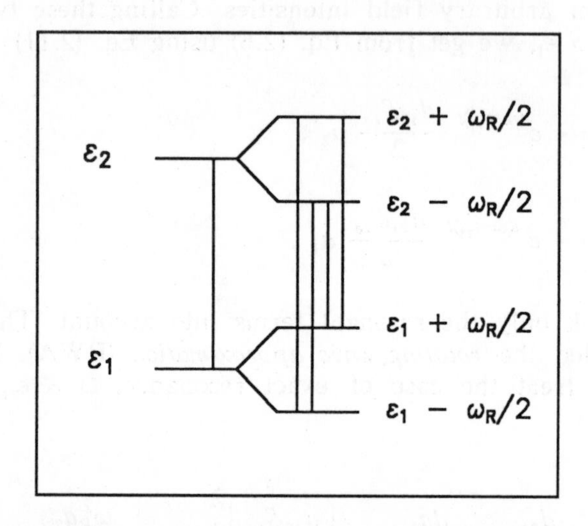

Fig. 2.1: Schematic drawing of the frequency scheme of a two-level system without the light field (left part of Figure) and the light-field induced level splitting (right part of Figure) for the case of a resonant field, i.e., zero detuning. The vertical lines indicate the possible optical transitions between the levels.

The two-level model can be solved also for the case of a finite detuning $\nu=\epsilon_{21}-\omega$. In this situation, Eqs. (2.27) and (2.28) can be written as

$$\frac{da_1}{dt} = ie^{-i\nu t} \frac{d_{12}\mathcal{E}_\omega}{2\hbar} a_2 \tag{2.32}$$

$$\frac{da_2}{dt} = ie^{i\nu t} \frac{d_{21}\mathcal{E}_\omega}{2\hbar} a_1 \ . \tag{2.33}$$

Taking the time derivative of Eq. (2.32)

$$\frac{d^2 a_1}{dt^2} = e^{-i\nu t}\, \frac{d_{21}\mathcal{E}_\omega}{2\hbar} \left[a_2 \nu + i \frac{da_2}{dt} \right] \tag{2.34}$$

and expressing $a_2(t)$ and $da_2(t)/dt$ in terms of $a_1(t)$ we get

$$\frac{d^2 a_1}{dt^2} = -i\nu \frac{da_1}{dt} - \frac{\omega_R^2 a_1}{4} \tag{2.35}$$

with the solution

$$a_1(t) = a_1(0)\, e^{i\Omega t} , \tag{2.36}$$

where

$$\Omega = -\frac{\nu}{2} \pm \frac{1}{2}\sqrt{\nu^2+\omega_R^2} . \tag{2.37}$$

Similarly, we obtain

$$a_2(t) = a_2(0)\, e^{-i\Omega t} . \tag{2.38}$$

Hence, we again get split and shifted levels

$$\epsilon_2 \to \Omega_2 \equiv \epsilon_2 + \Omega = \epsilon_2 - \frac{\nu}{2} \pm \frac{1}{2}\sqrt{\nu^2+\omega_R^2} \tag{2.39}$$

$$\epsilon_1 \to \Omega_1 \equiv \epsilon_1 - \Omega = \epsilon_1 + \frac{\nu}{2} \pm \frac{1}{2}\sqrt{\nu^2+\omega_R^2} . \tag{2.40}$$

The coherent modification of the atomic spectrum in the electric field of a light field resembles the Stark splitting and shifting in a static electric field. It is therefore called *optical Stark effect*. The modified or, as one also says, the renormalized states of the atom in the intense light field are those of a *dressed atom*. While the optical Stark effect is well-known in atoms, it has only recently been seen in semiconductors where the dephasing times are normally much shorter than in atoms, as will be discussed in more detail in later chapters of this book.

REFERENCES

For the basic quantum mechanical theory we recommend the textbooks:

A.S. Davydov, *Quantum Mechanics*, Pergamon, New York (1965)

L.I. Schiff, *Quantum Mechanics*, 3rd *ed.*, McGraw-Hill, New York (1968).

The quantum theory of two-level atoms in a light field is treated extensively, e.g., in:

L. Allen and J.H. Eberly, *Optical Resonance and Two-Level Atoms*, Wiley and Sons, New York (1975)

M. Sargent III, M.O. Scully, and W.E. Lamb, Jr., *Laser Physics*, Addison-Wesley, Reading, MA (1974).

PROBLEMS

Problem 2.1: To describe the dielectric relaxation in a dielectric medium, one often uses the *Debye model* where the polarization obeys the equation

$$\frac{d\mathcal{P}}{dt} = -\frac{1}{\tau}\left[\ \mathcal{P}(t) - \chi_0 \mathcal{E}(t)\ \right]. \tag{2.41}$$

Here τ is the relaxation time and χ_0 is the static dielectric susceptibility. The initial condition is

$$\mathcal{P}(t=-\infty) = 0\ .$$

Compute the optical susceptibility.

Problem 2.2: Compute the oscillator strength for the transitions between the states of a quantum mechanical harmonic oscillator. Verify the sum rule, Eq. (2.23).

Chapter 3
PERIODIC LATTICE OF ATOMS

After having introduced the basic optical properties of a single atom, we will now discuss how the atomic energy spectrum is modified in a solid, if the atoms form a periodic crystal lattice. In this book we will not discuss the bonding of the atoms or related problems, but simply ask ourselves how a single electron can move in the periodic potential of a crystal. In the spirit of a Hartree-Fock theory, we assume an effective periodic potential which contains already the mean field of the other electrons.

3-1. Reciprocal Lattice, Bloch Theorem

A) Bulk Semiconductors

We consider only simple crystals made of one sort of atoms. More complicated systems are treated in the textbooks referenced at the end of this chapter. The periodicity of the effective ionic potential is expressed by

$$V(\mathbf{r}) = V(\mathbf{r} + \mathbf{n}) \; , \tag{3.1}$$

where $\mathbf{n}$ is a lattice vector. It is convenient to expand the lattice vectors

$$\mathbf{n} = \sum_i \mathbf{a}_i n_i \; , \tag{3.2}$$

where the $\mathbf{a}_i$ are the basis vectors which span the unit cells. Note that the $\mathbf{a}_i$ are not unit vectors and they are generally not orthogonal. They point into the directions of the three axes of the unit cell, which may have a rhombic or more complicated shape. However, in the case of cubic lattices the $\mathbf{a}_i$ are parallel to the usual Cartesian unit vectors.

To make use of the fact that the potential acting on the electron has the periodicity of the lattice, we introduce a translation operator $T_\mathbf{n}$

$$T_\mathbf{n} f(\mathbf{r}) = f(\mathbf{r} + \mathbf{n}) , \tag{3.3}$$

where f is an arbitrary function. Applying T_n to the wavefunction ψ of an electron in the periodic potential $V(r)$, we obtain

$$T_\mathbf{n}\ \psi(\mathbf{r}) = \psi(\mathbf{r} + \mathbf{n}) = t_\mathbf{n}\psi(\mathbf{r}) , \tag{3.4}$$

where $t_\mathbf{n}$ is a phase factor, because $\psi(\mathbf{r})$ and $\psi(\mathbf{r+n})$ have to be identical up to a phase factor. Since the Hamilton operator

$$\mathcal{H} = \frac{p^2}{2m} + V(\mathbf{r}) \tag{3.5}$$

has the full lattice symmetry, the commutator of $\mathcal{H}$ and $T_\mathbf{n}$ vanishes:

$$[\mathcal{H}, T_\mathbf{n}] = 0 . \tag{3.6}$$

Under this condition a complete set of functions exists consisting of eigenfunctions to both $\mathcal{H}$ and $T_\mathbf{n}$:

$$\mathcal{H}\psi_{\lambda\mathbf{n}} = E_\lambda \psi_{\lambda\mathbf{n}} \tag{3.7}$$

and

$$T_\mathbf{n}\psi_{\lambda,\mathbf{n}}(\mathbf{r}) = \psi_{\lambda,\mathbf{n}}(\mathbf{r} + \mathbf{n}) = t_\mathbf{n}\psi_{\lambda,\mathbf{n}}(\mathbf{r}) . \tag{3.8}$$

As discussed above, $t_\mathbf{n}$ can only be a phase factor with the properties

$$|t_\mathbf{n}| = 1 , \tag{3.9}$$

and

$$t_\mathbf{n} t_\mathbf{m} = t_{\mathbf{n+m}} \tag{3.10}$$

since

$$T_{\mathbf{n+m}} = T_\mathbf{n} T_\mathbf{m} , \tag{3.11}$$

stating the obvious fact that a translation by $\mathbf{m} + \mathbf{n}$ is identical to a translation by $\mathbf{m}$ followed by another translation by $\mathbf{n}$. A possible choice to satisfy Eqs. (3.8) - (3.11) is

$$t_{\mathbf{n}} = e^{i(\mathbf{k}\cdot\mathbf{n} + 2\pi N)} , \tag{3.12}$$

where 2π is an allowed additional factor because

$$e^{i2\pi N} = 1 , \text{ for } N = \text{integer} . \tag{3.13}$$

We now define a *reciprocal lattice vector* $\mathbf{g}$ through the relation

$$e^{i\mathbf{k}\cdot\mathbf{n}} \equiv e^{i(\mathbf{k}+\mathbf{g})\cdot\mathbf{n}} \tag{3.14}$$

such that $\mathbf{g}\cdot\mathbf{n} = 2\pi N$. Expanding $\mathbf{g}$ in the basis vectors $\mathbf{b}_i$

$$\mathbf{g} = \sum_{i=1}^{3} m_i \mathbf{b}_i , \ m_i = \text{integer} , \tag{3.15}$$

we find

$$\mathbf{g}\cdot\mathbf{n} = \sum_{ij} m_i \ n_j \ \mathbf{b}_i\cdot\mathbf{a}_j = 2\pi N . \tag{3.16}$$

The last equality in this equation is satisfied if

$$\mathbf{b}_i\cdot\mathbf{a}_j = 2\pi \ \delta_{ij} , \ i,j = 1,2,3 . \tag{3.17}$$

Consequently $\mathbf{b}_i$ is perpendicular to both $\mathbf{a}_j$ and $\mathbf{a}_k$ with $i \neq k,j$ and can be written as a vector product

$$\mathbf{b}_i = c \ \mathbf{a}_j \times \mathbf{a}_k . \tag{3.18}$$

To determine the proportionality constant c, we use

$$\mathbf{a}_i\cdot\mathbf{b}_i = c \ \mathbf{a}_i\cdot(\mathbf{a}_j \times \mathbf{a}_k) = 2\pi , \tag{3.19}$$

which yields

$$c = \frac{2\pi}{\mathbf{a}_i\cdot(\mathbf{a}_j \times \mathbf{a}_k)} . \tag{3.20}$$

Hence

$$\mathbf{b}_i = 2\pi \frac{\mathbf{a}_j \times \mathbf{a}_k}{\mathbf{a}_i \cdot (\mathbf{a}_j \times \mathbf{a}_k)} \tag{3.21}$$

Similarly we can express the $\mathbf{a}_i$ in terms of the $\mathbf{b}_i$.

In summary, we have introduced two lattices characterized by the unit vectors $\mathbf{a}_i$ and $\mathbf{b}_i$, respectively, which are reciprocal to one another. Since the $\mathbf{a}_i$ are the vectors of the real crystal lattice, the lattice defined by the $\mathbf{b}_i$ is called the *reciprocal lattice*. The unit cells spanned by the $\mathbf{a}_i$ are called the *Wigner-Seitz cells*, while the unit cells spanned by the $\mathbf{b}_i$ are the *Brillouin zones*. One can think of a transformation between the real and reciprocal lattice spaces as a discrete three-dimensional Fourier transform.

For the example of a cubic lattice, we have

$$|\mathbf{b}_i| = \frac{2\pi}{|\mathbf{a}_i|} \, . \tag{3.22}$$

Consequently the smallest reciprocal lattice vector in this case has the magnitude

$$g_i = \frac{2\pi}{a_i} \, . \tag{3.23}$$

Going back to Eq. (3.14), it is clear that we can restrict the range of k values to the region

$$-\frac{g_i}{2} \leq k_i \leq \frac{g_i}{2} \, , \tag{3.24}$$

since all other values of k can be realized by adding (or subtracting) multiple reciprocal lattice vectors. This range of $\mathbf{k}$-values is called the first Brillouin zone.

Considerations along the lines of Eqs. (3.1) - (3.14) led F. Bloch to formulate the following theorem, which has to be fulfilled by the electronic wavefunctions in the lattice

$$\boxed{e^{i\mathbf{k}\cdot\mathbf{n}} \, \psi_{\lambda,\mathbf{k}}(\mathbf{r}) = \psi_{\lambda,\mathbf{k}}(\mathbf{r} + \mathbf{n})}$$

Bloch theorem (3.25)

To satisfy this relation, we make the ansatz

$$\boxed{\psi_{\lambda,\mathbf{k}}(\mathbf{r}) = \frac{e^{i\mathbf{k}\cdot\mathbf{r}}}{L^{3/2}}\, u_{\lambda,\mathbf{k}}(\mathbf{r})}\ ,$$

Bloch wavefunction (3.26)

where L^3 is the volume of the crystal, λ is the energy eigenvalue, and $\mathbf{k}$ is the eigenvalue for the periodicity (crystal momentum). The ansatz (3.26) fulfills the Bloch theorem (3.25), if the functions $u_{\lambda,\mathbf{k}}$ are periodic in real space

$$u_{\lambda,\mathbf{k}}(\mathbf{r}) = u_{\lambda,\mathbf{k}}(\mathbf{r} + \mathbf{n}) \ . \tag{3.27}$$

The $u_{\lambda,\mathbf{k}}$ are called Bloch functions. Inserting Eq. (3.26) into the Schrödinger equation

$$\mathscr{H}\psi_{\lambda,\mathbf{k}} = \left[\frac{p^2}{2m} + V(r)\right]\psi_{\lambda,\mathbf{k}} = E_\lambda \psi_{\lambda,\mathbf{k}} \tag{3.28}$$

and using the relation

$$\begin{aligned}\sum_j \frac{\partial^2}{\partial x_j^2}\,\psi_{\lambda,\mathbf{k}} &= -k^2\psi_{\lambda,\mathbf{k}} + 2i\sum k_j\,\frac{e^{i\mathbf{k}\cdot\mathbf{r}}}{L^{3/2}}\,\frac{\partial u_{\lambda,\mathbf{k}}}{\partial x_j} + \frac{e^{i\mathbf{k}\cdot\mathbf{r}}}{L^{3/2}}\sum_j \frac{\partial^2 u_{\lambda,\mathbf{k}}}{\partial x_j^2}\\ &= \frac{e^{i\mathbf{k}\cdot\mathbf{r}}}{L^{3/2}}\,(\nabla + i\mathbf{k})^2\, u_{\lambda,\mathbf{k}}\ ,\end{aligned} \tag{3.29}$$

we obtain

$$\left[-\frac{\hbar^2}{2m}\,(\nabla + i\mathbf{k})^2 + V(\mathbf{r})\right] u_{\lambda,\mathbf{k}} = E_{\lambda,\mathbf{k}}\, u_{\lambda,\mathbf{k}}\ . \tag{3.30}$$

Further treatment requires a specific form of the periodic potential $V(\mathbf{r})$. Before we discuss two simple approximations, namely that for tightly bound and nearly free electrons, (a) and (b), we derive the normalization condition for the Bloch functions. Since the wavefunctions (3.26) are normalized to the crystal volume L^3, we have

$$\int_{L^3} d^3r \, |\psi_{\lambda,\mathbf{k}}|^2 = 1 = \frac{1}{L^3} \int_{L^3} d^3r \, |u_{\lambda,\mathbf{k}}|^2 \, . \tag{3.31}$$

Since the Bloch functions have the periodicity of the lattice, the integral can be evaluated as

$$\int_{L^3} \rightarrow \sum_N \int_{\ell^3} ,$$

where ℓ^3 is the volume of of an elementary cell, such that $L^3 = N\ell^3$ when N is the number of elementary cells. Substituting Eq. (3.27) into Eq. (3.31) we obtain

$$\sum_{n=1}^{N} \frac{1}{N} \int_{\ell^3} \frac{d^3r}{\ell^3} \, |u_{\lambda,\mathbf{k}}|^2 = \frac{1}{\ell^3} \int_{\ell^3} d^3r \, |u_{\lambda,\mathbf{k}}|^2 = 1 \, , \tag{3.32}$$

showing that the Bloch functions have to be normalized within an elementary cell.

Sometimes it is useful to introduce localized functions as expansion set instead of the delocalized Bloch functions. An example of such localized functions are the *Wannier functions* $w_\lambda(\mathbf{r}-\mathbf{n})$. The Wannier functions are related to the Bloch wavefunctions, Eq. (3.26), via

$$w_\lambda(\mathbf{r}-\mathbf{n}) = \frac{1}{\sqrt{N}} \sum_{\mathbf{k}} e^{-i\mathbf{k}\cdot\mathbf{n}} \, \psi_{\lambda,\mathbf{k}}(\mathbf{r}) \tag{3.33}$$

and

$$\psi_{\lambda,\mathbf{k}}(\mathbf{r}) = \frac{1}{\sqrt{N}} \sum_{\mathbf{n}} e^{i\mathbf{k}\cdot\mathbf{n}} \, w_\lambda(\mathbf{r}-\mathbf{n}) \, . \tag{3.34}$$

Wannier functions are concentrated around a lattice point $\mathbf{n}$. They are orthogonal for different lattice points

$$\int d^3r \; w_\lambda^*(\mathbf{r}) \; w_{\lambda'}(\mathbf{r}-\mathbf{n}) = \delta_{\lambda,\lambda'} \; \delta_{n,0} \; , \tag{3.35}$$

see problem 3.2. Particularly in spatially inhomogeneous situations it is often advantageous to use Wannier functions as the expansion set.

B) Quantum Confined Semiconductors

In recent years, advances in crystal growth techniques made it possible to realize semiconductor microstructures, which are so small that their electronic properties deviate from those of bulk materials. For example, one can produce very thin semiconductor layers, so-called semiconductor *quantum wells*. The electrons in such a structure feel in addition to the periodic atomic potential in the direction of the layer another potential which approximately has the form of a square well. The walls of this well at $z = \pm L_c/2$ are just the surfaces of the layer.

Since usually L_c is still quite large in comparison to the lattice constant of the bulk semiconductor, it is a reasonable approximation to assume that the quantum confinement modifies only the plane-wave envelope part of the wavefunctions, Eq. (3.26), and does not change the lattice-periodic part $u_{\lambda,\mathbf{k}}$. For the case of a quantum well, one therefore obtains instead of Eq. (3.26)

$$\psi_{\lambda,\mathbf{k}}(\mathbf{r}) = \sqrt{\frac{2}{L_c}} \left\{ \begin{array}{c} \cos[(2n+1)\pi z/L_c] \\ \sin[2\pi n z/L_c] \end{array} \right\} \frac{e^{i\mathbf{k}_{||}\cdot\mathbf{r}}}{L^{3/2}} \, u_{\lambda,\mathbf{k}_{||}}(\mathbf{r}) \; , \tag{3.36}$$

where $\mathbf{k}_{||}$ is the wavenumber in the plane of the quantum well. Eq. (3.36) shows that the factor $\exp(ik_z z)/\sqrt{L}$ describing plane-wave propagation in a bulk semiconductor is replaced by a standing wave of even or odd symmetry. Correspondingly, one has discrete energy levels due to the quantization in z-direction. The separation of these energy levels is inversly proportional to L_c. If L_c is small enough, the excited states are well separated from the ground state and one can describe the electron gas essentially as a two-dimensional system. The translational degree of freedom in z-direction is removed by the confinement potential.

We call the layered structures quasi-two-dimensional, which reminds us that the z-extension, L_c, is still much larger than the lattice unit cell. Correspondingly, in a quasi-one-dimensional *quantum wire*, there is a confinement potential in two directions, say y and z, while the electron is still able to move freely in x-direction. Finally, in quasi-zero-dimensional semiconductor *quantum dots*, one has three-dimensional quantum confinement and the translational freedom is completely suppressed. In this case one has standing-wave solutions in all three space dimensions. Generally, quantum wells, wires and dots are therefore of mesoscopic nature, which means they are small enough to yield large quantum effects, but they are still large in comparison to the atomic unit cell.

3-2. Tight-Binding Approximation

Now, after these basic symmetry considerations, we discuss in this and in the following section two simple approximations for calculating the allowed energies and eigenfunctions of an electron in the crystal lattice. These two models are a) tightly bound electrons and b) nearly free electrons. They serve as elementary introduction into the wide field of band-structure calculations, for which we refer to textbook literature at the end of this chapter.

In the so-called tight-binding approximation, we treat those electrons, whose wavefunctions are not substantially changed when making a solid out of the atoms. We assume that these electrons stay close to the atomic sites and that the electronic wavefunctions centered around neighboring sites have little overlap. Consequently there is almost no overlap between wavefunctions for electrons that are separated by two or more atoms (next-nearest neighbors, next-next nearest neighbors, etc). We make use of this property by expanding the occurring integrals in terms of *overlap integrals* for the electronic wavefunctions at atoms m and ℓ. These overlap integrals become negligible for m substantially different from ℓ, allowing us to truncate this expansion.

The Schrödinger equation for a single atom located at the lattice point ℓ is

$$\mathcal{H}_0 \phi_\lambda(\mathbf{r} - \boldsymbol{\ell}) = \mathcal{E}_\lambda \phi_\lambda(\mathbf{r} - \boldsymbol{\ell}) \tag{3.37}$$

with the Hamiltonian

$$\mathcal{H}_0 = -\frac{\hbar^2 \nabla^2}{2m} + V(\mathbf{r} - \boldsymbol{\ell}) \ , \tag{3.38}$$

where $V(\mathbf{r} - \boldsymbol{\ell})$ is the potential of the ℓ-th atom. Eq. (3.37) is a good zero-order approximation for tightly bound electrons. The full problem of the periodic solid is then

$$\left[-\frac{\hbar^2\nabla^2}{2m} + \sum_{\ell} V(\mathbf{r} - \boldsymbol{\ell}) - E_\lambda\right]\psi_\lambda(\mathbf{r}) = 0 \; . \tag{3.39}$$

The total potential is the sum of the single-atom potentials. To solve Eq. (3.39) we make the ansatz

$$\boxed{\psi_\lambda(\mathbf{r}) = \sum_n \frac{e^{i\mathbf{k}\cdot\mathbf{n}}}{L^{3/2}}\,\phi_\lambda(\mathbf{r} - \mathbf{n})} \; ,$$

tight-binding wavefunction (3.40)

which obviously fulfills the Bloch theorem, Eq. (3.25).

In solid state physics one often wants to compute bulk properties without dealing with surface effects. To do this formally, one introduces the so-called periodic (Born-von Karman) boundary conditions

$$\psi(\mathbf{r}) \equiv \psi(\mathbf{r} + N_i\mathbf{a}_i), \tag{3.41}$$

where N_i is the number of atoms in i-direction. Inserting Eq. (3.41) into Eq. (3.40), we obtain

$$\begin{aligned}
\psi(\mathbf{r}) &= \frac{1}{L^{3/2}}\sum_n e^{i\mathbf{k}\cdot\mathbf{n}}\,\phi_\lambda(\mathbf{r} - \mathbf{n}) \\
&= \frac{1}{L^{3/2}}\sum_n e^{i\mathbf{k}\cdot\mathbf{n}}\,\phi_\lambda(\mathbf{r} + N_i\mathbf{a}_i - \mathbf{n}) \\
&= \frac{1}{L^{3/2}}\sum_m e^{i\mathbf{k}\cdot(\mathbf{m}+N_i\mathbf{a}_i)}\,\phi_\lambda(\mathbf{r} - \mathbf{m}) \; .
\end{aligned} \tag{3.42}$$

Since the first and last lines should be identical, we must have

$$e^{i\mathbf{k}\cdot(\mathbf{m}+N_i\mathbf{a}_i)} = e^{i\mathbf{k}\cdot\mathbf{m}} ,$$

which demands $\mathbf{k}N_i\mathbf{a}_i = 2\pi\nu$, where ν is an integer. In a cubic lattice,

$$-\frac{\pi}{a_i} \leq k \leq \frac{\pi}{a_i}$$

therefore ν runs over the following values in the first Brillouin zone

$$k = \frac{2\pi}{a_i}\frac{\nu}{N_i} , \text{ where } -\frac{N_i}{2} \leq \nu \leq \frac{N_i}{2} . \tag{3.43}$$

Now we proceed to compute the ground state energy

$$E_{\lambda,\mathbf{k}} = \frac{\int d^3r\psi^*_{\lambda,\mathbf{k}}(\mathbf{r})\ \mathscr{H}\psi_{\lambda,\mathbf{k}}(\mathbf{r})}{\int d^3r\psi^*_{\lambda,\mathbf{k}}(\mathbf{r})\ \psi_{\lambda,\mathbf{k}}(\mathbf{r})} = \frac{\mathscr{N}}{\mathscr{D}} , \tag{3.44}$$

where the numerator can be written as

$$\mathscr{N} = \sum_{n,m} e^{i\mathbf{k}\cdot(\mathbf{n}-\mathbf{m})} \int d^3r\ \phi^*_\lambda(\mathbf{r} - \mathbf{m})\ \mathscr{H}\ \phi_\lambda(\mathbf{r} - \mathbf{n}) \tag{3.45}$$

and the denominator is

$$\mathscr{D} = \sum_{n,m} e^{i\mathbf{k}\cdot(\mathbf{n}-\mathbf{m})} \int d^3r\ \phi^*_\lambda(\mathbf{r} - \mathbf{m})\ \phi_\lambda(\mathbf{r} - \mathbf{n}) . \tag{3.46}$$

Since we have assumed strongly localized electrons, it is a rapidly converging procedure to expand the integrals in terms of overlap integrals involving the atomic sites m and n. The leading contribution is $m = n$, then $n = m \pm 1$, etc. Since we only want to keep the leading order of the complete expression (3.44), it is sufficient to approximate in the denominator

$$\int d^3r\ \phi_\lambda^*(\mathbf{r} - \mathbf{m})\ \phi_\lambda(\mathbf{r} - \mathbf{n}) \cong \delta_{\mathbf{n},\mathbf{m}}\ , \tag{3.47}$$

yielding for the total denominator

$$\mathscr{D} = \sum_{n,m} \frac{\delta_{\mathbf{n},\mathbf{m}}}{L^3} = \sum_{n} \frac{1}{L^3} = \frac{N}{L^3}\ . \tag{3.48}$$

The integral in the numerator can be expanded as follows

$$\int d^3r\ \phi_\lambda^*(\mathbf{r} - \mathbf{m}) \left[-\frac{\hbar^2\nabla^2}{2m} + \sum_{\boldsymbol{\ell}} V(\mathbf{r} - \boldsymbol{\ell}) \right] \phi_\lambda(\mathbf{r} - \mathbf{n}) =$$

$$= \delta_{\mathbf{n},\mathbf{m}} \left[\sum_{\boldsymbol{\ell}} \delta_{\boldsymbol{\ell},\mathbf{n}}\ \mathscr{E}_\lambda + \sum_{\boldsymbol{\ell}\neq\mathbf{n}} \int d^3r\ \phi_\lambda^*(\mathbf{r} - \mathbf{n})\ V(\mathbf{r} - \boldsymbol{\ell})\ \phi_\lambda(\mathbf{r} - \mathbf{n}) \right]$$

$$+ \sum_{i} \delta_{\mathbf{n}\pm\mathbf{a}_i,\mathbf{m}} \sum_{\boldsymbol{\ell}} \int d^3r\ \phi_\lambda^*(\mathbf{r} \pm \mathbf{a}_i + \mathbf{n})\ V(\mathbf{r} - \boldsymbol{\ell})\ \phi_\lambda(r + n)$$

$$+ \ldots .$$

$$\equiv \delta_{\mathbf{n},\mathbf{m}}\ \mathscr{E}'_\lambda + \sum_{i} \delta_{\mathbf{n}\pm\mathbf{a}_i,\mathbf{m}} B_{\lambda,i} + \ldots\ , \tag{3.49}$$

where $B_{\lambda,i}$ is the overlap integral. The contributions of the next nearest neighbors are usually neglected, so that the total numerator can be written as

$$\mathcal{N}_{\lambda,\mathbf{k}} \cong \sum_{\mathbf{n},\mathbf{m}} e^{i\mathbf{k}\cdot(\mathbf{n}-\mathbf{m})} \left[\delta_{\mathbf{n},\mathbf{m}}\, \mathcal{E}'_\lambda + \sum_i \delta_{\mathbf{n}\pm\mathbf{a}_i,\mathbf{m}} B_{\lambda,i} \right] , \tag{3.50}$$

where $\mathcal{E}'_\lambda$ is the renormalized (shifted) atomic energy level.

Inserting Eqs. (3.45) - (3.50) into Eq. (3.44) and assuming a cubic lattice, where the $\mathbf{a}_i$ have equal magnitude and are proportional to the Cartesian unit vectors, we obtain $B_{\lambda,i} = B_\lambda$ and

$$E_{\lambda,\mathbf{k}} = \frac{1}{L^3} \sum_n \frac{\mathcal{E}'_\lambda + B_\lambda \sum_{i=1}^{3} (e^{ik_i a_i} + e^{-ik_i a_i})}{N/L^3}$$

$$= \mathcal{E}'_\lambda + 2B_\lambda \sum_i \cos(k_i a_i) , \tag{3.51}$$

which are the tight-binding cosine bands which are plotted schematically in Fig. 3.1.

To evaluate the detailed band structure, we have to compute B_λ. Without proof, we just want to mention at this point that for an attractive potential V (as in the case of electrons and ions) and p-type atomic functions ϕ, $B_\lambda > 0$, whereas for s-type atomic functions, $B_\lambda < 0$. Between the allowed energy levels, we have energy gaps, i.e., forbidden energy regions.

In summary, we have the following general results:

(i) The discrete atomic energy levels become quasi-continuous energy regions, called energy bands, with a certain band width.

(ii) There may be energy gaps between different bands.

(iii) Depending on the corresponding atomic functions, the bands $E_{\lambda,\mathbf{k}}$ may have positive or negative curvature around the band extrema.

(iv) In the vicinity of the band extrema one can often make a parabolic approximation

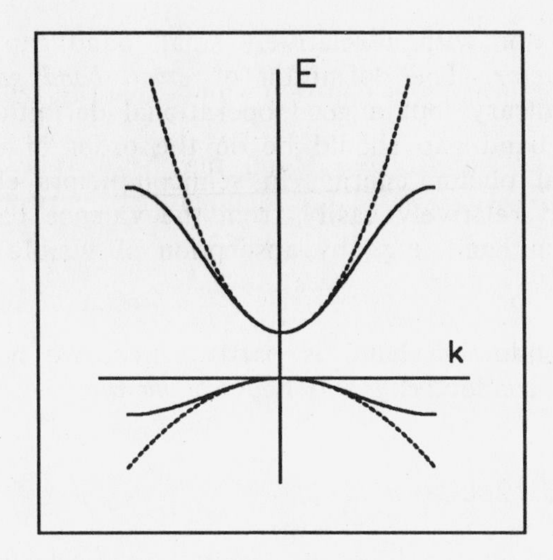

Fig. 3.1: Schematic drawing of the energy dispersion resulting from Eq. (3.51) for the cases of $B_\lambda > 0$ (lower band) and $B_\lambda < 0$ (upper band). The effective mass approximations, Eq. (3.52), are inserted as dashed lines.

$$E_k \cong E_0 + \frac{\hbar^2 k^2}{2m_{eff}}, \quad m_{eff} = \frac{\hbar^2}{\partial^2 E_{k=0} / \partial k^2}. \tag{3.52}$$

In the regimes where the parabolic approximation is valid, the electrons can be considered quasi-free electrons but with an effective mass m_{eff}, which may be positive or negative, as indicated in Fig. 3.1. A large value of the overlap integral B_λ results in a wide band and corresponding small effective mass m_{eff}.

(v) The states in the bands are filled according to the Pauli principle, beginning with the lowest states. The last completely filled band is called *valence band*. The next higher band is the *conduction band*.

There are three basic cases realized in nature:

(i) The conduction band is empty and separated by a large band gap from the valence band. This defines an *insula-*

tor. The electrons cannot be accelerated in an electric field since no empty states with slightly different E_k are available. Therefore we have no electrical conductivity.

(ii) An insulator with a relatively small band gap is called a *semiconductor*. The definition of *small band gap* is somewhat arbitrary, but a good operational definition is to say that the band gap should be on the order of or less than an optical photon energy. In semiconductors electrons can be moved relatively easily from the valence band into the conduction band, e.g., by absorption of visible or infrared light.

(iii) If the conduction band is partly filled, we have a finite electrical conductivity and hence a *metal*.

3-3. Nearly Free Electrons

Next we consider the limit of nearly free electrons, which in many respects is the opposite extreme of the tight-binding approximation. In spite of the very different nature of these two models, however, we will find a qualitatively similar band structure. In the free-electron model, we assume that the electrons move basically freely inside the crystal and the lattice potential of the ions acts only as a weak perturbation. This model is reasonable, e.g., for the conduction electrons in metals.

Here we use the free-electron wavefunctions as zero-order approximations

$$\boxed{\phi_{\mathbf{k}}(\mathbf{r}) = \frac{1}{L^{3/2}}\, e^{i\mathbf{k}\cdot\mathbf{r}}}$$

free-electron wavefunction (3.53)

and the unperturbed energies are

$$E_k = \frac{\hbar^2 k^2}{2m} \,. \tag{3.54}$$

When considering the electrons in a solid, we have to remember that the **k** values are not unique. By changing **k** to **k** + **g**, with **g** being a vector of the reciprocal lattice, we arrive at an identical situation. Hence

$$E_{|\mathbf{k}+\mathbf{g}|} = \frac{\hbar^2(\mathbf{k} + \mathbf{g})^2}{2m} \tag{3.55}$$

is another allowed unperturbed energy dispersion, as is $E_{\mathbf{k}+2\mathbf{g}}$, etc.

Since the ionic potential $V(\mathbf{r})$ has the full lattice periodicity, we can expand it as a discrete Fourier series

$$V(\mathbf{r}) = \sum_n V_n e^{i\mathbf{g}_n \cdot \mathbf{r}} \ . \tag{3.56}$$

In Fig. 3.2 we show schematically that the various possible energy eigenvalues E_k, E_{k+g}, E_{k+2g}, i.e., the energy dispersion of the free electron in the lattice has crossing points at the centers and boundaries of the Brillouin zones, indicating energy degeneracy.

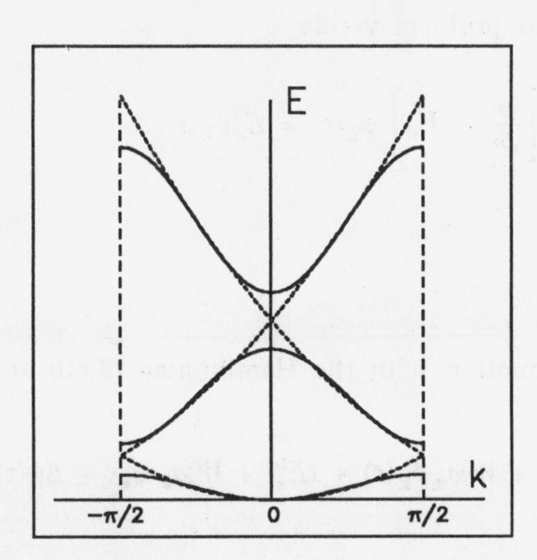

Fig. 3.2: Schematic drawing of the energy dispersion for nearly free electrons. The dashed lines show the unperturbed energy dispersion and the full lines show the modifications when the interaction with the lattice potential is included.

In order to investigate the effect of the ionic potential on the degenerate states we evaluate as an example the energy dispersion for k-values near the boundary of the Brillouin zone. Close to this boundary there are some nearly degenerate states $\mathbf{k}$ and $\mathbf{k}'$, e.g.

$$\mathbf{k}' = \mathbf{k} + k_x \mathbf{e}_x \ . \tag{3.57}$$

To analyze the influence of the ionic potential on these states, we use standard perturbation theory. A linear superposition of the nearly degenerate states is

$$\Phi(\mathbf{r}) = a_{\mathbf{k}} \phi_{\mathbf{k}}(\mathbf{r}) + a_{\mathbf{k}'} \phi_{\mathbf{k}'}(\mathbf{r}) \ , \tag{3.58}$$

where $\phi_{\mathbf{k}}(\mathbf{r})$ is given by Eq. (3.53). The Hamiltonian is

$$\begin{aligned} \mathcal{H} &= \frac{p^2}{2m} + V_0 + \sum_{m \neq 0} V_m e^{i\mathbf{g}_m \mathbf{r}} \\ &\equiv \mathcal{H}_0 + W \end{aligned} \tag{3.59}$$

and the unperturbed problem yields

$$\mathcal{H}_0 \ \phi_{\mathbf{k}}(\mathbf{r}) = \left[\frac{p^2}{2m} + V_0 \right] \phi_{\mathbf{k}}(\mathbf{r}) = E_k^0 \phi_{\mathbf{k}}(\mathbf{r}) \ , \tag{3.60}$$

where

$$E_k^0 = E_k + V_0 . \tag{3.61}$$

The Schrödinger equation with the Hamiltonian (3.60) and the ansatz (3.59) is written as

$$\mathcal{H} \Phi(\mathbf{r}) = (E_k^o + W) a_{\mathbf{k}} \phi_{\mathbf{k}}(\mathbf{r}) + (E_{k'}^o + W) a_{\mathbf{k}'} \phi_{\mathbf{k}'} = E \Phi(\mathbf{r}) \tag{3.62}$$

or

$$a_{\mathbf{k}} (E_k^o - E + W) \phi_{\mathbf{k}}(\mathbf{r}) + a_{\mathbf{k}'} (E'_k - E + W) \phi_{\mathbf{k}'}(\mathbf{r}) = 0 \tag{3.63}$$

Multiplying Eq. (3.63) by $\phi_{\mathbf{k}}^*(\mathbf{r})$, integrating $\int d^3 r$ ..., and using the orthogonality relation

$$\int d^3r \; \phi^*_{\mathbf{k}}(\mathbf{r}) \; \phi_{\mathbf{k}'}(\mathbf{r}) = \delta_{\mathbf{k}\mathbf{k}'},$$

we obtain

$$a_{\mathbf{k}}(E_k^o - E + W_{\mathbf{k}\mathbf{k}}) + a_{\mathbf{k}'}W_{\mathbf{k}\mathbf{k}'} = 0 \; . \tag{3.64}$$

The interaction matrix element is

$$W_{\mathbf{k}\mathbf{k}'} = \int d^3r \; \phi_{\mathbf{k}}^* \; W \; \phi_{\mathbf{k}'} = \sum_{m \neq 0} \frac{V_m}{L^3} \int d^3r \; e^{i(\mathbf{g}_m + k_x \mathbf{e}_x)\cdot\mathbf{r}}$$

$$\equiv V_x \; , \tag{3.65}$$

where Eq. (3.57) has been used. The diagonal element vanishes, $W_{\mathbf{k}\mathbf{k}}$=0, since $m \neq 0$. Inserting Eq. (3.65) into Eq. (3.64), we obtain

$$a_{\mathbf{k}}(E_k^o - E) + a_{\mathbf{k}'}V_x = 0 \; . \tag{3.66}$$

In the same way, by interchanging $k \leftrightarrow k'$ and noting that $W_{\mathbf{k}\mathbf{k}'} = W^*_{\mathbf{k}'\mathbf{k}}$ one finds

$$a_{\mathbf{k}'}(E_{k'}^o - E) + a_{\mathbf{k}}V_x^* = 0 \; . \tag{3.67}$$

The two coupled equations (3.66) and (3.67) have a solution only if the coefficient determinant vanishes

$$\begin{vmatrix} E_k^o - E & V_x \\ V_x^* & E_{k'}^o - E \end{vmatrix} = 0 \; , \tag{3.68}$$

i.e.,

$$(E_k^o - E)(E_{k'}^o - E) - |V_x|^2 = 0 \; , \tag{3.69}$$

or

$$E = \frac{1}{2}(E_k^o + E_{k'}^o) \pm \sqrt{\frac{1}{4}(E_k^o - E_{k'}^o)^2 + |V_x|^2} \ . \tag{3.70}$$

This result shows that the degeneracy at $k = k'$ is lifted through the finite interaction matrix element V_x. As usual for systems with crossing dispersions, the inclusion of the interaction removes the degeneracy at the crossing points and introduces a "repulsion" of the dispersion curves. As in the tight-binding calculations, also here we see the occurrence of energy bands with positive and negative curvature, which are separated by energy gaps. The qualitative features of these results are indicated as solid lines in Fig. 3.2.

In order to determine the band structure of a given crystal quantitatively, one has to use more sophisticated approximation schemes, which normally involve extensive numerical calculations.

REFERENCES

Most of the material discussed in this chapters, and much more, can be found in many solid state physics textbooks. We recommend, e.g.,

N. W. Ashcroft and N.D. Mermin, *Solid State Physics*, Saunders College (HRW), Philadelphia (1976)

C. Kittel, *Quantum Theory of Solids*, Wiley and Sons, New York (1967)

G. Bastard, *Wave Mechanics Applied to Semiconductor Heterostructures*, Les Editions de Physique, Paris (1988)

PROBLEMS

Problem 3.1: Energy bands in a solid may be discussed in the framework of a one-dimensional model with the potential

$$V(x) = -V_0 \sum_{n=-\infty}^{\infty} \delta(x+na) \ , \tag{3.71}$$

where $V_0 > 0$, n is an integer, and a is the lattice constant.

a) Determine the most general electron wavefunction ψ first in the region $0 < x < a$ and then, using Bloch's theorem, also in the region $a < x < 2a$.

b) Derive the relation between $d\psi/dx$ at $x_1 = a + \epsilon$ and $x_2 = a - \epsilon$ for $\epsilon \to 0$ by integrating the Schrödinger equation from x_1 to x_2.

c) Use the result of *b*) and $lim_{\epsilon\to 0}\ \psi(x_1) = lim_{\epsilon\to 0}\ \psi(x_2)$ to show that the energy eigenvalue $E(k)$ of an electron in the potential (3.71) satisfies the equation

$$\cos(ka) = \cos(\kappa a) - \frac{maV_0}{\hbar^2}\frac{\sin(\kappa a)}{\kappa a} \equiv f(\kappa a)\ , \tag{3.72}$$

where $\kappa = \sqrt{2me/\hbar^2}$.

d) Expand $f(\kappa a)$ around $\kappa a = n\pi - \delta$, $\delta > 0$, up to second order in δ assuming $maV_0/\hbar^2 \ll 1$ and determine the regimes in which Eq. (3.72) has no solutions (energy gaps).

e) Solve Eq. (3.72) graphically and discuss the qualitative features of the electron dispersion $E(k)$.

Problem 3.2: *a*) Prove the orthogonality relation, Eq. (3.35), for the Wannier functions.

b) Show that the Wannier function $w_\lambda(\mathbf{r}-\mathbf{n})$ is localized around the lattice point $\mathbf{n}$. Hint: Use the Bloch function, Eq. (3.26), around the band edge,

$$\psi_{\lambda,\mathbf{k}}(\mathbf{r}) \cong \frac{e^{i\mathbf{k}\cdot\mathbf{r}}}{L^{3/2}}\, u_{\lambda,0}(\mathbf{r})\ .$$

Problem 3.3: Use the effective mass approximation to calculate the electron energies

a) for a quantum well with infinitely high potential barriers in one dimension;

b) for a square quantum wire with infinitely high potential barriers in two dimensions;

c) for a square quantum dot (quantum box), in which the electrons are completely confined in all three dimensions.

Show that increasing quantum confinement causes an increasing zero-point energy due to the Heisenberg uncertainty principle.

Chapter 4
FREE CARRIER TRANSITIONS

In a typical semiconductor, the gap between the valence band and the conduction band corresponds to the energy $\hbar\omega$ of infrared or visible light. A photon with an energy $\hbar\omega > E_g$ can excite an electron from the valence band into the conduction band, leaving behind a hole in the valence band. The excited conduction-band electron and the valence-band hole carry opposite charges and interact via the mutually attractive Coulomb potential. This electron-hole Coulomb interaction will naturally influence the optical spectrum of a semiconductor. However, in order to obtain some qualitative insight, in a first approximation we disregard all the Coulomb effects and treat the electrons and holes as quasi-free particles.

4-1. Optical Susceptibility

Generally, electrons in the bands of a semiconductor are not in pure states but in so-called mixed states. Therefore, we have to extend the quantum mechanical method used to calculate the optical polarization in comparison to the treatment presented in Chap. 2. While pure states are described by wavefunctions, mixed states are described by a density matrix. In this chapter we again use the technique of Dirac state vectors $|n\mathbf{k}\rangle$ with the orthogonality relation

$$\langle n'\mathbf{k}' | n\mathbf{k}\rangle = \delta_{n',n}\, \delta_{\mathbf{k}',\mathbf{k}} \tag{4.1}$$

and the completeness relation

$$\sum_{n\mathbf{k}} |n\mathbf{k}\rangle\langle n\mathbf{k}| = 1 \ . \tag{4.2}$$

The state vectors $|n\mathbf{k}\rangle$ are eigenstates of the Hamiltonian $\mathcal{H}_0$ which now describes the electron in the periodic potential of a crystal

$$\mathcal{H}_0 |n\mathbf{k}\rangle = \hbar\epsilon_{nk} |n\mathbf{k}\rangle \ . \tag{4.3}$$

Here, the quantum number n labels the different energy bands and k denotes the quasi-momentum states. As usual, Eq. (4.3) is transformed into the Schrödinger equation in real-space representation by multiplying (4.3) from the left with the vector $\langle\mathbf{r}|$ and the Schrödinger wavefunction $\psi_{n\mathbf{k}}(\mathbf{r})$ for the state $|n\mathbf{k}\rangle$ is just the scalar product $\langle\mathbf{r}|n\mathbf{k}\rangle$.

The Hamiltonian of electrons in a crystal can be obtained in this representation by multiplying $\mathcal{H}_0$ from the left and right with the completeness relation (4.2)

$$\mathcal{H}_0 = \sum_{n'\mathbf{k}'} |n'\mathbf{k}'\rangle\langle n'\mathbf{k}'| \ \mathcal{H}_0 \sum_{n\mathbf{k}} |n\mathbf{k}\rangle\langle n\mathbf{k}| \tag{4.4}$$

Using Eqs. (4.3) and (4.1) we find the diagonal representation

$$\boxed{\mathcal{H}_0 = \hbar \sum_{n\mathbf{k}} \epsilon_{n,k} \ |n\mathbf{k}\rangle\langle n\mathbf{k}|}$$

electron Hamiltonian (4.5)

The action of this Hamiltonian on an arbitrary state vector can easily be understood. The "bra-vector" $\langle n\mathbf{k}|$ projects out that part which contains the state with the quantum numbers $n,\mathbf{k}$ represented by the "ket-vector" $|n\mathbf{k}\rangle$. The energy of this state is $\hbar\epsilon_{n,k}$.

As discussed in Chap. 2, Eq. (2.4), the dipole interaction with the light is described by

$$\mathcal{H}_I = - ex \ \mathcal{E}(t) = - d \ \mathcal{E}(t). \tag{4.6}$$

Using the completeness relation twice yields

$$\mathcal{H}_I = - \mathcal{E}(t) \sum_{n,n',\mathbf{k},\mathbf{k}'} |n'\mathbf{k}'\rangle\langle n'\mathbf{k}'| d |n\mathbf{k}\rangle\langle n\mathbf{k}| \ . \tag{4.7}$$

To simplify the problem we consider only one valence band v and one conduction band c out of the many bands of a real semiconductor, i.e., the index n can be c or v. In many cases, this two-band approximation is sufficient to calculate the optical response of real material because most of the energetically lower or higher bands are not optically active in the frequency region of interest. For the dipole matrix element we get

$$\langle c\mathbf{k}'|\, d\, |v\mathbf{k}\rangle = \delta_{\mathbf{k}',\mathbf{k}}\, d_{cv} \ . \tag{4.8}$$

Eq. (4.8) is valid in the so-called dipole approximation, where the momentum of the absorbed or emitted photon is neglected because it is small in comparison to typical values $\hbar k$ of the electron momentum. The $\mathbf{k}$-dependence of d_{cv} can often be neglected in the spectral region around the semiconductor band edge. Only if the variation of d_{cv} over the whole first Brillouin zone is needed, one has to include the $\mathbf{k}$-dependence of the dipole matrix elements.

Within these approximations the interaction Hamiltonian has only nondiagonal elements in the band index, and Eq. (4.7) reduces to

$$\mathcal{H}_I = -\,\mathcal{E}(t) \sum_{\mathbf{k},\{n\neq n'\}=\{c,v\}} d_{n',n}\, |n'\mathbf{k}\rangle\langle n\mathbf{k}| \ . \tag{4.9}$$

The optical properties of semiconductors in the band edge region are determined by the optically induced interband transitions of the electrons between valence and conduction band.

In the calculation of the optical susceptibility we will again restrict ourselves to the resonant terms. Using the rotating wave approximation we get

$$\boxed{\mathcal{H}_I = -\,\frac{\mathcal{E}_\omega}{2} e^{-i(\omega+i\delta)t} d_{cv}\, |c\mathbf{k}\rangle\langle v\mathbf{k}| \;-\; \frac{\mathcal{E}_\omega}{2} e^{i(\omega-i\delta)t} d_{vc}\, |v\mathbf{k}\rangle\langle c\mathbf{k}|}$$

interaction Hamiltonian (4.10)

where the field is assumed to be switched-on adiabatically at $t=-\infty$, as introduced in Eq. (2.11). It becomes clear that Eq. (4.10) contains only the resonant terms when we transform this expression into the interaction representation

$$\tilde{\mathscr{H}}_I(t) = e^{\frac{i}{\hbar} H_0 t} \mathscr{H}_I e^{-\frac{i}{\hbar} H_0 t}$$

$$= - \frac{\mathscr{E}_\omega}{2} \left[e^{i(\epsilon_{c,k} - \epsilon_{v,k} - (\omega + i\delta))t} d_{cv} \; |c\mathbf{k}\rangle \langle v\mathbf{k}| + \text{h.c.} \right] , \tag{4.11}$$

where h.c. denotes the Hermitian conjugate of the preceding term. The rapidly oscillating exponential term does not average out the influence of the interaction only if $\omega + \epsilon_{v,k} = \epsilon_{c,k}$.

The single-particle density matrix $\rho(t)$ can be expanded into the eigenstates $|n\mathbf{k}\rangle$

$$\boxed{\rho(t) = \sum_{m,n} \rho_{m,n}(t) \; |m\mathbf{k}\rangle \langle n\mathbf{k}|}$$

density matrix (4.12)

The equation of motion for the density matrix is the Liouville equation

$$\frac{d}{dt} \rho(t) = - \frac{i}{\hbar} [\mathscr{H}, \rho(t)] , \tag{4.13}$$

which is written in the interaction representation as

$$\frac{d}{dt} \tilde{\rho}(t) = - \frac{i}{\hbar} [\tilde{\mathscr{H}}_I, \tilde{\rho}(t)] \tag{4.14}$$

where

$$\tilde{\rho}(t) = e^{\frac{i}{\hbar} H_0 t} \rho e^{-\frac{i}{\hbar} H_0 t} . \tag{4.15}$$

Using Eq. (4.11) we get

$$\frac{d}{dt}\tilde{\rho}(t) = \frac{i}{\hbar}\frac{\mathcal{E}_\omega}{2} e^{i(\epsilon_{c,k} - \epsilon_{v,k} - (\omega + i\delta))t} d_{cv}$$

$$\times \sum_{m,n} \Big[|c\mathbf{k}\rangle\langle v\mathbf{k}|m\mathbf{k}\rangle\langle n\mathbf{k}| - |m\mathbf{k}\rangle\langle n\mathbf{k}|c\mathbf{k}\rangle\langle v\mathbf{k}| \Big] \tilde{\rho}_{mn}(t)$$

$$+ \frac{i}{\hbar}\frac{\mathcal{E}_\omega}{2} e^{-i(\epsilon_{c,k} - \epsilon_{v,k} - (\omega - i\delta))t} d^*_{cv}$$

$$\times \sum_{m,n} \Big[|v\mathbf{k}\rangle\langle c\mathbf{k}|m\mathbf{k}\rangle\langle n\mathbf{k}| - |m\mathbf{k}\rangle\langle n\mathbf{k}|v\mathbf{k}\rangle\langle c\mathbf{k}| \Big] \tilde{\rho}_{mn}(t) \ . \tag{4.16}$$

Taking the matrix element

$$\tilde{\rho}_{cv}(t) = \langle c\mathbf{k}|\, \tilde{\rho}(t) \,|v\mathbf{k}\rangle \tag{4.17}$$

of Eq. (4.16) yields

$$\frac{d}{dt}\tilde{\rho}_{cv}(t) = \frac{i}{\hbar} d_{cv} \frac{\mathcal{E}_\omega}{2} e^{i[\epsilon_{c,k} - \epsilon_{v,k} - (\omega + i\delta)]t} [\tilde{\rho}_{vv}(t) - \tilde{\rho}_{cc}(t)]. \tag{4.18}$$

Hence, the off-diagonal elements $\tilde{\rho}_{cv}$ of the density matrix for the momentum state k couple to the diagonal elements $\tilde{\rho}_{cc}$, $\tilde{\rho}_{vv}$, of the same state. As discussed in Chap. 12, a coupling between different k-values is introduced as soon as we consider also the Coulomb interaction among the carriers.

The diagonal elements of the density matrix $\tilde{\rho}_{nn} = \rho_{nn}$ give the probability to find an electron in the state $|nk\rangle$, i.e., ρ_{nn} is the population distribution of the electrons in band n. We could now proceed in our calculation and obtain dynamic equations for ρ_{cc} and ρ_{vv} by taking the respective matrix elements of Eq. (4.16). The resulting equations couple back to ρ_{cv} and we would obtain the *Bloch equations* for the free carrier system. These, and more general Bloch equations, will be discussed in later chapters of this book. For the present treatment of free carrier transitions we take a shortcut and do not compute the diagonal elements of the density matrix, but rather assume that the electrons in each band are in thermal equilibrium.

As is well known, the thermal equilibrium distribution for electrons is the Fermi distribution

$$\rho^0_{nn} = \frac{1}{e^{(\epsilon_{n,k}-\mu_n)\beta}+1} \equiv f_{n,k} \ , \tag{4.19}$$

where $\beta = 1/(k_B T)$ is the inverse thermal energy and k_B is the Boltzmann constant. The Fermi distribution and its properties are discussed in more detail in Chap. 6 of this book. For the present chapter, it is sufficient to note that the chemical potential μ_n is determined by the condition that the sum $\Sigma_{\mathbf{k}} f_{n,k}$ yields the total number of electrons N_n in band n, i.e.,

$$\sum_{\mathbf{k}} f_{n,k} = N_n \quad \rightarrow \quad \mu_n = \mu_n(N_n,T) \ , \tag{4.20}$$

where we assume that the summation over the two spin directions is included with the $\mathbf{k}$-summation. In total equilibrium and for thermal energies which are small in comparison to the band gap, the valence band is completely filled and the conduction band is empty, i.e.,

$$N_v = N, \ N_c = 0 \ \text{for} \ 1/\beta << E_g,$$

where N is the number of atoms.

It is worthwhile to mention at this point, that one can generate situations in semiconductors where $N_v < N$ and $N_c = N\text{-}N_v \neq 0$, e.g., through injection of electrons and holes in a p-n junction or through optical pumping. At the same time radiative or non-radiative recombination processes occur, such that the steady-state concentration of electrons in the conduction and valence bands is determined by the balance between the generation and recombination processes. Such a situation is called a *quasi-equilibrium state*, in which equilibrium is established only within each band but not between different bands. Usually, quasi-equilibrium is realized on a time scale which is long in comparison to the carrier thermalization times, which are typically on the order of hundreds of femtoseconds, but short in comparison to the carrier recombination time, which is of the order of nanoseconds in most direct-gap semiconductors.

Restricting ourselves to the quasi-equilibrium regime, inserting the distribution functions (4.19) into the RHS of Eq. (4.18), and integrating over time, we obtain

$$\tilde{\rho}_{cv}(t) = \frac{1}{2\hbar} \frac{d_{cv} \mathcal{E}_\omega e^{i(\epsilon_{c,k} - \epsilon_{v,k} - \omega)t}}{\epsilon_{c,k} - \epsilon_{v,k} - \omega - i\delta} (f_{v,k} - f_{c,k}) . \tag{4.21}$$

The optical polarization is given by

$$P(t) = \text{tr} \left[\rho(t) d \right] = \text{tr} \left[\tilde{\rho}(t) \tilde{d}(t) \right] , \tag{4.22}$$

where tr stands for trace, i.e., the sum over all diagonal matrix elements

$$P = \sum_{\mathbf{k}} (\tilde{\rho}_{cv} \tilde{d}_{vc} + \tilde{\rho}_{vc} \tilde{d}_{cv})$$

$$= \sum_{\mathbf{k}} \frac{|d_{cv}|^2 (f_{v,k} - f_{c,k})}{\hbar(\epsilon_{c,k} - \epsilon_{v,k} - \omega - i\delta)} \frac{\mathcal{E}_\omega}{2} e^{-i\omega t} + \text{c.c.} . \tag{4.23}$$

Using the Fourier transformation (1.13) of Eq. (1.14), we see that

$$\mathcal{P}(t) = \frac{P(t)}{L^3} = \int \frac{d\omega'}{2\pi} \chi(\omega') \mathcal{E}(\omega') e^{-i\omega' t} . \tag{4.24}$$

Since the Fourier-transform $\mathcal{E}(\omega')$ of $\mathcal{E}(t)$, given by Eq. (2.11), is

$$\mathcal{E}(\omega') = [\delta(\omega - \omega') + \delta(\omega + \omega')] \pi \mathcal{E}_\omega / 2 , \tag{4.25}$$

Eq. (4.24) becomes

$$\mathcal{P}(t) = \chi(\omega) \frac{\mathcal{E}_\omega}{2} e^{-i\omega t} + \text{c.c.} , \tag{4.26}$$

showing that the coefficient between $\mathcal{P}(t)$ and $e^{-i\omega t} \mathcal{E}_\omega / 2$ in Eq. (4.23), divided by L^3, is the optical susceptibility

$$\chi(\omega) = \frac{1}{L^3}\sum_{\mathbf{k}} \frac{|d_{cv}|^2(f_{v,k} - f_{c,k})}{\hbar(\epsilon_{c,k}-\epsilon_{v,k}-\omega-i\delta)}$$

optical susceptibility for free carriers (4.27)

4-2. Free Carrier Absorption

According to Eq. (1.53), the absorption spectrum is determined by the imaginary part of $\chi(\omega)$

$$\alpha(\omega) = \frac{4\pi\omega}{n_b c}\chi''(\omega)$$

$$= \frac{4\pi^2\omega}{L^3 n_b c}\sum_{\mathbf{k}}|d_{cv}|^2\,(f_{v,k}-f_{c,k})\;\delta(\hbar(\epsilon_{c,k}-\epsilon_{v,k}-\omega))\ . \tag{4.28}$$

Since it is possible to evaluate Eq. (4.28) independent of the dimensionality d of the electron system, we will give the result for the d-dimensional general case. The situation of d=3 describes the usual bulk semiconductor, whereas d=2 and d=1 are the idealized situations in narrow quantum-wells and quantum wires, respectively.

As discussed in Chap. 3, it is often possible to approximate the band energies $\epsilon_{c,k}$ and $\epsilon_{v,k}$ by quadratic functions around the band extrema. Unless noted otherwise, we always assume that the extrema of both bands occur at the center of the Brillouin zone, i.e. at k=0. Such semiconductors are called direct-gap semiconductors. Introducing the effective masses m_c and m_v for electrons in the conduction band and valence band, respectively, we write the energy difference as

$$\hbar(\epsilon_{c,k} - \epsilon_{v,k}) = \frac{\hbar^2k^2}{2m_c} - \frac{\hbar^2k^2}{2m_v} + E_g \tag{4.29}$$

Since the valence-band curvature is negative, we have a negative mass for the electrons in the valence band, $m_v < 0$. To avoid dealing with negative masses, one often prefers to introduce *holes* as new quasi-particles with a positive effective mass

$$m_h = - m_v . \tag{4.30}$$

In the *electron-hole representation*, one discusses electrons in the conduction band and holes in the valence band. The probability $f_{h,k}$ to have a hole at state k is given as

$$f_{h,k} = 1 - f_{v,k} \, . \tag{4.31}$$

The charge of the hole is opposite to that of the electron, i.e. $+e$. Eq. (4.30) implies that the energy of a hole is counted in the opposite way of the electron energy, i.e., the hole has minimum energy when it is at the top of the valence band. To emphasize the symmetry in our results, we rename the conduction-band mass $m_c \to m_e$, and understand from now on that the terms electron is used for conduction-band electrons and hole for valence-band holes, respectively.

In the electron-hole representation we write the energy difference as

$$\hbar(\epsilon_{c,k} - \epsilon_{v,k}) = \hbar(\epsilon_{c,k} + \epsilon_{h,k}) = \frac{\hbar^2 k^2}{2m_r} + E_g \, , \tag{4.32}$$

where

$$m_r = \frac{m_e m_h}{m_e + m_h} \tag{4.33}$$

is the reduced electron-hole mass.

In order to proceed with our evaluation of the absorption coefficient for electrons with d translational degrees of freedom, it is useful to convert the sum over $\mathbf{k}$ into an integral. Formally we can do this conversion by introducing a density of k states, since except in one dimension the k values are not distributed uniformly along the k axis. We start from the identity

$$\sum_{\mathbf{k}} \to \sum_{\mathbf{k}} \frac{(\Delta k)^d}{(\Delta k)^d} \, . \tag{4.34}$$

Periodic boundary conditions for a medium of length L imply that the allowed k-values are $k_m = 2\pi m/L$, i.e., $\Delta k_m = 2\pi/L$, and

$$(\Delta k)^d = \left[\frac{2\pi}{L}\right]^d . \tag{4.35}$$

For a sufficiently large system with dense k-values, we can make the replacement

$$\left[\frac{L}{2\pi}\right]^d \sum_{\mathbf{k}} (\Delta k)^d \to 2\left[\frac{L}{2\pi}\right]^d \int d^d k = 2\left[\frac{L}{2\pi}\right]^d \int d\Omega_d \int dk\, k^{d-1} , \tag{4.36}$$

where the integral over $d\Omega_d$ denotes the d-dimensional angle integration and the factor 2 comes from the spin summation implicitly incluced in the k-summation. If the integrand has no angle-dependence, as in Eq. (4.28), the angle integration simply yields Ω_d, where $\Omega_3 = 4\pi$, $\Omega_2 = 2\pi$, and $\Omega_1 = 1$, respectively. With these transformations, we obtain from Eq. (4.28)

$$\alpha(\omega) = \frac{8\pi^2\omega|d_{cv}|^2}{n_b c L_c^{3-d}} \frac{1}{(2\pi)^d} \Omega_d\, S(\omega) . \tag{4.37}$$

In Eq. (4.37), we have replaced the ratio L^d/L^3 by $1/L_c^{3-d}$, where L_c denotes again the length of the system in the confined space dimensions, see Chap. 3. Furthermore, we introduced

$$S(\omega) = \int_0^\infty dk\, k^{d-1} \delta\left[\frac{\hbar^2 k^2}{2m_r} + E_g + E_0^{(d)} - \hbar\omega\right] (1-f_{e,k}-f_{h,k}) . \tag{4.38}$$

In the energy-conserving δ-function we included also the zero-point energy

$$E_0^{(d)} = \frac{\hbar^2}{2m_r}\left[\frac{\pi}{L_c}\right]^2 (3 - d) \tag{4.39}$$

of the (3-d) confined directions (see problem 4.2). Taking the electron-hole-pair energy

$$\frac{\hbar^2 k^2}{2m_r} = x \tag{4.40}$$

as the integration variable, we can evaluate the integral in Eq. (4.38) with

$$k = \left[\frac{2m_r}{\hbar^2}\right]^{1/2} x^{1/2} \text{ and } dk = \frac{1}{2}\left[\frac{2m_r}{\hbar^2}\right]^{1/2} \frac{dx}{x^{1/2}} \tag{4.41}$$

as

$$S(\omega) = \frac{1}{2}\left[\frac{2m_r}{\hbar^2}\right]^{\frac{d}{2}} \int_0^\infty dx\; x^{\frac{d-2}{2}}\; \delta(x+E_g+E_0^{(d)}-\hbar\omega)\; [1-f_e(x)-f_h(x)] \;. \tag{4.42}$$

Here, we have defined

$$f_i(x) = \frac{1}{e^{\beta[(x-E_g+E_0^{(d)})m_r/m_i \;-\; \mu_i]} + 1} \quad \text{for} \quad i = e, h \;. \tag{4.43}$$

The final integral in Eq. (4.42) is easily evaluated yielding

$$S(\omega) = \frac{1}{2}\left[\frac{2m_r}{\hbar^2}\right]^{\frac{d}{2}} (\hbar\omega-E_g-E_0^{(d)})^{\frac{d}{2}-1}\; \Theta(\hbar\omega - E_g-E_0^{(d)})\; A(\omega) \;, \tag{4.44}$$

where $\Theta(x)$ is again the Heavyside unit-step function and

$$A(\omega) = 1 - f_e(\hbar\omega) - f_h(\hbar\omega) \;. \tag{4.45}$$

The factor $A(\omega)$ in (4.44) is often referred to as *band-filling factor*. Inserting the result for $S(\omega)$ into Eq. (4.37), we obtain for the absorption coefficient

$$\alpha(\omega) = \alpha_0^d\; \frac{\hbar\omega}{E_0} \left[\frac{\hbar\omega-E_g-E_0^{(d)}}{E_0}\right]^{\frac{d}{2}-1} \Theta(\hbar\omega-E_g-E_0^{(d)})\; A(\omega) \;, \tag{4.46}$$

where we have introduced the energy $E_0=\hbar^2/(2m_r a_0^2)$ and the length $a_0=\hbar^2\epsilon_0/(e^2 m_r)$ as scaling parameters, see Eqs. (10.70) and (10.71), and

$$\alpha_0^d = \frac{4\pi^2 |d_{cv}|^2}{\hbar n_b c} \frac{1}{(2\pi a_0)^d} \Omega_d \frac{1}{L_c^{3-d}} \tag{4.47}$$

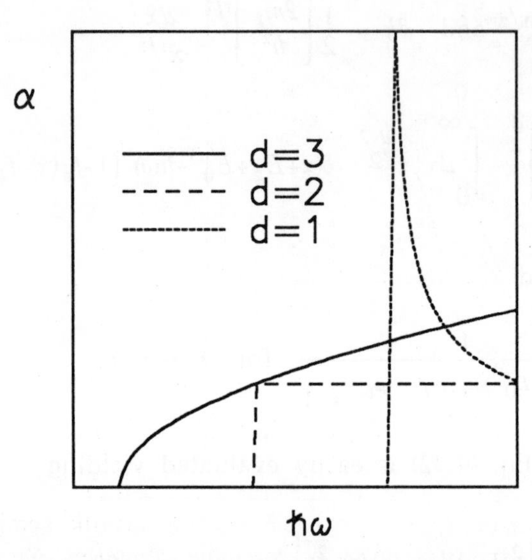

Fig. 4.1: Free electron absorption spectra for semiconductors, where the electrons can move freely in one, two, and three space dimensions.

To discuss the resulting semiconductor absorption, we first consider the case of unexcited material, where $f_e(\omega) = f_h(\omega) = 0$, i.e., $A(\omega) = 1$. The absorption spectra obtained from Eq. (4.46) for this case are plotted in Fig. 4.1. The figure shows that in two-dimensional materials the absorption sets in at $E_g + E_0^{(2)}$ like a step function, while it starts like a square root $\sqrt{\hbar\omega - E_g}$ in bulk material with d=3, and it diverges like $1/\sqrt{\hbar\omega - E_g - E_0^{(1)}}$ for d=1. The function $S(\omega)$ is just the density of states. If we would have considered not strictly two- or one-dimensional conditions, but a quantum well or quantum wire with a finite thickness, the density of states would exhibit steps corresponding to the quantization of the electron motion in the confined space dimensions. The first step, which is all what we have taken into account, belongs to the lowest eigenvalue.

Further steps occur at the energies of the correspondingly higher energy values of the excited states.

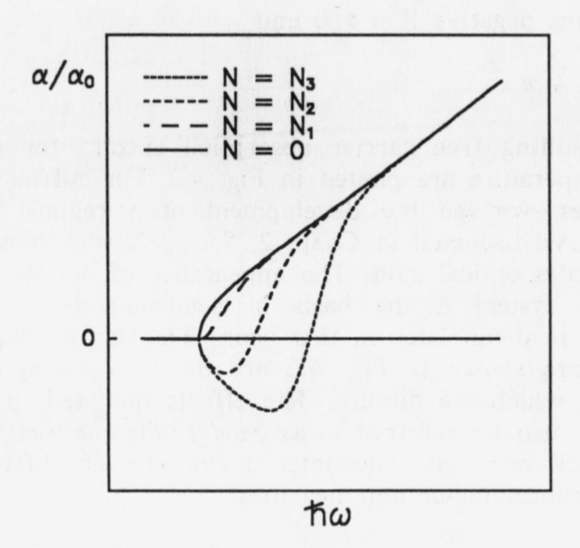

Fig. 4.2: Free electron absorption spectra for a bulk semiconductor, $d = 3$, for different electron-hole densities $N_3 > N_2 > N_1 > 0$.

As mentioned earlier, through optical pumping or injection of carriers, one may realize a situation with a finite number of electrons and holes. In this case one speaks about an excited semiconductor, where the band-filling factor $A(\omega)$, Eq. (4.45), differs from one. Using the properties of the Fermi functions, one can rewrite $A(\omega)$ as (see problem 4.1)

$$A(\omega) = \Big[(1-f_e(\omega))\ (1-f_h(\omega)) + f_e(\omega)f_h(\omega) \Big] \tanh\left[\frac{\beta}{2}(\hbar\omega - E_g - \mu)\right] , \tag{4.48}$$

where we introduced the total chemical potential μ as

$$\mu = \mu_e + \mu_h \ . \tag{4.49}$$

Since

$$0 \leq f_{e/h} \leq 1 \ , \tag{4.50}$$

we see that the prefractor of the tanh term in Eq. (4.48) is strictly positive, varying between 0.5 and 1. However, tanh(x) changes its sign at $x = 0$. The band-filling factor, and therefore the optical absorption, can become negative if $\mu > 0$ and

$$E_g < \hbar\omega < E_g + \mu \ . \tag{4.51}$$

Examples of the resulting free carrier absorption spectra for bulk material at room temperature are plotted in Fig. 4.2. For sufficiently high carrier densities, we see the development of a regime with negative absorption. As discussed in Chap. 2, Sec. 2-2, such a negative absorption indicates optical gain. The appearance of optical gain in the electron-hole system is the basis of semicondcutor lasers, which are discussed in detail later in this book. The density-dependent absorption spectra shown in Fig. 4.2 are the first example of optical nonlinearities which we discuss. The effects included in our present treatment are usually referred to as *band-filling nonlinearities*. Throughout this book we will encounter a variety of different sources for optical semiconductor nonlinearities.

REFERENCES

J.L. Pankove, *Optical Processes in Semiconductors*, Dover Publ., New York (1971)

PROBLEMS

Problem 4.1: Show that

$$1-f_e(\omega)-f_h(\omega) = \Big[(1-f_e(\omega))(1-f_h(\omega)) + f_e(\omega)f_h(\omega)\Big] \tanh\left[\frac{\beta}{2}(\hbar\omega-E_g-\mu)\right] \ .$$

Hint: Use $\tanh(x) = (e^x - e^{-x})/(e^x + e^{-x})$.

Problem 4.2: Calculate the onset of the absorption due to the second subband in a quasi-two-dimensional semiconductor quantum well.

Chapter 5
FIELD QUANTIZATION

Since we want to progress beyond the approximation of non-interacting particles, we have to resort to methods of quantum field theory. For this purpose, we review the general quantization procedure of a given classical field. As a result we obtain a formulation of quantum mechanics in terms of generation and annihilation operators, which is well suited to describe many-body systems. In this language, the transition from state ℓ to state m is described as annihilation of the respective quasi-particle in state ℓ and simultaneous creation of that quasi-particle in state m.

Formally one can also apply the quantization procedure to the one-particle Schrödinger field, which is the origin of the name *second quantization* which is often used instead of *field quantization.* In this chapter we discuss the general formalism of field quantization and illustrate it with three examples: i) the electromagnetic field, ii) the displacement field, iii) the Schrödinger field. In the process we introduce the appropriate quanta, namely, *photons* for the electromagnetic field, *phonons* for the displacement field, and *electrons* for the Schrödinger field. Anticipating the introduction of these quanta, we use their names as headings for our subsections. For the electromagnetic and displacement fields we use the ordinary commutator (with a minus sign), while for the electrons we use anti commutators. These simple choices automatically yield the correct statistics, namely Bose-Einstein statistics for the electromagnetic and displacement fields and Fermi-Dirac statistics for the electrons.

5-1. Lagrange Functional

We denote by $\phi_j(\mathbf{r},t)$ the general field variable, which may be the vector potential A_j, the displacement field ξ_j, or any other field. Generalizing Lagrangian and Hamiltonian point mechanics to the continuum case, we introduce a Lagrange functional which depends on fields according to

$$L[\phi_j] = \int d^3r \; \mathcal{L}(\phi_j, \partial\phi_j/\partial t, \partial\phi_j/\partial x_i, \mathbf{r}, t) \; . \tag{5.1}$$

The Lagrange equations (field equations) are obtained from Hamilton's principle

$$\delta \int_{t_0}^{t_1} L \; dt = 0 = \delta \int_{t_0}^{t_1} dt \int d^3r \; \mathcal{L}(\phi_j, \partial\phi_j/\partial t, \partial\phi_j/\partial x_i, \mathbf{r}, t) \; . \tag{5.2}$$

In Eq. (5.2), the variation δ acts on the fields (not on the coordinates)

$$\delta\mathcal{L} = \sum_j \left[\frac{\delta\mathcal{L}}{\delta\phi_j} \delta\phi_j + \frac{\delta\mathcal{L}}{\delta \frac{\partial\phi_j}{\partial t}} \delta\dot{\phi}_j + \sum_i \frac{\delta\mathcal{L}}{\delta\phi_j^i} \delta\phi_j^i \right] , \tag{5.3}$$

where we introduced the abbreviations $\dot{\phi} = \partial\phi/\delta t$, $\phi_j^i = \partial\phi_j/\partial x_i$. $\delta\mathcal{L}/\delta\phi_j$ denotes the functional derivative of $\mathcal{L}$ with respect to the j-th component of the vector field $\boldsymbol{\phi}$. Assuming as usual that variations and derivatives commute, we can use partial integration. Since $\delta\phi(t_0) = \delta\phi(t_1) = 0$ and since $\delta\phi$ also vanishes at the boundaries of the system, we obtain

$$\delta \int_{t_0}^{t_1} dt L = \int_{t_0}^{t_1} dt \int d^3r \left[\frac{\delta\mathcal{L}}{\delta\phi_j} - \frac{\partial}{\partial t} \frac{\delta\mathcal{L}}{\delta \frac{\partial\phi_j}{\partial t}} - \sum_i \frac{\partial}{\partial x_i} \frac{\delta\mathcal{L}}{\delta \frac{\partial\phi_j}{\partial x_i}} \right] \delta\phi_j = 0. \tag{5.4}$$

The variations $\delta\phi_j$ are arbitrary, so that Eq. (5.4) can be satisfied only if

$$\frac{\delta \mathscr{L}}{\delta \phi_j} - \frac{\partial}{\partial t} \frac{\delta \mathscr{L}}{\delta \dfrac{\partial \phi_j}{\partial t}} - \sum_i \frac{\partial}{\partial x_i} \frac{\delta \mathscr{L}}{\delta \dfrac{\partial \phi_j}{\partial x_i}} = 0 \ . \tag{5.5}$$

These equations are the Lagrange equations (field equations).

i) **photons**

Photons are the quanta of the electromagnetic field. To keep things as simple as possible, we restrict our discussion to the electromagnetic field in vacuum. For this we have

$$\int dt \ L = \int dt \int d^3r \ \frac{1}{8\pi} (\mathscr{E}^2 - H^2) \ , \tag{5.6}$$

where

$$\mathbf{H} = \nabla \times \mathbf{A} = \text{curl } \mathbf{A} \ , \tag{5.7}$$

$\mathbf{A}$ is the vector potential,

$$\mathscr{E} = -\nabla\phi - \frac{1}{c} \frac{\partial \mathbf{A}}{\partial t} \ , \tag{5.8}$$

and ϕ is the scalar potential. Using Eqs. (5.7) and (5.8) in Eq. (5.6), we write the Lagrangian density as

$$\mathscr{L} = \frac{1}{8\pi} \left[\left[\nabla\phi + \frac{1}{c} \frac{\partial \mathbf{A}}{\partial t} \right]^2 - (\text{curl } \mathbf{A})^2 \right] . \tag{5.9}$$

The components of the vector potential A_i and the scalar potential ϕ are the elements of the general field variable $\phi_i(\mathbf{r},t)$ discussed above, i.e.,

$$\phi_i = A_i \quad \text{for} \quad i\text{=1, 2, 3} \quad \text{and} \quad \phi_4 = \phi \ . \tag{5.10}$$

To obtain the field equations, we need the results

$$\frac{\delta \mathscr{L}}{\delta A_j} = \frac{\delta \mathscr{L}}{\delta \phi} = \frac{\delta \mathscr{L}}{\delta \dfrac{\partial \phi}{\partial t}} = 0 \, . \tag{5.11}$$

$$\frac{\delta \mathscr{L}}{\delta \dfrac{\partial A_j}{\partial t}} = \frac{1}{4\pi c} \left[\frac{\partial \phi}{\partial x_j} + \frac{1}{c} \frac{\partial A_j}{\partial t} \right] \tag{5.12}$$

$$\frac{\delta \mathscr{L}}{\delta \dfrac{\partial \phi}{\partial x_i}} = \frac{1}{4\pi} \left[\frac{\partial \phi}{\partial x_i} + \frac{1}{c} \frac{\partial A_i}{\partial t} \right] \tag{5.13}$$

$$\begin{aligned}
\frac{\delta \mathscr{L}}{\delta \dfrac{\partial A_j}{\partial x_\ell}} &= - \frac{\delta}{\delta \dfrac{\partial A_j}{\partial x_\ell}} \sum \epsilon_{imk} \frac{\partial A_k}{\partial x_m} \epsilon_{iqn} \frac{\partial A_n}{\partial x_q} \frac{1}{8\pi} \\
&= - \frac{1}{8\pi} \sum \epsilon_{i\ell j} \, \epsilon_{iqn} \frac{\partial A_n}{\partial x_q} - \frac{1}{8\pi} \sum \epsilon_{imk} \frac{\partial A_k}{\partial x_m} \epsilon_{i\ell j} \\
&= - \frac{1}{4\pi} \sum \epsilon_{i\ell j} \epsilon_{iqn} \frac{\partial A_n}{\partial x_q} \\
&= - \frac{1}{4\pi} \left[\frac{\partial A_j}{\partial x_\ell} - \frac{\partial A_\ell}{\partial x_j} \right] .
\end{aligned} \tag{5.14}$$

In Eq. (5.14) we have used the so-called Levicivita tensor ϵ_{ijk} which has the following properties

$$\epsilon_{123} = 1 \, , \; \epsilon_{213} = -1 \, , \; \epsilon_{iij} = 0 \, , \tag{5.15}$$

where *cyclic permutations* of the indices do not change the result, i.e., $\epsilon_{123} = \epsilon_{312} = \epsilon_{231}$, etc. For the product of the tensors we have the rule

$$\sum_i \epsilon_{ij\ell} \epsilon_{imn} = \delta_{jm} \delta_{\ell n} - \delta_{jn} \delta_{\ell m} \, . \tag{5.16}$$

The Levicivita tensor is very useful when evaluating vector products, curls and the like. For example, the curl of a vector is

$$\text{curl } \mathbf{A} = \sum \mathbf{e}_i \frac{\partial}{\partial x_j} A_\ell \, \epsilon_{ij\ell} \ . \tag{5.17}$$

Combining Eqs. (5.9) - (5.14) with the Lagrange equations, we get

$$-\frac{1}{4\pi}\left[\sum \frac{\partial}{\partial x_i}\left[\frac{\partial A_j}{\partial x_i} - \frac{\partial A_i}{\partial x_j}\right] - \frac{1}{c}\frac{\partial}{\partial t}\left[\frac{\partial \phi}{\partial x_j} + \frac{1}{c}\frac{\partial A_j}{\partial t}\right]\right] = 0 \tag{5.18}$$

or, using Eqs. (5.7) and (5.8)

$$\text{curl } \mathbf{H} = \frac{1}{c}\frac{\partial}{\partial t}\mathcal{E} \ , \tag{5.19}$$

which is one of Maxwell's equations. Analogously we obtain

$$\sum \frac{\partial}{\partial x_i}\left[\frac{\partial \phi}{\partial x_i} + \frac{1}{c}\frac{\partial A_i}{\partial t}\right] = 0 \ , \tag{5.20}$$

which is nothing but

$$\text{div } \mathcal{E} = 0 \ . \tag{5.21}$$

The other two Maxwell equations are just the definitions of the potentials. Equation (5.8) yields

$$\text{curl } \mathcal{E} = -\frac{1}{c}\frac{\partial \mathrm{H}}{\partial t} \ , \tag{5.22}$$

since

$$\nabla \times \mathcal{E} = \nabla \times \left[-\nabla\phi - \frac{1}{c}\frac{\nabla \mathbf{A}}{\partial t}\right] = -\frac{1}{c}\frac{\partial}{\partial t}\nabla \times \mathbf{A} \tag{5.23}$$

and from Eq. (5.7) we have

$$\text{div } \mathbf{H} = 0 \tag{5.24}$$

since

$$\nabla \cdot (\nabla \times \mathbf{A}) = 0 \ . \tag{5.25}$$

Equations (5.19), (5.21), (5.22) and (5.24) are the complete Maxwell equations in vacuum, which occur here as our field equations.

ii) **phonons**

As second example we treat waves in an elastic medium. The quanta of these waves will turn out to be the phonons. We denote by $\xi_i(\mathbf{r},t)$ the i-th component of the displacement field at point $\mathbf{r}$ and time t. The displacement field satisfies the classical wave equation

$$\frac{\partial^2}{\partial t^2}\xi_i - c^2\nabla^2\xi_i = 0 \ . \tag{5.26}$$

The correct Lagrangian, whose field equation yields (5.26), is

$$\mathscr{L} = \frac{1}{2}\rho\left[\sum_i\left[\frac{\partial\xi_i}{\partial t}\right]^2 - \sum_{ij}c^2\left[\frac{\partial\xi_i}{\partial x_j}\right]^2\right], \tag{5.27}$$

where the density ρ appears so that $\mathscr{L}$ has the correct unit of an energy density. We arrived at Eq. (5.27) by an educated guess and by verifying that the field equations are indeed just Eq. (5.26).

iii) **electrons**

As we see below, the electrons appear through the quantization of the Schrödinger equation. This is the origin of the name *second quantization*, although the name *field quantization* is generally more appropriate. Here we consider the single-particle Schrödinger equation as our classical wave equation

$$i\hbar\frac{\partial\psi}{\partial t} - \mathscr{H}_{sch}\,\psi = 0 \ , \tag{5.28}$$

where

$$\mathcal{H}_{sch} = -\frac{\hbar^2\nabla^2}{2m} + V(\mathbf{r}) \tag{5.29}$$

and ψ is the complex wavefunction. We treat ψ and ψ^* as the quantities that have to be varied independently. Our educated guess for the Lagrangian is

$$\mathcal{L} = \psi^* \, (i\hbar\dot{\psi} - \mathcal{H}_{sch}\psi) \,, \tag{5.30}$$

which can be verified by evaluating Eq. (5.5). The variation with respect to ψ^* yields

$$\frac{\delta\mathcal{L}}{\delta\psi^*} - \frac{\partial}{\partial t}\frac{\delta\mathcal{L}}{\delta\dfrac{\partial\psi^*}{\partial t}} - \sum\frac{\partial}{\partial x_i}\frac{\delta\mathcal{L}}{\delta\dfrac{\partial\psi^*}{\partial x_i}} = 0 \,. \tag{5.31}$$

The last two terms on the LHS of this equation are identically zero and

$$\frac{\delta\mathcal{L}}{\delta\psi^*} = 0 = i\hbar\,\frac{\partial\psi}{\partial t} - \mathcal{H}_{sch}\psi \,, \tag{5.32}$$

which is the Schrödinger equation (5.28). The variation with respect to ψ is more lengthy, but it can be shown to yield the conjugate complex Schrödinger equation

$$-i\hbar\,\frac{\partial\psi^*}{\partial t} - \mathcal{H}_{sch}\psi^* = 0 \,. \tag{5.33}$$

5-2. Canonical Momentum and Hamilton Function

The next step in our general procedure is to introduce the canonical momentum and the Hamilton functional. As in point mechanics, the canonical momentum $\Pi_i(\mathbf{r},t)$ for $\phi_i(\mathbf{r},t)$ is defined as

$$\Pi_i = \frac{\delta\mathcal{L}}{\delta\dfrac{\partial\phi_i}{\partial t}} \tag{5.34}$$

and the Hamilton density is defined as

$$\hbar = \sum_i \Pi_i \frac{\partial \phi_i}{\partial t} - \mathscr{L} , \tag{5.35}$$

where all variables have to be expressed in terms of Π and ϕ.

i) **photons**

Using the Lagrangian (5.9) in Eq. (5.34), we obtain

$$\Pi_i = \frac{\delta \mathscr{L}}{\delta \dfrac{\partial A_i}{\partial t}} = \frac{1}{4\pi c}\left[\frac{\partial \phi}{\partial x_i} + \frac{1}{c}\frac{\partial A_i}{\partial t} \right], \tag{5.36}$$

so that $\Pi_i = - \mathscr{E}/4\pi c$. Solving (5.36) for $\partial A_i/\partial t$ yields

$$\frac{\partial A_i}{\partial t} = 4\pi c^2 \Pi_i - c\,\frac{\partial \phi}{\partial x_i} . \tag{5.37}$$

The momentum density canonical to the scalar potential ϕ vanishes identically because $\mathscr{L}$ does not depend on $\partial\phi/\partial t$. Therefore, we cannot treat ϕ in the same way as $\mathbf{A}$. Actually, as we will show below, the Hamilton density is independent of ϕ. Therefore, we can completely eliminate ϕ from the discussion by choosing the gauge ϕ=0.

The Hamilton density is

$$\hbar = \sum \Pi_i \left[4\pi c^2 \Pi_i - c\,\frac{\partial \phi}{\partial x_i} \right] - \frac{1}{8\pi}\left[(4\pi c\Pi)^2 - (\text{curl } \mathbf{A})^2\right]$$

$$= 2\pi c^2 \Pi^2 + \frac{1}{8\pi}(\text{curl } \mathbf{A})^2 , \tag{5.38}$$

since the term $c\Pi\cdot\nabla\phi$ vanishes. To prove this, we make a partial integration of the total Hamiltonian. This transforms the $c\Pi\cdot\nabla\phi$ into $-c\phi\nabla\cdot\Pi$. We can drop this term if we limit the treatment to solutions for which $\nabla\cdot\Pi = \nabla\cdot\mathscr{E} = 0$. To see that this is possible we consider the Hamilton equation for Π,

$$\frac{\partial \Pi}{\partial t} = - \frac{1}{4\pi} \nabla \times \nabla \times \mathbf{A} = 0. \tag{5.39}$$

This equation shows that $\nabla \cdot \Pi$ always remains zero, if we choose it to be zero initially.

ii) **phonons**

Using the Lagrangian (5.27) in Eq. (5.34), we obtain the canonical momentum

$$\Pi_i = \frac{\delta \mathscr{L}}{\delta \dfrac{\partial \xi_i}{\partial t}} = \rho \frac{\partial \xi_i}{\partial t} . \tag{5.40}$$

This result is the continuum version of the well-known relation from point mechanics, where $p = m\dot{x}$ The Hamiltonian is computed as

$$h = \frac{\Pi^2}{2\rho} + \frac{\rho c^2}{2} \sum_{ik} \left[\frac{\partial \xi_i}{\partial x_k} \right]^2 , \tag{5.41}$$

where the first term on the RHS is again just the continuum version of $p^2/2m$ and the second term is the potential.

iii) **electrons**

Here we use the Lagrangian (5.30) to obtain

$$\Pi = i\hbar\psi^* \tag{5.42}$$

and

$$\begin{aligned} h &= \Pi \dot{\psi} - \frac{1}{i\hbar} \Pi i\hbar\dot{\psi} + \frac{1}{i\hbar} \Pi \mathscr{H}_{sch}\psi \\ &= \frac{1}{i\hbar} \Pi \mathscr{H}_{sch}\psi . \end{aligned} \tag{5.43}$$

5-3. Quantization of the Fields

Now we have prepared everything to come to the crucial step of field quantization. First a reminder: the transition from classical mechanics to quantum mechanics can be done formally by replacing the classical variables $\mathbf{r}$ and $\mathbf{p}$ by operators $\hat{\mathbf{r}}$ and $\hat{\mathbf{p}}$ that fulfill the commutation relations

$$\begin{aligned}
[\hat{r}_\ell, \hat{p}_j] &= i\hbar\delta_{\ell j} \\
[\hat{r}_\ell, \hat{r}_j] &= 0 \\
[\hat{p}_\ell, \hat{p}_j] &= 0 \ ,
\end{aligned} \tag{5.44}$$

where here and in the following a commutator without a subscript always denotes the *minus commutator*

$$[A,B] \equiv [A,B]_- = AB - BA \ . \tag{5.45}$$

In this chapter, we use hats on top of the variables to denote operators. This notation is used throughout the following chapters. In later parts of this book we keep the hats only when it is necessary to distinguish between operators and the corresponding classical quantities or expectation values. Otherwise it is obvious from the context when we are discussing operators.

The field quantization is now done by replacing the fields $\phi_i \to \hat{\phi}_i$, $\Pi_i \to \hat{\Pi}_i$ and the Hamiltonian $\mathscr{H} \to \hat{\mathscr{H}}$. For the field operators, we demand the commutation relations

$$\begin{aligned}
[\hat{\phi}_\ell(\mathbf{r},t), \hat{\Pi}_j(\mathbf{r}',t)]_\pm &= i\hbar\delta_{\ell j}\delta(\mathbf{r} - \mathbf{r}') \\
[\hat{\phi}_\ell(\mathbf{r},t), \hat{\phi}_j(\mathbf{r}',t)]_\pm &= 0 \\
[\hat{\Pi}_\ell(\mathbf{r},t), \hat{\Pi}_j(\mathbf{r}',t)]_\pm &= 0 \ ,
\end{aligned} \tag{5.46}$$

where

$$[A,B]_+ = AB + BA \tag{5.47}$$

and $[A,B]_-$ is given by Eq. (5.45). Both commutators occur in nature, the minus commutator for Bosons like photons and phonons, and the plus-commutator, usually called the *anti-commutator*, for Fermions, e.g., electrons. By choosing the correct type of commutator, we automatically get the right quantized field theory.

i) **photons**

In the ϕ=0 gauge we have $\Pi_i(r,t) = 1/4\pi c^2\ \partial A_i/\partial t$ and $A_i(\mathbf{r},t)$ as canonical variables. After quantization, the commutation relations are

$$[\hat{A}_\ell(\mathbf{r},t),\ \hat{\Pi}_j(\mathbf{r}',t)]_- = i\hbar\delta_{\ell j}\delta(\mathbf{r}-\mathbf{r}') \ , \tag{5.48}$$

and the Hamilton operator becomes

$$\hat{\mathscr{H}} = \int d^3r \left[2\pi c^2\hat{\Pi}^2 + \frac{1}{8\pi}\,(\mathrm{curl}\ \hat{A})^2\right] . \tag{5.49}$$

For many applications we want to work in the Coulomb gauge

$$\nabla\cdot\mathbf{A} = 0 \ , \tag{5.50}$$

where the electromagnetic field is transverse. In this gauge, the commutator between $\mathbf{A}$ and Π is no longer given by Eq. (5.48) because the individual components of these operators are connected by $\nabla\cdot\mathbf{A} = 0$ and $\nabla\cdot\Pi = 0$. In the Coulomb gauge, the commutator (5.48) has to be replaced by

$$[\hat{A}_\ell(\mathbf{r},t),\ \hat{\Pi}_j(\mathbf{r}',t)]_- = i\hbar\delta^t_{\ell j}(\mathbf{r}-\mathbf{r}') \ , \tag{5.51}$$

see Schiff (1968). Here, the transverse δ-function is defined as

$$\delta^t_{\ell j}(\mathbf{r}-\mathbf{r}') = \delta_{\ell j}\delta(\mathbf{r}-\mathbf{r}') - \frac{1}{4\pi}\frac{\partial}{\partial x_\ell}\frac{\partial}{\partial x_j}\frac{1}{|\mathbf{r}-\mathbf{r}'|} \ . \tag{5.52}$$

In order to introduce photon operators (in vacuum), we expand the field operators in terms of plane waves

$$\mathbf{u}_{\lambda\mathbf{k}} = \frac{e^{i\mathbf{k}\cdot\mathbf{r}}}{L^{3/2}}\,\epsilon_{\lambda\mathbf{k}} \ , \tag{5.53}$$

where $\epsilon_{\lambda\mathbf{k}}$ is the polarization unit vector, such that

$$\int d^3r \; \mathbf{u}^*_{\lambda\mathbf{k}}(\mathbf{r}) \; \mathbf{u}_{\lambda'\mathbf{k}'}(\mathbf{r}) = \delta_{\lambda\lambda'} \; \delta_{\mathbf{k},\mathbf{k}'} \tag{5.54}$$

and

$$\sum_{\lambda=1,2} \epsilon_{\lambda\mathbf{k},\ell} \; \epsilon_{\lambda\mathbf{k},j} + \frac{k_\ell k_j}{k^2} = \delta_{\ell,j}. \tag{5.55}$$

Since the electromagnetic field in Coulomb gauge is transverse, we know that $\mathbf{k}$ is perpendicular to $\epsilon_{\lambda k}$, implying that $\nabla\cdot\mathbf{u} = 0$. Note that if we want to introduce photons in a resonator we use resonator eigenmodes for the expansion instead of plane waves. The expansion of the field operators is

$$\hat{A}(\mathbf{r},t) = \sum_{\mathbf{k},\lambda} [\hat{b}_{\lambda\mathbf{k}} \; \mathbf{u}_{\lambda\mathbf{k}}(\mathbf{r}) + \hat{b}^\dagger_{\lambda\mathbf{k}} \; \mathbf{u}^*_{\lambda\mathbf{k}}(\mathbf{r})] B_{\lambda k} \tag{5.56}$$

$$\hat{\Pi}(\mathbf{r},t) = \frac{1}{4\pi c^2} \frac{\partial \hat{A}}{\partial t} = \sum_{\mathbf{k},\lambda} \left[\frac{\partial \hat{b}_{\mathbf{k}\lambda}}{\partial t} \mathbf{u}_{\lambda\mathbf{k}} + \frac{\partial \hat{b}^\dagger_{\mathbf{k}\lambda}}{\partial t} \mathbf{u}^*_{\lambda\mathbf{k}} \right] \frac{B_{\lambda k}}{4\pi c^2} , \tag{5.57}$$

where the expansion coefficients b, $b^\dagger$ are the photon operators. We determine the quantity $B_{\lambda k}$ such that the photon operators fulfill simple commutation relations.

To evaluate the time derivative of the photon operators, we make use of the fact that the vector potential $\mathbf{A}$ satisfies the wave equation

$$\left[\frac{\partial^2}{\partial t^2} - c^2\nabla^2 \right] \hat{\mathbf{A}} = 0 . \tag{5.58}$$

Inserting the expansion (5.56), we obtain

$$\frac{\partial^2}{\partial t^2} \hat{b}_{\lambda\mathbf{k}} = -\omega_k^2 \hat{b}_{\lambda\mathbf{k}}$$

$$\frac{\partial^2}{\partial t^2}\,\hat{b}^\dagger_{\lambda\mathbf{k}} = -\omega_k^2\hat{b}^\dagger_{\lambda\mathbf{k}}\,, \tag{5.59}$$

where $\omega_k = ck$. Equation (5.59) is fulfilled if

$$\frac{\partial}{\partial t}\,\hat{b}_{\lambda\mathbf{k}} = -i\omega_k\hat{b}_{\lambda\mathbf{k}}$$

$$\frac{\partial}{\partial t}\,\hat{b}^\dagger_{\lambda\mathbf{k}} = i\omega_k\hat{b}^\dagger_{\lambda\mathbf{k}}\,. \tag{5.60}$$

Inserting Eq. (5.60) into Eq. (5.57) yields

$$\hat{\Pi}(\mathbf{r},t) = -\frac{1}{4\pi c^2}\sum i\omega_k B_{\lambda k}(\hat{b}_{\lambda\mathbf{k}}\mathbf{u}_{\lambda\mathbf{k}} - \hat{b}^\dagger_{\lambda\mathbf{k}}\mathbf{u}^*_{\lambda\mathbf{k}})\,. \tag{5.61}$$

To determine the commutation relations of the photon operators, we insert the expansions (5.56) and (5.61) into the commutator (5.48) to obtain

$$\sum_{\substack{\lambda,k\\ \lambda',k'}}\frac{-iB_{\lambda k}B_{\lambda' k'}\omega_{k'}}{4\pi c^2}\left[\left(\hat{b}_{\lambda\mathbf{k}}u_{\lambda\mathbf{k},\ell}(\mathbf{r}) + \hat{b}^\dagger_{\lambda\mathbf{k}}u^*_{\lambda\mathbf{k},\ell}(\mathbf{r})\right),\right.$$

$$\left.\left(\hat{b}_{\lambda'\mathbf{k}'}u_{\lambda'\mathbf{k}',j}(\mathbf{r}') - \hat{b}^\dagger_{\lambda'\mathbf{k}'}u^*_{\lambda'k,'j}(\mathbf{r})\right)\right] = i\hbar\delta_{\ell j}\delta(\mathbf{r}-\mathbf{r}')\,. \tag{5.62}$$

This commutator is satisfied if

$$[\hat{b}_{\lambda\mathbf{k}},\,\hat{b}^\dagger_{\lambda'\mathbf{k}'}] = \delta_{\lambda\lambda'}\delta_{\mathbf{k}\mathbf{k}'}$$

$$[\hat{b}_{\lambda\mathbf{k}},\,\hat{b}_{\lambda'\mathbf{k}'}] = 0 = [\hat{b}^\dagger_{\lambda\mathbf{k}},\,\hat{b}^\dagger_{\lambda'\mathbf{k}'}] \tag{5.63}$$

and the normalization constant is choosen as

$$B_{\lambda k} = \left[\frac{2\pi c^2\hbar}{\omega_k}\right]^{1/2}\,. \tag{5.64}$$

Summarizing these results, we have

$$\hat{\mathbf{A}}(\mathbf{r},t) = \sum_{\lambda\mathbf{k}} \sqrt{\frac{2\pi c^2\hbar}{\omega_k}}\,(\hat{b}_{\lambda\mathbf{k}}\,\mathbf{u}_{\lambda\mathbf{k}}(\mathbf{r}) + \hat{b}^{\dagger}_{\lambda\mathbf{k}}\,\mathbf{u}_{\lambda\mathbf{k}}{}^{*}(\mathbf{r}))\;, \tag{5.65}$$

$$\hat{\Pi}(\mathbf{r},t) = -\,i\sum_{\lambda\mathbf{k}} \sqrt{\frac{\hbar\omega_k}{8\pi c^2}}\,(\hat{b}_{\lambda\mathbf{k}}\,\mathbf{u}_{\lambda\mathbf{k}}(\mathbf{r}) - \hat{b}^{\dagger}_{\lambda\mathbf{k}}\,\mathbf{u}_{\lambda\mathbf{k}}{}^{*}(\mathbf{r}))\;, \tag{5.66}$$

and the Hamiltonian

$$\hat{\mathscr{H}} = \sum \frac{\hbar\omega_k}{2}\,(\hat{b}^{\dagger}_{\lambda\mathbf{k}}\hat{b}_{\lambda\mathbf{k}} + \hat{b}_{\lambda\mathbf{k}}\hat{b}^{\dagger}_{\lambda\mathbf{k}}) = \sum \hbar\omega_k(\hat{b}^{\dagger}_{\lambda\mathbf{k}}\hat{b}_{\lambda\mathbf{k}} + 1/2)\;. \tag{5.67}$$

ii) **phonons**

Here we quantize the canonical variables into the field operators $\hat{\xi}_i(\mathbf{r},t)$ and $\hat{\Pi}_i(\mathbf{r},t) = \rho\,\partial\hat{\xi}_i/\partial t$ with the commutations relations

$$[\hat{\xi}_j(\mathbf{r},t),\,\hat{\Pi}_\ell(\mathbf{r}',t)]_- = i\hbar\delta_{j\ell}\delta(\mathbf{r}-\mathbf{r}')\;. \tag{5.68}$$

Again we expand the field operators

$$\hat{\xi}(\mathbf{r},t) = \sum [\hat{b}_{\mathbf{k}}\mathbf{u}_{\mathbf{k}}(\mathbf{r}) + \hat{b}^{\dagger}_{\mathbf{k}}\mathbf{u}_k{}^{*}(\mathbf{r}')]B_k \tag{5.69}$$

and similarly for $\hat{\Pi}$. Choosing the functions $\mathbf{u}$ as plane waves

$$\mathbf{u}_{\mathbf{k}}(\mathbf{r}) = \epsilon_k\,\frac{e^{i\mathbf{kr}}}{\sqrt{V}}\;,$$

where $\epsilon_k = \mathbf{k}/|\mathbf{k}|$ and the normalization constant as $B_k = \sqrt{\hbar/2\rho\omega_k}$, we find the commutation relation between the phonon operators

$$[\hat{b}_{\mathbf{k}}, \hat{b}^{\dagger}_{\mathbf{k}'}] = \delta_{\mathbf{k},\mathbf{k}'} , \tag{5.70}$$

and the Hamilton operator becomes

$$\hat{\mathcal{H}} = \sum \hbar\omega_k \, (\hat{b}^{\dagger}_{\mathbf{k}}\hat{b}_{\mathbf{k}} + 1/2) . \tag{5.71}$$

iii) **electrons**

Here we introduce the field operators $\hat{\psi}$ and $\hat{\Pi} = i\hbar\hat{\psi}^*$, and we use the Fermi anti-commutation relations

$$[\hat{\psi}, \hat{\psi}]_+ = 0 = [\hat{\psi}^{\dagger}, \hat{\psi}^{\dagger}]_+ \tag{5.72}$$

and

$$[\hat{\psi}(\mathbf{r},t), \hat{\Pi}(\mathbf{r}',t)]_+ = i\hbar\delta(\mathbf{r} - \mathbf{r}') \tag{5.73}$$

which is equivalent to

$$[\hat{\psi}(\mathbf{r},t), \hat{\psi}^{\dagger}(\mathbf{r}',t)]_+ = \delta(\mathbf{r} - \mathbf{r}') . \tag{5.74}$$

For the Hamilton density we had

$$\hat{h} = \frac{1}{i\hbar} \hat{\Pi}(\mathbf{r}) \, \mathcal{H}_{sch}\hat{\psi}(\mathbf{r}) = \hat{\psi}^{\dagger}(\mathbf{r}) \, \mathcal{H}_{sch}\hat{\psi}(\mathbf{r}) , \tag{5.75}$$

which yields the Hamilton operator of a noninteracting electron system as

$$\hat{\mathcal{H}} = \int d^3r \, \hat{\psi}^{\dagger}(\mathbf{r}) \, \mathcal{H}_{sch}\hat{\psi}(\mathbf{r}) . \tag{5.76}$$

The derivation of the Hamiltonian for an interacting electron system is discussed, e.g., in the textbook of Davidov. Here we only want to mention that one has to start from the N-particle Schrödinger Hamiltonian which is then transformed into the Fock representation. One finally obtains

$$\hat{\mathcal{H}} = \int d^3r \; \hat{\psi}^\dagger(\mathbf{r}) \; \mathcal{H}_{sch} \hat{\psi}(\mathbf{r})$$

$$+ \frac{1}{2} \int d^3r \int d^3r' \; \hat{\psi}^\dagger(\mathbf{r}) \; \hat{\psi}^\dagger(\mathbf{r}') \; V(\mathbf{r},\mathbf{r}') \; \hat{\psi}(\mathbf{r}') \; \hat{\psi}(\mathbf{r}) \; . \tag{5.77}$$

This expression shows that the interaction term has a similar structure as the single-particle term, but instead of $\mathcal{H}_{sch}$ the electron density operator $\hat{\psi}^\dagger(\mathbf{r}) \; \hat{\psi}(\mathbf{r})$ times the potential V appears. The whole interaction term is thus the product of the density operators at $\mathbf{r}$ and $\mathbf{r}'$ multiplied by the pair interaction potential. The factor 1/2 appears to avoid double counting.

We now expand the field operators into the eigenfunctions ϕ_n of the single-particle Schrödinger Hamiltonian

$$\mathcal{H}_{sch} \; \phi_n = E_n \; \phi_n \; , \tag{5.78}$$

such that

$$\hat{\psi}(\mathbf{r},t) = \sum \hat{a}_n(t) \; \phi_n(\mathbf{r}) \; . \tag{5.79}$$

These eigenfunctions could be any complete set, such as the Bloch or Wannier functions. It is important to choose the appropriate set for the problem at hand. The electron operators obey the anti-commutation relation

$$[\hat{a}_n, \, \hat{a}_n^\dagger]_+ = \delta_{n,m} \; , \tag{5.80}$$

and we obtain the single-particle Hamilton operator

$$\hat{\mathcal{H}} = \sum_n E_n \; \hat{a}_n^\dagger \; \hat{a}_n \; . \tag{5.81}$$

Similarly, from Eq. (5.77) we get the many-body Hamiltonian

$$\hat{\mathcal{H}} = \sum_n E_n \hat{a}_n^\dagger \hat{a}_n + \frac{1}{2} \sum V_{m'n'mn} \hat{a}_{m'}^\dagger \hat{a}_{n'}^\dagger \hat{a}_m \hat{a}_n , \tag{5.82}$$

where

$$V_{m'n'mn} = \langle m'n' | V | nm \rangle .$$

is the matrix element of the interaction potential.

REFERENCES

Field quantization is treated in most quantum mechanics textbooks. See, e.g.,

C. Cohen-Tannoudji, J. Dapont-Roc, and G. Grynberg, *Photons and Atoms*, Wiley, New York (1989)

A.S. Davydov, *Quantum Mechanics*, Pergamon, New York (1965)

L.I. Schiff, *Quantum Mechanics*, McGraw-Hill, New York (1968)

PROBLEMS

Problem 5.1: Consider a linear chain of atoms with masses M and interatomic distance a. The coupling between the atoms is given by a harmonic force with the force constant K.

a) Show that the Lagrange function is

$$L = \frac{1}{2} M \sum_{r=1}^{N} \left[\frac{dq_r}{dt}\right]^2 - \frac{1}{2} K \sum_{r=1}^{N} (q_r - q_{r+1})^2 \tag{5.83}$$

where q_r is the displacement of the r-th atom from its equilibrium position.

b) Compute the canonical momentum p_r.

c) The displacements can be expanded into *normal coordinates* Q_k

$$q_r(t) = \frac{1}{\sqrt{N}} \sum_k Q_k(t)\, e^{ikar}. \tag{5.84}$$

Use

$$\sum_r e^{iar(k-k')} = N\, \delta_{kk'} \tag{5.85}$$

and the periodic boundary conditions $q_{r+N} = q_r$ to determine the allowed k-values.

d) The displacements are quantized by introducing the commutation relations

$$[\, \hat{p}_r \,,\, \hat{q}_s \,] = \frac{\hbar}{i}\, \delta_{rs}$$

$$[\, \hat{p}_r \,,\, \hat{p}_s \,] = 0 = [\, \hat{q}_r \,,\, \hat{q}_s \,]\,. \tag{5.86}$$

Use the fact that the displacement is a Hermitian operator

$$\hat{q}_r^\dagger = \hat{q}_r \tag{5.87}$$

to show the relations

$$\hat{Q}_k^\dagger = \hat{Q}_{-k}\ ; \tag{5.88}$$

$$\hat{P}_k \equiv \frac{\partial \hat{\mathscr{L}}}{\partial\, (\partial \hat{Q}_k / \partial t)} = \hat{P}_{-k}^\dagger\ ; \tag{5.89}$$

$$[\, \hat{Q}_k \,,\, \hat{P}_{k'} \,] = -\,\frac{\hbar}{i}\, \delta_{kk'}\ ; \tag{5.90}$$

and

$$\hat{\mathcal{H}} = \frac{1}{2M} \sum_k \hat{P}_k \hat{P}_{-k} + K \sum_k [1-\cos(ak)] \hat{Q}_k \hat{Q}_{-k}$$

$$= \sum_k \hbar\omega_k \left[\frac{1}{2M\hbar\omega_k} \hat{P}_k \hat{P}_{-k} + \frac{M\omega_k}{2\hbar} \hat{Q}_k \hat{Q}_{-k} \right], \tag{5.91}$$

where $\omega_k^2 = 2 \frac{K}{M} [1-\cos(ak)]$.

e) Introduce the phonon operators $\hat{a}_k$ and $\hat{a}_k^\dagger$ through the linear transformations

$$\hat{a}_k^\dagger = \sqrt{\frac{M\omega_k}{2\hbar}} \hat{Q}_{-k} - i \sqrt{\frac{1}{2\hbar M\omega_k}} \hat{P}_k \tag{5.92}$$

$$\hat{a}_k = \sqrt{\frac{M\omega_k}{2\hbar}} \hat{Q}_k + i \sqrt{\frac{1}{2\hbar M\omega_k}} \hat{P}_{-k} . \tag{5.93}$$

Verify the phonon commutation relation

$$[\hat{a}_k , \hat{a}_{k'}^\dagger] = \delta_{kk'} \tag{5.94}$$

and show that the Hamiltonian becomes

$$\hat{\mathcal{H}} = \sum_k \hbar\omega_k \left[\hat{a}_k^\dagger \hat{a}_k + \frac{1}{2} \right]. \tag{5.95}$$

Problem 5.2: The Fourier expansion of $\hat{\mathbf{A}}$ is given by

$$\hat{\mathbf{A}}(\mathbf{r},t) = \sum_{\mathbf{k}} e^{i\mathbf{k}\cdot\mathbf{r}} \hat{\mathbf{A}}(\mathbf{k},t) , \tag{5.96}$$

and correspondingly for $\hat{\Pi}(\mathbf{r},t)$. Prove, that in the Coulomb gauge the commutator of $\hat{\mathbf{A}}(\mathbf{k},t)$ and $\hat{\Pi}(\mathbf{k},t)$ is given by

$$[\ \hat{A}_\ell(\mathbf{k},t)\ ,\ \hat{\Pi}_j(\mathbf{k},t)\ ] = i\hbar\left[\delta_{\ell,j} - \frac{k_\ell k_j}{k^2}\right]. \tag{5.97}$$

Chapter 6
IDEAL QUANTUM GASES

In this chapter we discuss ideal quantum gases. An ideal gas is a system of non-interacting particles that is nevertheless in thermodynamic equilibrium. We analyze these systems in some detail to get experience in working with creation and destruction operators and also because we need several of the obtained results in later parts of this book.

In order to describe quantum mechanical systems at finite temperatures, we need the concept of *ensemble averages*. Such averages are computed using the statistical operator $\hat{\rho}$ which is defined as

$$\boxed{\hat{\rho} = \frac{e^{-\beta(\hat{H}-\mu\hat{N})}}{\mathrm{tr}\{e^{-\beta(\hat{H}-\mu\hat{N})}\}}} \; .$$

statistical operator for grand-canonical ensemble (6.1)

Here, we have given the statistical operator for a *grand-canonical ensemble* with variable number of particles. The expectation value $\langle\hat{Q}\rangle$ of an arbitrary operator $\hat{Q}$ is computed as

$$\langle\hat{Q}\rangle = \mathrm{tr}\,\{\,\hat{\rho}\,\hat{Q}\,\}\;. \tag{6.2}$$

The trace of an operator $\hat{Q}$ can be evaluated using any complete orthonormal set of functions $|n\rangle$ or $|\ell\rangle$, since

$$\begin{aligned}\operatorname{tr}\{\hat{Q}\} &= \sum_{\ell} \langle \ell | \hat{Q} | \ell \rangle \\ &= \sum_{\ell, n} \langle \ell | n \rangle \langle n | \hat{Q} | \ell \rangle \\ &= \sum_{n} \langle n | \hat{Q} | n \rangle \ . \end{aligned} \tag{6.3}$$

For practical calculations, it is most convenient to choose the functions as eigenfunctions to the operator $\hat{Q}$. If this is not possible, we want to choose the functions at least as eigenfuctions to some part of $\hat{Q}$, so that the part which cannot be evaluated exactly is *small* in some sense. The precise meaning of *small* and how to choose the most appropriate functions to evaluate the respective traces will be discussed for special cases in later chapters of this book.

6-1. The Ideal Fermi Gas

For didactic purposes we write the spin index explicitly in this and in the following chapters. The Hamiltonian for a system of non-interacting Fermions is

$$\hat{\mathscr{H}} = \sum_{\mathbf{k}, s} E_k \ \hat{a}^{\dagger}_{\mathbf{k},s} \ \hat{a}_{\mathbf{k},s} \ , \tag{6.4}$$

where $E_k = \hbar^2 k^2 / 2m$ is the kinetic energy. Consequently we have

$$\hat{\mathscr{H}} - \mu \hat{N} = \sum_{\mathbf{k}, s} (E_k - \mu) \ \hat{a}^{\dagger}_{\mathbf{k},s} \ \hat{a}_{\mathbf{k},s} \ . \tag{6.5}$$

To obtain the probability distribution function for Fermions, we compute the expectation value of the particle number operator in the state $(\mathbf{k}, s)$, i.e., we compute the mean occupation number $f_{\mathbf{k},s}$

$$f_{\mathbf{k},s} = \langle \hat{a}^\dagger_{\mathbf{k},s}\, \hat{a}_{\mathbf{k},s}\rangle = \frac{\mathrm{tr}\, e^{-\beta \sum\limits_{\mathbf{k}',s'} (E_{k'}-\mu)\hat{a}^\dagger_{\mathbf{k}',s'}\hat{a}_{\mathbf{k}',s'}}\, \hat{a}^\dagger_{\mathbf{k},s}\, \hat{a}_{\mathbf{k},s}}{\mathrm{tr}\, e^{-\beta \sum\limits_{\mathbf{k}',s'} (E_{k'}-\mu)\hat{a}^\dagger_{\mathbf{k}',s'}\hat{a}_{\mathbf{k}',s'}}} \; . \tag{6.6}$$

The evaluation of these expressions is greatly simplified since we can use

$$e^{-\beta \sum\limits_{\mathbf{k}',s'} (E_{k'}-\mu)\hat{a}^\dagger_{\mathbf{k}',s'}\hat{a}_{\mathbf{k}',s'}} = \prod_{\mathbf{k}',s'} e^{-\beta(E_{k'}-\mu)\hat{a}^\dagger_{\mathbf{k}',s'}\hat{a}_{\mathbf{k}',s'}} \; . \tag{6.7}$$

It should be noted that such a replacement is only possible if the operators in the different terms of the sum in the exponent on the LHS of (6.7) commute. This is trivially correct in the present case since the number operators

$$\hat{n}_{\mathbf{k},s} = \hat{a}^\dagger_{\mathbf{k},s}\hat{a}_{\mathbf{k},s} \tag{6.8}$$

for different states $(\mathbf{k}, s)$ commute. Hence, Eq. (6.6) can be written as

$$f_{\mathbf{k},s} = \frac{\mathrm{tr}\left\{\prod\limits_{\mathbf{k}'s'} e^{-\beta(E_{k'}-\mu)\hat{n}_{\mathbf{k}',s'}}\; \hat{n}_{\mathbf{k},s}\right\}}{\mathrm{tr}\left\{\prod\limits_{\mathbf{k}'s'} e^{-\beta(E_{k'}-\mu)\hat{n}_{\mathbf{k}',s'}}\right\}} \; . \tag{6.9}$$

We use

$$\mathrm{tr} \prod_{\mathbf{k}'s'} \ldots = \prod_{\mathbf{k}'s'} \mathrm{tr}\, \ldots \; ,$$

and evaluate the trace with the eigenfunctions of the particle number operator

$$\hat{n}_{\mathbf{k},s} | n_{\mathbf{k},s} \rangle = n_{\mathbf{k},s} | n_{\mathbf{k},s} \rangle . \tag{6.10}$$

All factors $(\mathbf{k}',s')$ in numerator and denominator of Eq. (6.9) cancel, except for the term with $\mathbf{k}'=\mathbf{k}$ and $s'=s$. Therefore, Eq. (6.9) yields

$$\boxed{f_{\mathbf{k},s} = \frac{1}{e^{\beta(E_k-\mu)} + 1}}$$

Fermi-Dirac distribution (6.11)

This is the Fermi-Dirac distribution which has already been used in Chap. 4. Eq. (6.11) shows that the distribution function depends only on the magnitude of **k** and not on the spin. Therefore, we often denote the Fermi-Dirac distribution simply by f_k. Examples for the Fermi-Dirac distribution function are plotted in Fig. 6.1 for three different temperatures.

As discussed in Eq. (4.20), where the spin summation was implicitly included in the **k**-summation, we obtain the total number of particles N by summing the distribution function f_k over all quantum states $\mathbf{k},s$

$$N = \sum_{\mathbf{k},s} f_k . \tag{6.12}$$

This relation determines the chemical potential $\mu = \mu(n,T)$ as a function of particle density n and temperature T. In order to evaluate Eq. (6.12) it is again useful to convert the sum over **k** into an integral over the energy ϵ

$$\sum_{\mathbf{k}} \to \int d\epsilon \; \rho^d(\epsilon), \tag{6.12a}$$

where, according to Eqs. (4.36) - (4.41) the d-dimensional density of states is given by

$$\rho^d(\epsilon) = \Omega_d \left[\frac{L}{2\pi}\right]^d \frac{1}{2} \left[\frac{2m}{\hbar^2}\right]^{d/2} \epsilon^{\frac{d-2}{2}} . \tag{6.13}$$

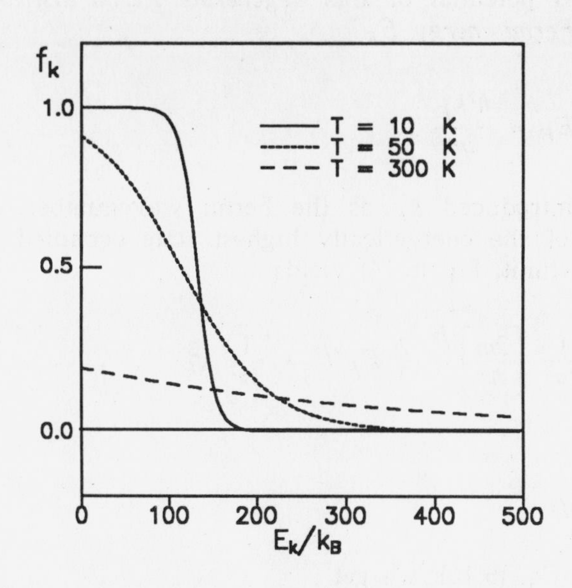

Fig. 6.1: Fermi-Dirac distribution function f_k as function of E_k/k_B for the particle density n=$1\cdot10^{18}$ cm^{-3} and three different temperatures.

6-1.1 The Ideal Fermi Gas in Three Dimensions

For a system with three dimensions, Eq. (6.12) yields

$$N = \frac{L^3}{2\pi^2} \left[\frac{2m}{\hbar^2}\right]^{\frac{3}{2}} \int_0^\infty d\epsilon\sqrt{\epsilon}\ \frac{1}{e^{\beta(\epsilon-\mu)}+1} . \tag{6.14}$$

Unfortunately this integral cannot be evaluated analytically. We will therefore consider first the low-temperature limit $T\to0$ or $\beta\to\infty$. If

$\beta \to \infty$

$$f_k = \begin{bmatrix} 1 \\ 0 \end{bmatrix} \text{ for } \begin{bmatrix} \epsilon < \mu \\ \epsilon > \mu \end{bmatrix} \quad \text{or} \quad f_k = \theta(\mu - \epsilon) \,, \tag{6.15}$$

showing that the Fermi function degenerates into the unit-step function. The chemical potential of this *degenerate Fermi distribution* is often denoted as *Fermi energy* E_F

$$\mu(n, T{=}0) = E_F = \frac{\hbar^2 k_F^2}{2m} \,, \tag{6.16}$$

where we have introduced k_F as the Fermi wavenumber. This is the wavenumber of the energetically highest state occupied at T=0. In this degenerate limit, Eq. (6.14) yields

$$n = \frac{N}{L^3} = \frac{1}{2\pi^2} \left[\frac{2m}{\hbar^2}\right]^{3/2} \frac{2}{3} \, E_F{}^{3/2} = \frac{1}{3\pi^2} \, k_F^3 \tag{6.17}$$

and thus

$$k_F = (3\pi^2 n)^{1/3}. \tag{6.18}$$

Inserting this into Eq. (6.16), we get

$$E_F = \frac{\hbar^2}{2m} (3\pi^2 n)^{2/3}. \tag{6.19}$$

The following discussion shows that the Fermi energy is basically the zero-point energy E_0 of a particle due to its localization in the volume

$$(\Delta x)^3 \propto 1/n = L^3/N. \tag{6.20}$$

The uncertainty relation for space and momentum uncertainty, Δx and Δp, can be written as

$$(\Delta p)^2 (\Delta x)^2 \geq \frac{\hbar^2}{4} \,, \tag{6.21}$$

or, using Eq. (6.20),

$$(\Delta p)^2 \propto \hbar^2 n^{2/3} . \tag{6.22}$$

The zero point energy is therefore given as

$$E_0 = \frac{\Delta p^2}{2m} \propto \frac{\hbar^2}{2m} n^{2/3} \propto E_F . \tag{6.23}$$

In the high-temperature limit where $\beta \to 0$, the chemical potential must grow fast to large negative values

$$\lim_{\beta \to 0} (-\mu\beta) = \infty \tag{6.24}$$

in order to keep the integral in Eq. (6.14) finite. The quantity $\exp(\beta\mu)$, called the *virial*, is thus a small quantity for $\beta E_F \ll 1$ and can be used as expansion parameter. In lowest approximation the Fermi function can be approximated by

$$f_k = \frac{e^{\beta(\mu - E_k)}}{1 + e^{\beta(\mu - E_k)}} \cong e^{\beta\mu} e^{-\beta E_k} . \tag{6.25}$$

In this case, Eq. (6.14) yields

$$n = \frac{e^{\beta\mu}}{2\pi^2} \left[\frac{2m}{\hbar^2\beta}\right]^{3/2} \int_0^\infty dx \sqrt{x}\, e^{-x} . \tag{6.26}$$

The integral is $1/2 \sqrt{\pi}$, so that

$$n = n_0 e^{\beta\mu}, \tag{6.27}$$

where

$$n_0 = \frac{1}{4} \left[\frac{2m}{\hbar^2\pi\beta}\right]^{3/2} \tag{6.28}$$

or, using Eqs. (6.26) and (6.27),

$$e^{\beta\mu} = 4n \left[\frac{\hbar^2\pi\beta}{2m}\right]^{3/2} . \tag{6.29}$$

Inserting this result into Eq. (6.25) yields the classical non-degenerate, or *Boltzmann distribution*

$$\boxed{f_k = 4n \left[\frac{\hbar^2 \pi \beta}{2m}\right]^{3/2} e^{-\beta E_k}}$$

Boltzmann distribution (6.30)

For the parameters used in Fig. 6.1, the distribution function at T=300K is practically undistinguishable from the Boltzmann distribution function (6.30) for the same conditions.

At this point, we will briefly describe how one can obtain an analytic approximation for $\mu(n,T)$ which is good for all except very strongly degenerate situations. Here we follow the work of Joyce and Dixon (1977) and Aguilera-Navaro *et al.* (1988). According to Eq. (6.14) the normalized density $\nu = n/n_0$ can be written as

$$\nu = \frac{n}{n_0} = \frac{2}{\sqrt{\pi}} \int_0^\infty dx \, \sqrt{x} \, \frac{z\, e^{-x}}{1 + ze^{-x}}, \tag{6.31}$$

where $z = \exp(\beta\mu)$. The integral can be evaluated using the series representation

$$\nu = \frac{2}{\sqrt{\pi}} \int_0^\infty dx \, \sqrt{x} \sum_{n=0}^{\infty} (-1)^n \, z^{n+1} \, e^{-x(n+1)}$$

$$= \sum_{n=1}^{\infty} (-1)^{n+1} \frac{z^n}{n^{3/2}}, \tag{6.32}$$

Clearly, this expansion converges only for μ<0 or z<1. However, the convergence range can be extended using the following resummation. First we invert Eq. (6.32) to express z in terms of ν

$$z = \sum_{n=1}^{\infty} b_n \nu^n \tag{6.33}$$

where the comparison with Eq. (6.32) shows that $b_1 = 1$. Taking the logarithm of Eq. (6.33), we can write $\beta\mu$ as

$$\beta\mu = ln\nu + \sum_{n=1}^{\infty} B_n \nu^n . \tag{6.34}$$

The logarithmic derivative of Eq. (6.34) yields

$$\nu \frac{d\beta\mu}{d\nu} = 1 + \sum_{n=1}^{\infty} B_n \, n \, \nu^n . \tag{6.35}$$

Now we make a *Padé* approximation by writing the infinite sum on the RHS of Eq. (6.34) as the ratio of two polynomials of order L and M

$$\nu \frac{d\beta\mu}{d\nu} \cong \frac{\sum_{i=0}^{L} p_i \nu^i}{\sum_{i=0}^{M} q_i \nu^i} = [L/M](\nu) . \tag{6.36}$$

This approximation is called the *L/M-Padé* approximation. The approximation with L=2 and M=1 already gives very accurate results. A final integration yields

$$\beta\mu \cong ln\nu + K_1 \ln(K_2\nu + 1) + K_3\nu \tag{6.37}$$

with K_1=4.8966851, K_2=0.04496457, and K_3=0.1333760, see Ell *et al.* (1989).

The comparison of Eqs. (6.37) and (6.29) shows that the logarithmic term in Eq. (6.37) is exactly the classical result. The chemical potential according to Eq. (6.37) is plotted in Fig. 6.2. Within drawing accuracy, the result is indistinguishable from the exact chemical potential obtained as numerical solution of Eq. (6.14) showing that Eq. (6.37) yields an excellent approximation for the range $-\infty < \mu\beta \leq 30$.

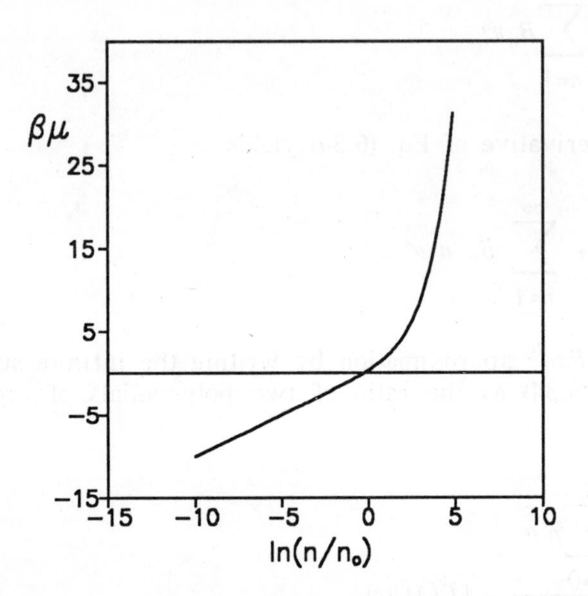

Fig. 6.2: Chemical potential μ for a three-dimensional Fermi gas as function of n/n_0, where n_0 is defined in Eq. (6.28).

6-1.2 The Ideal Fermi Gas in Two Dimensions

For a two-dimensional system, Eq. (6.12) yields

$$n = \frac{N}{L^2} = \frac{1}{2\pi}\left[\frac{2m}{\hbar^2\beta}\right]\int_0^\infty dx\ \frac{1}{e^x\ e^{-\beta\mu}+1}\ , \tag{6.38}$$

where $n=N/L^2$ now is the 2-dimensional particle density and L^2 is the area. Using $\exp(x)=t$ as a new integration variable, the integral in Eq. (6.38) becomes

$$\int_1^\infty dt\ \frac{1}{t(te^{-\mu\beta}+1)} = \int_1^\infty dt\ \left[\frac{1}{t} - \frac{1}{t+e^{\mu\beta}}\right] = \ln(1+e^{\mu\beta}). \tag{6.39}$$

Hence, we find the analytical result

$$n = \frac{m}{\hbar^2 \beta \pi} \ln(1+e^{\beta\mu}) \tag{6.40}$$

or

$$\boxed{\beta\mu(n,T) = \ln\left[e^{\hbar^2 \beta \pi n/m} - 1\right]}$$

2d Fermion chemical potential (6.41)

6-2. Ideal Bose Gas

Our discussion of the ideal Bose gas with spin s=0 proceeds similar to the analysis of the ideal Fermi gas. The Hamiltonian is

$$\hat{\mathscr{H}} = \sum_{\mathbf{k}} E_k \, \hat{b}_{\mathbf{k}}^{\dagger} \hat{b}_{\mathbf{k}} = \sum_{k} E_k \, \hat{n}_{\mathbf{k}} \, . \tag{6.42}$$

The expectation value of the particle number operator is

$$g_k \equiv \langle \hat{n}_{\mathbf{k}} \rangle = \frac{\mathrm{tr}\{e^{-\beta(E_k - \mu)\hat{n}_{\mathbf{k}}} \, \hat{n}_{\mathbf{k}}\}}{\mathrm{tr}\{e^{-\beta(E_k - \mu)\hat{n}_{\mathbf{k}}}\}} \, . \tag{6.43}$$

As in the Fermi case, the traces in Eq. (6.43) are evaluated choosing the eigenfunctions $|n_{\mathbf{k}}\rangle$ of the particle number operator

$$\hat{n}_{\mathbf{k}} | n_{\mathbf{k}} \rangle = n_{\mathbf{k}} | n_{\mathbf{k}} \rangle \, , \text{ where } n_{\mathbf{k}} = 0, 1, 2, \ldots, N, \ldots \infty \, . \tag{6.44}$$

In contrast to the Fermi gas, where the Pauli principle allows all quantum states to be occupied only once, each state can be populated arbitrarily often in the Bose system. We obtain

$$\mathrm{tr}\{e^{-\beta(E_k-\mu)\hat{n}_\mathbf{k}}\} = \sum_{n_\mathbf{k}=0}^{\infty} e^{-\beta(E_k-\mu)n_\mathbf{k}}$$

$$= \sum_{n=0}^{\infty} a^n = \frac{1}{1-a}, \tag{6.45}$$

where $a = e^{-\beta(E_k-\mu)}$. It is straightforward to evaluate the numerator in Eq. (6.43) as derivative of the denominator, showing that Eq. (6.43) yields the Bose-Einstein distribution function

$$\boxed{g_k = \frac{1}{e^{\beta(E_k-\mu)} - 1}}$$

Bose-Einstein distribution (6.46)

Generally for Bosons we have two possible cases:

i) Not conserved particle number, i.e., $N = \Sigma_\mathbf{k} g_k \neq$ constant. In this case μ cannot be determined from this relation, it has to be equal to zero: $\mu \equiv 0$. Examples for this class of Bosons are thermal photons and phonons.

ii) Conserved particle number $N = \Sigma_\mathbf{k} g_k =$ constant. Then $\mu = \mu(N,T)$ is determined from Eq. (6.12) as in the Fermi system. Due to the minus sign in the denominator of the Bose-Einstein distribution, the sum in Eq. (6.45) converges only for $\mu \leq 0$ since the smallest value of E_k is zero. Examples for this class of Bosons are He atoms.

In the remainder of this chapter we discuss some properties of the Bose system with conserved particle number, case ii). As in the Fermi case $\beta\mu$ takes on large negative values for high temperatures. Thus for $T \to \infty$, we can neglect the -1 in the denominator of Eq. (6.46) as compared to $e^{-\beta\mu}$ showing that the Bose-Einstein distribution also converges toward the Boltzmann distribution for high temperatures.

6-2.1 The Ideal Bose Gas in Three Dimensions

If we study the chemical potential of the ideal Bose gas for decreasing temperatures, we find that μ is negative and that its absolute value decreases toward zero. We denote the critical temperature at which μ becomes zero as T_c

$$\mu(n,T=T_c) = 0 \rightarrow T_c \ . \tag{6.47}$$

To determine the value of T_c we use Eq. (6.46) with $\mu = 0$ and compute the total number of Bosons first for the three-dimensional system

$$N = \sum_{\mathbf{k}} g_k = \int_0^\infty d\epsilon \ \rho^3(\epsilon) \ \frac{1}{e^{\beta\epsilon} - 1}$$

$$= \frac{L^3}{4\pi^2} \left[\frac{2m}{\hbar^2}\right]^{3/2} \int_0^\infty d\epsilon\sqrt{\epsilon} \ \frac{1}{e^{\beta\epsilon} - 1} \ . \tag{6.48}$$

The series representation

$$\frac{1}{e^x - 1} = \sum_{n=1}^\infty e^{-nx} \tag{6.49}$$

allows to rewrite Eq. (6.48) as

$$N = \frac{L^3}{(2\pi)^2} \left[\frac{2m}{\beta\hbar^2}\right]^{3/2} \sum_{n=1}^\infty \int_0^\infty dx \ e^{-nx} x^{1/2} \tag{6.50}$$

and the substitution $nx = t$ yields

$$N = \frac{L^3}{(2\pi)^2}\left[\frac{2m}{\beta\hbar^2}\right]^{3/2}\left[\sum_{n=1}^{\infty}\frac{1}{n^{3/2}}\right]\int_0^{\infty} dt\ e^{-t}t^{1/2}\ , \tag{6.51}$$

where the sum is the ζ function and the integral is the Γ function, both with the argument 3/2,

$$N = \frac{L^3}{(2\pi)^2}\left[\frac{2m}{\beta\hbar^2}\right]^{3/2}\zeta(3/2)\ \Gamma(3/2)\ . \tag{6.52}$$

Setting $T = T_c$, i.e., $\beta = \beta_c$, we find from Eq. (6.52) that

$$k_B T_c = \frac{\hbar^2}{2m}\, n^{2/3}\left[\frac{2\pi^2}{\Gamma(3/2)\zeta(3/2)}\right]^{2/3}\ . \tag{6.53}$$

The result (6.53) shows that T_c is a finite temperature ≥ 0. Now we know that $\mu = 0$ at $T = T_c$, but what happens if T falls below T_c? The chemical potential has to remain zero, since otherwise the Bose-Einstein distribution function would diverge and the calculations leading to Eq. (6.52) are therefore also valid for $T < T_c$. However, from the result (6.52) we see that N decreases with temperature like $T^{3/2}$ yielding the apparent contradiction

$$N(T{<}T_c) < N(T{=}T_c) = N\ . \tag{6.54}$$

The solution for the missing particles came from Einstein. He realized that the apparently missing particles are in fact condensed into the state $k = 0$, which has zero weight in the transformation from the sum to the integral in Eq. (6.12*a*). Therefore the term with $k = 0$ has to be treated separately for Bose systems at $T < T_c$. This can be done by writing

$$N = \sum_{k \neq 0} n_k + N_o$$

$$= \frac{L^3}{(2\pi)^2} \left[\frac{2mk_B T}{\hbar^2} \right]^{3/2} \zeta(3/2)\ \Gamma(3/2) + N_o\ . \qquad (6.55)$$

This equation shows that all particles are condensed into the state $k = 0$ at $T = 0$. This condensation in k-space is called the *Bose-Einstein condensation.* It corresponds to a real-space correlation effect and leads to superconductivity and superfluidity. For temperatures between $T = 0K$ and T_c the Bose system consists of a mixture of superfluid and normal state.

6-1.2 The Ideal Bose Gas in Two Dimensions

Using the two-dimensional density of states, we get for the total number of Bosons

$$N = \frac{L^2}{4\pi} \left[\frac{2m}{\hbar^2 \beta} \right] \int_0^\infty dx\ \frac{1}{e^{x-\beta\mu} - 1}\ . \qquad (6.56)$$

Now the two-dimensional particle density $n = N/L^2$ can be evaluated in the same way as the corresponding expression for Fermions, yielding

$$n = -\frac{1}{4\pi} \left[\frac{2m}{\hbar^2 \beta} \right] \ln(1 - e^{\beta\mu}) \qquad (6.57)$$

The argument of the logarithmic term has to be larger than 0, i.e. $e^{\beta\mu} < 1$ and $\mu < 0$ for any finite β value. Therefore, the chemical potential approaches zero only asymptotically as $T \to 0$ and there is no Bose-Einstein condensation in an ideal two-dimensional Bose system.

REFERENCES

J.B. Joyce and R.W. Dixon, Appl. Phys. Lett. **31**, 354 (1977)

V.C. Aguilera-Navaro, G.A. Esterez, and A. Kostecki, J. Appl. Phys. **63**, 2848 (1988)

C. Ell, R. Blank, S. Benner, and H. Haug, Journ. Opt. Soc. Am. **B6**, 2006 (1989)

PROBLEMS

Problem 6.1: Expand the chemical potential of a nearly degenerate *3d* Fermion system for low temperatures (Sommerfeld expansion).

Problem 6.2: Determine the temperature at which a *2d* Fermion system of a given density has zero chemical potential.

Problem 6.3: Calculate the energy and specific heat of a nearly degenerate *2d* Fermion system.

Chapter 7
THE INTERACTING ELECTRON GAS

In this chapter we discuss a model for the interacting electron gas in a solid. To keep the analysis as simple as possible, we neglect the discrete lattice structure of the ions in the solid and treat the positive charges as a smooth background, called jellium – like jelly.

This jellium model has been originally designed to describe the conduction characteristics in simple metals. However, as we will see in later chapters of this book, this model is also useful to compute some of the intraband properties of an excited semiconductor. In an excited semiconductor we have to deal with an electron-hole gas, which consists of the excited electrons in the conduction band and the corresponding holes, i.e., missing electrons, in the valence band. In this case, one again has total charge neutrality, since the negative charges of the electrons are compensated by the positive charges of the holes.

In the following sections, we will discuss the jellium model in such a way that only very minor changes are required when we want to apply the results to the case of an excited semiconductor.

7-1. Three-Dimensional Electron Gas

The Hamiltonian for the three-dimensional electron system is

$$\hat{\mathscr{H}} = \sum_{s} \int d^3r\ \hat{\psi}_s^{\dagger}(\mathbf{r}) \left[-\frac{\hbar^2\nabla^2}{2m}\right]\hat{\psi}_s(\mathbf{r})$$

$$+ \frac{1}{2} \sum_{s,s'} \int d^3r\ d^3r' \hat{\psi}_s^{\dagger}(\mathbf{r})\hat{\psi}_{s'}^{\dagger}(\mathbf{r}') \frac{e^2}{\epsilon_0 |\mathbf{r}-\mathbf{r}'|} \hat{\psi}_{s'}(\mathbf{r}')\ \hat{\psi}_s(\mathbf{r})$$

$$- \sum_{s} \int d^3r\ d^3r' \hat{\psi}_s^{\dagger}(\mathbf{r}) \frac{e^2 n(\mathbf{r}')}{\epsilon_0 |\mathbf{r}-\mathbf{r}'|} \hat{\psi}_s(\mathbf{r}) + \frac{e^2}{2} \int d^3r\ d^3r' \frac{n(\mathbf{r})n(\mathbf{r}')}{\epsilon_0 |\mathbf{r}-\mathbf{r}'|}\ , \qquad (7.1)$$

where the ionic charge distribution $n(\mathbf{r})$ in the jellium model is not an operator but the c-number

$$n(r) = N/L^3 = n\ . \qquad (7.2)$$

The first term in Eq. (7.1) is the kinetic energy of the electrons, the second term describes the electron-electron repulsion, the third term is the electron-ion attraction and the last term describes the repulsive ion-ion interaction. ϵ_0 in Eq. (7.1) is a background dielectric constant which takes into account the polarizability of the valence electrons and of the lattice.

The assumption (7.2) introduces the full translational symmetry into our problem, and we can simplify the treatment by expanding $\hat{\psi}_s(\mathbf{r})$ into plane waves $\phi_{\mathbf{k}}(\mathbf{r}) = \exp(i\mathbf{k}\cdot\mathbf{r})/L^{3/2}$

$$\hat{\psi}_s(\mathbf{r}) = \sum \hat{a}_{\mathbf{k},s} \frac{e^{i\mathbf{k}\cdot\mathbf{r}}}{L^{3/2}}\ . \qquad (7.3)$$

Using the plane-wave normalization

$$\int d^3r\ \phi_{\mathbf{k}}^*(\mathbf{r})\ \phi_{\mathbf{k}'}(\mathbf{r}) = \delta_{\mathbf{k},\mathbf{k}'} \qquad (7.4)$$

we find

$$\hat{a}_{\mathbf{k},s} = \frac{1}{L^{3/2}} \int d^3r \; \hat{\psi}_s(\mathbf{r}) e^{-i\mathbf{k}\cdot\mathbf{r}} \quad . \tag{7.5}$$

Inserting the expansion (7.3) into the Hamiltonian (7.1) yields

$$\begin{aligned}
\hat{\mathcal{H}} &= \sum_{s,\mathbf{k}} E_k \hat{a}^\dagger_{\mathbf{k},s} \hat{a}_{\mathbf{k},s} \\
&+ \frac{1}{2} \sum_{\substack{\mathbf{k}_1,\mathbf{k}_1' \\ \mathbf{k}_2,\mathbf{k}_2' \\ s,s'}} \int \frac{d^3r d^3r'}{L^6} \hat{a}^\dagger_{\mathbf{k}_1',s} \hat{a}^\dagger_{\mathbf{k}_2',s'} \hat{a}_{\mathbf{k}_2 s'} \hat{a}_{\mathbf{k}_1,s} e^{i(\mathbf{k}_1-\mathbf{k}_1')\cdot\mathbf{r}} \frac{e^2}{\epsilon_0 |\mathbf{r}-\mathbf{r}'|} e^{i(\mathbf{k}_2-\mathbf{k}_2')\cdot\mathbf{r}'} \\
&- \frac{e^2 n}{L^3} \sum_{\mathbf{k},\mathbf{k}',s} \int d^3r d^3r' e^{i(\mathbf{k}-\mathbf{k}')\cdot\mathbf{r}} \frac{1}{\epsilon_0 |\mathbf{r}-\mathbf{r}'|} \hat{a}^\dagger_{\mathbf{k},s} \hat{a}_{\mathbf{k}',s} \\
&+ \frac{1}{2} \int \frac{d^3r \; d^3r'}{L^6} \frac{e^2 N^2}{\epsilon_0 |\mathbf{r}-\mathbf{r}'|} \quad .
\end{aligned} \tag{7.6}$$

It is advantageous to introduce relative and center-of-mass coordinates

$$\boldsymbol{\rho} \equiv \mathbf{r} - \mathbf{r}' \quad \text{and} \quad \mathbf{R} \equiv \frac{\mathbf{r} + \mathbf{r}'}{2} \, , \tag{7.7}$$

which transform the double integrals in Eq. (7.6) according to

$$\int d^3r d^3r' \cdots \to \int d^3R d^3\rho \cdots \; . \tag{7.8}$$

Then we obtain integrals of the kind:

$$I = \int \frac{d^3R \; d^3\rho}{L^6} \; e^{i(\mathbf{k}_1+\mathbf{k}_2-\mathbf{k}_1'-\mathbf{k}_2')\cdot\mathbf{R}} \; e^{i\rho\cdot(\mathbf{k}_1-\mathbf{k}_1'-\mathbf{k}_2+\mathbf{k}_2')/2} \; \frac{1}{\rho}$$

$$= \delta_{\mathbf{k}_1+\mathbf{k}_2,\mathbf{k}_1'+\mathbf{k}_2'} \frac{1}{L^3} \int d^3\rho \; e^{i\rho\cdot(\mathbf{k}_1-\mathbf{k}_1')} \frac{1}{\rho} \; . \tag{7.9}$$

Denoting the momentum transfer $\mathbf{k}_1 - \mathbf{k}_1' = \mathbf{q}$, we can rewrite the last integral in Eq. (7.9) as

$$\rho_q = \int d^3\rho \; e^{i\boldsymbol{\rho}\cdot\mathbf{q}} \frac{1}{\rho} = 2\pi \int_0^\infty d\rho \; \rho \int_{-1}^{1} d\cos\theta \; e^{iq\rho\cos\theta}$$

$$= -\frac{2\pi i}{q} \int_0^\infty d\rho \; (e^{iq\rho} - e^{-iq\rho}) \; . \tag{7.10}$$

To evaluate the remaining integral in Eq. (7.10), we introduce the convergence generating factor $\exp(-\epsilon\rho)$ under the integral and take the limit of $\epsilon \to 0$ after the evaluation. This yields

$$\rho_q = -\frac{2\pi i}{q} \lim_{\epsilon\to 0} \int_0^\infty d\rho \; (e^{iq\rho} - e^{-iq\rho})e^{-\epsilon\rho} = \frac{4\pi}{q^2} \; . \tag{7.11}$$

Up to prefactors, the final result in Eq. (7.11) is just the Fourier-transform of the statically screened Coulomb interaction potential V_q

$$V_q = \frac{1}{L^3} \int d^3r \; e^{-i\mathbf{q}\cdot\mathbf{r}} \; V(r) \tag{7.12}$$

yielding

$$\boxed{V_q = \frac{4\pi e^2}{\epsilon_0 L^3 q^2}}$$

3d Coulomb potential (7.13)

Using Eq. (7.11) in the electron-electron (*e-e*) interaction term in Eq. (7.6), the *e-e* interaction becomes ($\mathbf{k}_1 = \mathbf{k}$, $\mathbf{k}_2 = \mathbf{k}'$)

$$\frac{1}{2} \sum_{\substack{\mathbf{k},\mathbf{k}' \\ \mathbf{q},s,s'}} \hat{a}^{\dagger}_{\mathbf{k}-\mathbf{q},s} \hat{a}^{\dagger}_{\mathbf{k}'+\mathbf{q},s'} \hat{a}_{\mathbf{k}',s'} \hat{a}_{\mathbf{k},s} V_q \,. \tag{7.14}$$

Similar manipulations can be used to show that the interaction of the electrons with the positive background can be transformed into

$$- \frac{e^2 n}{L^3} \sum_{\mathbf{k},\mathbf{k}',s} \int d^3r d^3r' e^{i(\mathbf{k}-\mathbf{k}')\cdot\mathbf{r}} \frac{1}{\epsilon_0 |\mathbf{r}-\mathbf{r}'|} \hat{a}^{\dagger}_{\mathbf{k},s} \hat{a}_{\mathbf{k}',s}$$

$$= - \frac{e^2 n}{\epsilon_0} \sum_{\mathbf{k},\mathbf{k}',s} \delta_{\mathbf{k},\mathbf{k}'} \hat{a}^{\dagger}_{\mathbf{k},s} \hat{a}_{\mathbf{k},s} \int d^3\rho \, \frac{1}{\rho} \, e^{i(\mathbf{k}-\mathbf{k}')\cdot\boldsymbol{\rho}/2} \,. \tag{7.15}$$

Comparison with Eqs. (7.9) - (7.11) shows us that

$$\delta_{\mathbf{k},\mathbf{k}'} \int d^3\rho \, \frac{1}{\rho} \, e^{i(\mathbf{k}-\mathbf{k}')\cdot\boldsymbol{\rho}/2} = \lim_{q\to 0} \rho_q \,, \tag{7.16}$$

which has a quadratic divergence. However the $q = 0$ term in Eq. (7.12) has the same divergence as does the ion-ion interaction (last term in Eq. (7.1)). The *e-e* and *ion-ion* interaction terms contribute each

$$1/2 \lim_{q\to 0} \rho_q$$

and the *e*-ion attraction contributes

$$- \lim_{q\to 0} \rho_q .$$

Replacing the electron density operators in the q=0 terms by their expectation values, we see that all the diverging terms cancel.

The resulting Hamiltonian reduces to

$$\hat{\mathcal{H}} = \sum E_k \hat{a}^\dagger_{\mathbf{k},s}\hat{a}_{\mathbf{k},s} + \frac{1}{2} \sum_{\substack{\mathbf{k},\mathbf{k}' \\ q\neq 0 \\ s,s'}} \hat{a}^\dagger_{\mathbf{k}+\mathbf{q},s}\hat{a}^\dagger_{\mathbf{k}'-\mathbf{q},s'}\hat{a}_{\mathbf{k}',s'}\hat{a}_{\mathbf{k},s}\, V_q\,, \tag{7.17}$$

The calculations leading to the electron gas Hamiltonian (7.17) show that the only, but extremely important effect resulting from the interaction of the electrons with the homogeneous positive charge background is to eliminate the diverging term $q = 0$ from the sum in the electron-electron interaction Hamiltonian.

Now we use the electron gas Hamiltonian to compute the ground-state energy ($T = 0$) in Hartree-Fock approximation. Since we know that at $T = 0$ all particles are in states with $|\mathbf{k}| \leq k_F$, the Hartree-Fock ground-state wavefunction is

$$|0\rangle_{HF} = \hat{a}^\dagger_{\mathbf{k}_1,s_1}\hat{a}^\dagger_{\mathbf{k}_2,s_2} \cdots \hat{a}^\dagger_{\mathbf{k}_N,s_N} |0\rangle$$

$$= \prod_{k_i \leq k_F} \hat{a}^\dagger_{\mathbf{k}_i,s_i} |0\rangle\,. \tag{7.18}$$

Due to the anti-commutation relations between the Fermi operators, Eq. (7.18) automatically has the correct symmetry. The Hartree-Fock ground state energy is

$$E_0^{HF} = {}_{HF}\langle 0|\hat{H}|0\rangle_{HF}$$

$$= E_{kin}^{HF} + E_{pot}^{HF}\,. \tag{7.19}$$

First we evaluate the kinetic energy

$$E_{kin}^{HF} = \sum_{\mathbf{k},s} E_k\; {}_{HF}\langle 0|\hat{a}^\dagger_{\mathbf{k},s}\hat{a}_{\mathbf{k},s}|0\rangle_{HF}\,. \tag{7.20}$$

We simply get

$$ {}_{HF}\langle 0|\hat{a}^{\dagger}_{\mathbf{k},s}\hat{a}_{\mathbf{k},s}|0\rangle_{HF} = \Theta(E_F - E_k) = f_k \,. \tag{7.21} $$

which is just the Fermi distribution at $T = 0$. Therefore,

$$ E^{HF}_{kin} = 2\sum_{\mathbf{k}} E_k \,\Theta(E_F - E_k) = \frac{L^3}{\pi^2}\,\frac{\hbar^2}{2m}\,\frac{k_F^5}{5} \,. \tag{7.22} $$

With the Fermi wavenumber $k_F = (3\pi^2 n)^{1/3}$, Eq. (6.18), we find

$$ E^{HF}_{kin} = \frac{\hbar^2 L^3}{10 m\pi^2}\,(3\pi^2 n)^{5/3} \,. \tag{7.23} $$

According to Eqs. (6.17) - (6.20) the zero-point energy is proportional to $n^{2/3}$. Thus, in Hartree-Fock approximation the kinetic energy per electron is essentially the zero-point energy of the electron.

For the potential energy we obtain

$$ E^{HF}_{pot} = \frac{1}{2}\sum_{\substack{\mathbf{q}\neq 0\\ \mathbf{k},\mathbf{k}',s,s'}} V_q \;{}_{HF}\langle 0|\hat{a}^{\dagger}_{\mathbf{k}+\mathbf{q},s}\hat{a}^{\dagger}_{\mathbf{k}'-\mathbf{q},s'}\hat{a}_{\mathbf{k}',s'}\hat{a}_{\mathbf{k},s}|0\rangle_{HF} \,. \tag{7.24} $$

This term is nonzero only if $|\mathbf{k}|, |\mathbf{k}'| \leq k_F$ and $|\mathbf{k}+\mathbf{q}|, |\mathbf{k}'-\mathbf{q}| \leq k_F$. Using these conditions and the commutation relations repeatedly, we obtain

$$ \begin{aligned} E^{HF}_{pot} &= -\frac{1}{2}\sum_{\substack{\mathbf{q}\neq 0\\ \mathbf{k},\mathbf{k}',s}} V_q\, \delta_{\mathbf{k}+\mathbf{q},\mathbf{k}'}\,\Theta(E_F - E_k)\,\Theta(E_F - E_{k'}) \\ &= -\frac{1}{2}\sum_{\mathbf{k},\mathbf{k}',s} V_{|\mathbf{k}-\mathbf{k}'|}\, f_k\, f_{k'} \,. \end{aligned} \tag{7.25} $$

Explicit evaluation of the sum in the last line and use of Eq. (6.18) yields (see problem 7.1)

$$E_{pot}^{HF} \equiv E_{exc} = -\frac{e^2 L^3}{4\pi^3 \epsilon_0} (3\pi^2 n)^{4/3} . \tag{7.26}$$

The Hartree-Fock result for the potential energy due to electron-electron repulsion is just the *exchange energy*, which increases with density with a slightly smaller power than the kinetic energy.

In order to understand why the exchange effect yields an energy reduction, calculate the conditional probability to find in the Hartree-Fock ground state an electron at the position $\mathbf{r}$ with spin s and simultaneously at $\mathbf{r}'$ with s'. This conditional probability is just the correlation function

$$R_{ss'}(\mathbf{r},\mathbf{r}') = {}_{HF}\langle 0|\, \hat{\psi}_s^\dagger(\mathbf{r})\, \hat{\psi}_{s'}^\dagger(\mathbf{r}')\, \hat{\psi}_{s'}(\mathbf{r}')\, \hat{\psi}_s(\mathbf{r})|\, 0\rangle_{HF} . \tag{7.27}$$

Obviously, this correlation function is only finite if the annihilation operators simultaneously find an electron in $(\mathbf{r}, s)$ and $(\mathbf{r}', s')$. The creation operators simply put the annihilated electrons back into their previous states. To evaluate Eq. (7.27), we again expand the field operators into plane waves

$$\hat{\psi}_s(\mathbf{r}) = \sum_{\mathbf{k}} \hat{a}_{\mathbf{k},s} \frac{e^{i\mathbf{k}\cdot\mathbf{r}}}{L^{3/2}} . \tag{7.28}$$

Using this expansion in Eq. (7.27), we find

$$R_{ss'}(\mathbf{r},\mathbf{r}') = \frac{1}{L^6} \sum e^{i(\boldsymbol{\ell}'-\mathbf{k}')\cdot\mathbf{r}'+i(\boldsymbol{\ell}-\mathbf{k})\cdot\mathbf{r}}\, {}_{HF}\langle 0|\, \hat{a}_{\mathbf{k},s}^\dagger \hat{a}_{\mathbf{k}',s'}^\dagger \hat{a}_{\boldsymbol{\ell}',s'} \hat{a}_{\boldsymbol{\ell},s}\,|\, 0\rangle_{HF} . \tag{7.29}$$

The sum runs over $(\boldsymbol{\ell}, \boldsymbol{\ell}', \mathbf{k}, \mathbf{k}')$ with $(|\boldsymbol{\ell}|, |\boldsymbol{\ell}'|, |\mathbf{k}|, |\mathbf{k}'|) < k_F$. Again we commute all creation operators to the right to use $\hat{a}_{\mathbf{k}',s'}^\dagger |\, 0\rangle_{HF} = 0$ for $k' < k_F$ and obtain

$$R_{ss'}(\mathbf{r},\mathbf{r}') = \frac{1}{L^6} \left[\frac{N^2}{4} - \delta_{ss'}\, |F(\mathbf{r} - \mathbf{r}')|^2 \right] , \tag{7.30}$$

where

$$F(\rho) = \sum_{\substack{\mathbf{k} \\ |k| < k_F}} e^{i\mathbf{k}\cdot\boldsymbol{\rho}} = n\,\frac{3}{2}\,\frac{\sin k_F\rho - \rho k_F \cos k_F\rho}{(k_F\rho)^3}\,. \qquad (7.31)$$

In Eq. (7.31) we again use Eq. (6.18) to express k_F in terms of the density n. Inserting Eq. (7.31) into Eq. (7.30), we obtain

$$R_{ss'}(\mathbf{r},\mathbf{r}') = \frac{n^2}{4}\left\{1-\delta_{ss'}9\left[\frac{\sin(k_F|\mathbf{r}-\mathbf{r}'|)-|\mathbf{r}-\mathbf{r}'|\,k_F\cos(k_F|\mathbf{r}-\mathbf{r}'|)}{(k_F|\mathbf{r}-\mathbf{r}'|)^3}\right]^2\right\}. \qquad (7.32)$$

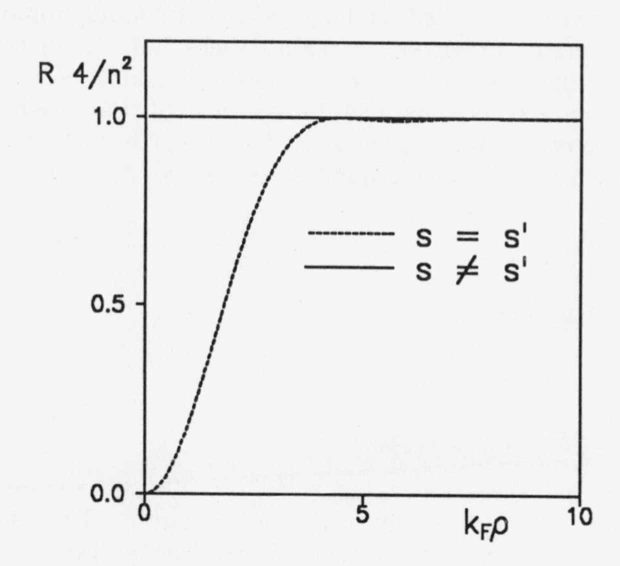

Fig. 7.1: Pair correlation function $R_{ss'}$ for the three-dimensional electron plasma, Eq. (7.32), as function of the dimensionless particle distance $k_F\rho$, where $\rho=|\mathbf{r}-\mathbf{r}'|$ and k_F is given by Eq. (6.18).

This result shows that the conditional probability to find an electron at $\mathbf{r}'$ with spin s', given that there is an electron at $\mathbf{r}$ with spin s, depends only on the separation $|\mathbf{r} - \mathbf{r}'|$ between the two electrons. Furthermore if s and s' are different, the second term on the RHS of Eq. (7.32) vanishes, and we find that the correlation function is constant. However for electrons with equal spin, $s = s'$, we can convince ourselves by a Taylor expansion that $R_{ss}(\rho \to 0) \to 0$. This result shows that the electrons with equal spin avoid each other as a consequence of the Pauli exclusion principle (exchange repulsion). We can say that each electron is surrounded by an *exchange hole*. The exchange hole indicates that the mean separation between electrons with equal spin is larger that it would be without the Pauli principle. Hence, the existence of the exchange hole *reduces the overall Coulomb repulsion* explaining the energy reduction described by Eq. (7.26).

According to the Hartree-Fock theory, electrons with different spin do not avoid each other, since the states are chosen to satisfy the exchange principle but they do not include Coulomb correlations. The exchange principle is satisfied as long as one quantum number, here the spin, is different. However, in reality there will be an additional correlation, which leads to the so-called *Coulomb hole*. To treat these correlation effects, one has to go beyond Hartree-Fock theory, e.g., using screened Hartree-Fock (RPA), see Chap. 9. Generally, one can write for the exact ground state energy E_0

$$E_0 = E_0^{HF} + E_{cor}$$

$$= E_{kin}^{HF} + E_{exc} + E_{cor} \ , \tag{7.33}$$

where the correlation energy is defined as

$$E_{cor} = E_0 - E_0^{HF} \ . \tag{7.34}$$

An exact calculation of E_{cor} is generally not possible. To obtain good estimates for E_{cor} is one of the tasks of many-body theory.

7-2. Two-Dimensional Electron Gas

Even in those semiconductor structures which we consider as low-dimensional, such as quantum wells, quantum wires, or quantum dots, the Coulomb potential varies as $1/r$. The reason is that the electric field lines between two charges are not confined within these structures. The field lines also pass through the surrounding material, which often is another semiconductor material with very similar dielectric constant. Exceptions are semiconductor quantum dots in glass or in colloids. The corresponding modifications of the Coulomb potential will be discussed in Chap. 20.

In this section we discuss the situation that the electron motion is confined to two dimensions, but the Coulomb interaction has its three-dimensional space dependence. For simplicity we disregard here all modifications which occur for different dielectric constants in the confinement layer and the embedding material. We call such a situation quasi-two dimensional, and it is always this case which we consider in this book when we analyze modifications due to dimensional confinement.

We assume that the carriers are confined to an x,y layer and we put the transverse coordinate z=0. For this case we need only the two dimensional Fourier transform of the Coulomb potential. One gets in polar coordinates

$$V_q = \frac{e^2}{\epsilon_0 L^2}\int_0^\infty dr\ r \int_0^{2\pi} d\phi\ \frac{e^{iqr\cos\phi}}{r}$$

$$= \frac{2\pi e^2}{\epsilon_0 q L^2}\int_0^\infty d(qr)\ J_0(qr), \tag{7.35}$$

where $J_0(x)$ is the zero-order Bessel function of first kind. Using

$$\int_0^\infty dx\ J_0(x) = 1$$

we find the result

$$\boxed{V_q = \frac{2\pi e^2}{\epsilon_0 q L^2}}$$

quasi-$2d$ Coulomb potential (7.36)

Eq. (7.36) shows that the Coulomb potential in two dimensions exhibits a $1/q$ dependence instead of the $1/q^2$ dependence in $3d$.

Using the quasi-*2d* Coulomb potential (7.36) in Eq. (7.25), we obtain the exchange energy as (see problem 7.1)

$$E_{exc}^{2d} = -\frac{L^2 e^2 C}{3\pi\epsilon_0}(2\pi n)^{3/2}, \tag{7.37}$$

where

$$C = \sum_{\ell=0,2,\ldots,\infty} \frac{2}{\ell+2}\left[\frac{1}{2\ell}\binom{\ell}{\ell/2}\right]^2. \tag{7.38}$$

is a numerical constant.

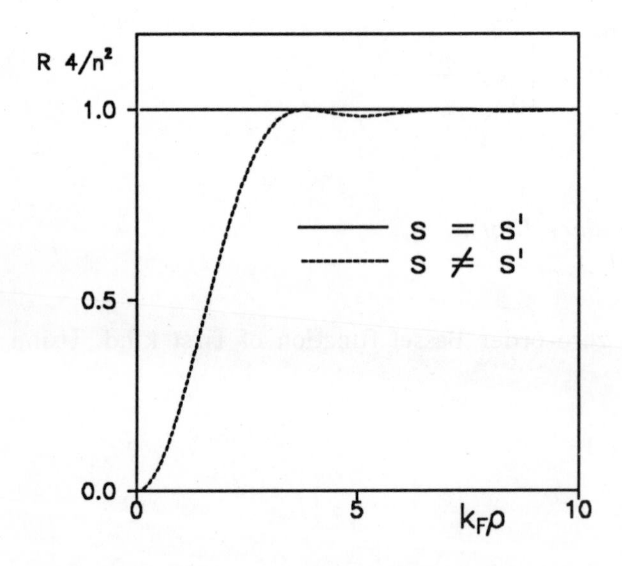

Fig. 7.2: Pair correlation function $R_{ss'}$ for the quasi-two-dimensional electron plasma, Eqs. (7.30) and (7.44), as function of the dimensionless particle distance $k_F\rho$, where $\rho=|\mathbf{r}-\mathbf{r}'|$ and k_F is given by Eq. (7.46)

In order to compare the exchange energy in *2d* and *3d*, we introduce the normalized distance r_s between particles through the relation

$$\frac{4\pi}{3} r_s^3 = \frac{1}{n a_0^3} = \frac{L^3}{N a_0^3} \tag{7.39}$$

in *3d*. Here, a_0 is a characteristic length given by the Bohr radius of the bound electron-hole pairs, i.e., excitons (see Chap. 10 for details). At this point we use the definition of a_0 as

$$a_0 = \frac{\hbar^2 \epsilon_0}{m e^2} \tag{7.40}$$

in three dimensions. In two dimensions we have

$$\pi r_s^2 = \frac{1}{n a_0^2} = \frac{L^2}{N\, a_0^2} , \tag{7.41}$$

where a_0 is the *2d* Bohr radius, which is half of the *3d* Bohr radius. The *3d* exchange energy is seen to vary with the particle distance r_s as

$$E_{exc}^{3d} \propto - r_s^{-4} , \tag{7.42}$$

whereas

$$E_{exc}^{2d} \propto - r_s^{-3} , \tag{7.43}$$

i.e., in three dimensions the exchange energy increases more strongly with decreasing particle distance than in two dimensions.

In the evaluation of the pair correlation function $R_{ss'}(\mathbf{r}, \mathbf{r}')$ for the *2d* electron gas, we obtain formally the same result as in Eq. (7.30), where however now

$$F(\mathbf{r} - \mathbf{r}') = n \, \frac{J_1(k_F |\mathbf{r}-\mathbf{r}'|)}{k_F |\mathbf{r}-\mathbf{r}'|} \tag{7.44}$$

with J_1 being the first-order Bessel function of the first kind. Here we have introduced the two-dimensional Fermi wavenumber k_F through the relation

$$2\sum_{\mathbf{k}} f_k = 2\,\frac{2\pi F}{(2\pi)^2}\int_0^{k_F} kdk \tag{7.45}$$

yielding

$$k_F^2 = 2\pi n = 2\pi\,\frac{L^2}{F} \tag{7.46}$$

The resulting pair correlation function for *2d* is plotted in Fig. 7.2. Schematically, the variations of the correlation function in two-dimensions resemble those of the three-dimensional result shown in Fig. 7.1. A more detailed comparison between Figs. 7.1 and 7.2 however shows that the oscillatory structures are somewhat more pronounced in the *2d* system indicating that the exchange correlations of the electrons are stronger in *2d* than in *3d*.

REFERENCES

C. Kittel, *Quantum Theory of Solids*, Wiley and Sons, New York (1967)

D. Pines and P. Nozieres, *The Theory of Quantum Liquids*, Benjamin, Reading, Mass. (1966)

PROBLEMS

Problem 7.1: Use the Hamiltonian (7.17) and the Hartree-Fock wavefunction (7.18) to compute the ground-state energy with the *3d* and the *2d* Coulomb interaction potentials, respectively. Hint: Use the expansions

$$\frac{1}{|\mathbf{k}-\mathbf{k}'|^2} = \frac{1}{k^2}\sum_{\ell,\ell'}\left[\frac{k'}{k}\right]^{\ell+\ell'} P_\ell(\cos\theta)P_{\ell'}(\cos\theta) \quad \text{for} \quad k'<k$$

$$= \frac{1}{k'^2} \sum_{\ell,\ell'} \left[\frac{k}{k'}\right]^{\ell+\ell'} P_\ell(\cos\theta) P_{\ell'}(\cos\theta) \quad \text{for} \quad k<k'$$

$$\frac{1}{|\mathbf{k}-\mathbf{k}'|} = \frac{1}{k} \sum_{\ell} \left[\frac{k'}{k}\right]^{\ell} P_\ell(\cos\theta) \quad \text{for} \quad k'<k$$

$$= \frac{1}{k'} \sum_{\ell} \left[\frac{k}{k'}\right]^{\ell} P_\ell(\cos\theta) \quad \text{for} \quad k<k'$$

and the integrals

$$\int_{-1}^{1} d\cos\theta \; P_\ell(\cos\theta) P_{\ell'}(\cos\theta) = \frac{2}{2\ell+1} \delta_{\ell,\ell'}$$

$$\int_0^{2\pi} d\theta \; P_\ell(\cos\theta) = 2\pi \left[\frac{1}{2^\ell} \binom{\ell}{\ell/2} \right]^2 \quad \text{for } \ell \text{ even}$$

$$= 0 \quad \text{for } \ell \text{ odd} .$$

Problem 7.2: Compute the pair correlation function (7.27) for the *3d* and the *2d* case. Prove that $R_{ss}(\mathbf{r}-\mathbf{r}') \to 0$ for $\mathbf{r} \to \mathbf{r}'$.

Chapter 8
PLASMONS AND PLASMA SCREENING

In this chapter we discuss collective excitations in the electron gas. As mentioned earlier, *collective excitations* are excitations that belong to the entire system. The collective excitations of the electron gas (= plasma) are called *plasmons*. These excitations and their effect on the dielectric constant have been discussed in Chap. 1 in the framework of classical electrodynamics. In this chapter we now develop the corresponding second-quantized formalism, which reveals that also electron-electron pair excitations occur which influence the dielectric constant and other properties in fundamental ways. The excitations in the electron plasma are responsible for screening of the Coulomb potential, effectively reducing it to a potential whose interaction range is reduced with increasing plasma density. A simplified description of the screening is developed in terms of an effective collective excitation, called plasmon pole approximation.

8-1. Plasmons and Pair Excitations

In the following we discuss the elementary plasma excitations using second quantization. First we compute the equation of motion for the particle density operator. We simplify this equation making a *mean field approximation* for the local density with respect to the surrounding density. The mean field approximation in the equation of motion is usually called *random phase approximation* (RPA) and allows us to factor four-operator averages into products of two-operator averages.

To simplify the notation in the remainder of this book we suppress from now on the superscript ^ for operators, unless this is needed to avoid misunderstandings. Furthermore, the spin index is only given where necessary. In all other cases it can be assumed to be included in the quasi-momentum subscript.

The operator for the particle density in real space is

$$n(\mathbf{r}) = \psi^\dagger(\mathbf{r})\psi(\mathbf{r}) \ , \tag{8.1}$$

or, using the plane-wave expansion

$$\psi(\mathbf{r}) = \frac{1}{L^{3/2}} \sum_{\mathbf{k}} e^{i\mathbf{k}\cdot\mathbf{r}} a_{\mathbf{k}} \ , \tag{8.2}$$

$$n(\mathbf{r}) = \sum_{\mathbf{k},\mathbf{k}'} a^\dagger_{\mathbf{k}'} a_{\mathbf{k}} \frac{1}{L^3} e^{i(\mathbf{k}-\mathbf{k}')\cdot\mathbf{r}} \ . \tag{8.3}$$

Next, we take the Fourier transformation of the density operator. For the spatial *3d* Fourier transformation we use the following conventions

$$f_{\mathbf{q}} = \int \frac{d^3r}{L^3} f(\mathbf{r}) \ e^{-i\mathbf{q}\cdot\mathbf{r}} \tag{8.4}$$

$$f(\mathbf{r}) = \sum_{\mathbf{q}} f_{\mathbf{q}} \ e^{i\mathbf{q}\cdot\mathbf{r}} \tag{8.5}$$

and

$$\sum_{\mathbf{q}} e^{i\mathbf{q}\cdot(\mathbf{r}-\mathbf{r}')} = L^3 \ \delta(\mathbf{r}-\mathbf{r}') \tag{8.6}$$

$$\int \frac{d^3r}{L^3} e^{i(\mathbf{q}-\mathbf{q}')\cdot\mathbf{r}} = \delta_{\mathbf{q},\mathbf{q}'} \ . \tag{8.7}$$

With these definitions, we obtain from Eq. (8.3)

$$n_{\mathbf{q}} = \frac{1}{L^3} \sum_{\mathbf{k}} a^{\dagger}_{\mathbf{k}-\mathbf{q}} \, a_{\mathbf{k}} \ . \tag{8.8}$$

To compute $n_{\mathbf{q}}$ we use the Heisenberg equation for $a^{\dagger}_{\mathbf{k}-\mathbf{q}} a_{\mathbf{k}}$

$$\frac{d}{dt} \, a^{\dagger}_{\mathbf{k}-\mathbf{q}} \, a_{\mathbf{k}} = \frac{i}{\hbar} \, [\mathcal{H}, \, a^{\dagger}_{\mathbf{k}-\mathbf{q}} \, a_{\mathbf{k}}] \tag{8.9}$$

with the electron gas Hamiltonian

$$\mathcal{H} = \sum_{\mathbf{k}} E_k a^{\dagger}_{\mathbf{k}} a_{\mathbf{k}} + \frac{1}{2} \sum_{\substack{\mathbf{k},\mathbf{k}',\mathbf{q} \\ q \neq 0}} V_q \; a^{\dagger}_{\mathbf{k}-\mathbf{q}} a^{\dagger}_{\mathbf{k}'+\mathbf{q}} a_{\mathbf{k}'} a_{\mathbf{k}} \ . \tag{8.10}$$

Evaluating the commutators on the RHS of Eq. (8.9), we get for the kinetic term

$$\frac{i}{\hbar} \sum_{\boldsymbol{\ell}} E_{\ell} \; [a^{\dagger}_{\boldsymbol{\ell}} a_{\boldsymbol{\ell}}, \; a^{\dagger}_{\mathbf{k}-\mathbf{q}} a_{\mathbf{k}}] = i \, (\epsilon_{k-q} - \epsilon_k) a^{\dagger}_{\mathbf{k}-\mathbf{q}} a_{\mathbf{k}} \ , \tag{8.11}$$

where we have again introduced the frequencies

$$\epsilon_k = \frac{E_k}{\hbar} \ . \tag{8.12}$$

For the Coulomb term we obtain

$$\frac{1}{2} \, \frac{i}{\hbar} \sum V_p \; [\, a^{\dagger}_{\boldsymbol{\ell}-\mathbf{p}} a^{\dagger}_{\mathbf{n}+\mathbf{p}} a_{\mathbf{n}} a_{\boldsymbol{\ell}}, \; a^{\dagger}_{\mathbf{k}-\mathbf{q}} a_{\mathbf{k}}]$$

$$= \sum \frac{iV_p}{2\hbar} \Big(a^{\dagger}_{\mathbf{k}-\mathbf{q}-\mathbf{p}} a^{\dagger}_{\mathbf{n}+\mathbf{p}} a_{\mathbf{n}} a_{\mathbf{k}} - a^{\dagger}_{\boldsymbol{\ell}-\mathbf{p}} a^{\dagger}_{\mathbf{k}-\mathbf{q}+\mathbf{p}} a_{\boldsymbol{\ell}} a_{\mathbf{k}}$$

$$+ \, a^{\dagger}_{\mathbf{k}-\mathbf{q}} a^{\dagger}_{\boldsymbol{\ell}-\mathbf{p}} a_{\mathbf{k}-\mathbf{p}} a_{\boldsymbol{\ell}} - a^{\dagger}_{\mathbf{k}-\mathbf{q}} a^{\dagger}_{\mathbf{n}+\mathbf{p}} a_{\mathbf{n}} a_{\mathbf{k}+\mathbf{p}} \Big) \tag{8.13}$$

After renaming **p** to -**p** and using $V_{-p} = V_p \propto 1/p^2$, we see that the first and second and the third and fourth terms become identical. Collecting all contributions of the commutator, we obtain

$$\frac{d}{dt} a^{\dagger}_{\mathbf{k-q}} a_{\mathbf{k}} = i(\epsilon_{k-q} - \epsilon_k) a^{\dagger}_{\mathbf{k-q}} a_{\mathbf{k}}$$

$$+ \frac{i}{\hbar} \sum_{\mathbf{n},\mathbf{p}} V_p \, a^{\dagger}_{\mathbf{k-q-p}} a^{\dagger}_{\mathbf{n+p}} a_{\mathbf{n}} a_{\mathbf{k}} + \frac{i}{\hbar} \sum_{\mathbf{n},\mathbf{p}} V_p \, a^{\dagger}_{\mathbf{k-q}} a^{\dagger}_{\mathbf{n-p}} a_{\mathbf{k-p}} a_{\mathbf{n}} \; . \tag{8.14}$$

Since we are interested in $n_{\mathbf{q}}$, we have to solve Eq. (8.14) and sum over **k**. However we see from Eq. (8.14) that the two-operator dynamics is coupled to four-operator terms. One way to proceed would therefore be to compute as a next step the equation of motion for the four-operator term. Doing this we construct a hierarchy of equations, since the four-operator equation couples to six-operator terms, etc. If we would follow this approach we then would have to truncate the hierarchy at one stage and solve the coupled set of differential equations.

Instead of deriving such a hierarchy of equations, we make a *mean field approximation* (RPA) for the local density with respect to the surrounding density in Eq. (8.14). We pick such combinations of wavenumbers from the sums on the RHS of Eq. (8.14) that two operators $a^{\dagger}_{\mathbf{k}} a_{\mathbf{k}'}$ can be combined to become the density operator $a^{\dagger}_{\mathbf{k}} a_{\mathbf{k}}$, which is then replaced by its expectation value. A hand-waving argument for the RPA approximation is to say that an operator has a time-dependent phase associated with it, such as

$$a_{\mathbf{k}} \propto e^{-i\omega_{\mathbf{k}} t} \; , \; a^{\dagger}_{\mathbf{k}'} \propto e^{i\omega_{\mathbf{k}'} t} \; . \tag{8.15}$$

Since these operators occur under sums, we generate expressions like

$$\sum e^{i(\omega_{\mathbf{k}} - \omega_{\mathbf{k}'})t} \; . \tag{8.16}$$

When we evaluate such sums over all **k**-values we realize that terms with $\mathbf{k} \neq \mathbf{k}'$ oscillate rapidly, averaging out more or less, while the term with $\mathbf{k} = \mathbf{k}'$ gives the dominant contribution.

In the term

$$\frac{i}{\hbar}\sum_{\mathbf{n},\mathbf{p}} V_p\, a^\dagger_{\mathbf{k-q-p}} a^\dagger_{\mathbf{n+p}} a_{\mathbf{n}} a_{\mathbf{k}} \tag{8.17}$$

we choose $\mathbf{p} = -\mathbf{q}$ and obtain

$$\frac{i}{\hbar}\sum_{\mathbf{n},\mathbf{p}} V_p\, a^\dagger_{\mathbf{k-q-p}} a^\dagger_{\mathbf{n+p}} a_{\mathbf{n}} a_{\mathbf{k}} \cong \sum_{\mathbf{n}} \frac{iV_q}{\hbar}\, a^\dagger_{\mathbf{k}} a^\dagger_{\mathbf{n-q}} a_{\mathbf{n}} a_{\mathbf{k}} \,. \tag{8.18}$$

Now we replace the number operator by its expectation value, i.e.,

$$a^\dagger_{\mathbf{k}} a_{\mathbf{k}} \rightarrow \langle a^\dagger_{\mathbf{k}} a_{\mathbf{k}} \rangle = f_k \,, \tag{8.19}$$

where f_k is the Fermi distribution. This leads to

$$\frac{i}{\hbar}\sum_{\mathbf{n},\mathbf{p}} V_p\, a^\dagger_{\mathbf{k-q-p}} a^\dagger_{\mathbf{n+p}} a_{\mathbf{n}} a_{\mathbf{k}} \cong \frac{\mathrm{i}V_q}{\hbar} f_k \sum_{\mathbf{n}} a^\dagger_{\mathbf{n-q}} a_{\mathbf{n}} \,. \tag{8.20}$$

In the term

$$\frac{i}{\hbar}\sum_{\mathbf{n},\mathbf{p}} V_p\, a^\dagger_{\mathbf{k-q}} a^\dagger_{\mathbf{n-p}} a_{\mathbf{k-p}} a_{\mathbf{n}} \,,$$

we select $\mathbf{p} = \mathbf{q}$ and replace $a^\dagger_{\mathbf{k-q}} a_{\mathbf{k-q}} \rightarrow f_{k-q}$ to obtain

$$\frac{i}{\hbar}\sum_{\mathbf{n},\mathbf{p}} V_p\, a^\dagger_{\mathbf{k-q}} a^\dagger_{\mathbf{n-p}} a_{\mathbf{k-p}} a_{\mathbf{n}} \cong \frac{\mathrm{i}V_q}{\hbar} f_{k-q} \sum_{\mathbf{n}} a^\dagger_{\mathbf{n-q}} a_{\mathbf{n}} \,. \tag{8.21}$$

Summarizing the different contributions, we approximate Eq. (8.14) by

$$\frac{d}{dt}\, a^\dagger_{\mathbf{k}-\mathbf{q}} a_\mathbf{k} \cong i(\epsilon_{k-q} - \epsilon_k) a^\dagger_{\mathbf{k}-\mathbf{q}} a_\mathbf{k} - \frac{i}{\hbar} \sum_\mathbf{n} V_q (f_{k-q} - f_k)\, a^\dagger_{\mathbf{n}-\mathbf{q}} a_\mathbf{n} . \tag{8.22}$$

In order to find the eigenfrequencies of the plasma density, we use the ansatz

$$a^\dagger_{\mathbf{k}-\mathbf{q}}(t)\; a_\mathbf{k}(t) = e^{-i(\omega+i\delta)t} a^\dagger_{\mathbf{k}-\mathbf{q}}(0)\; a_\mathbf{k}(0) \;, \tag{8.23}$$

in Eq. (8.22) and get

$$\hbar(\omega + i\delta + \epsilon_{k-q} - \epsilon_k)\; a^\dagger_{\mathbf{k}-\mathbf{q}} a_\mathbf{k} = V_q\; (f_{k-q} - f_k) \sum_\mathbf{n} a^\dagger_{\mathbf{n}-\mathbf{q}} a_\mathbf{n} . \tag{8.24}$$

Dividing both sides by $\hbar(\omega+i\delta+\epsilon_{k-q}-\epsilon_k)$ and summing the resulting equation over $\mathbf{k}$, we find

$$n_\mathbf{q} = V_q\; n_\mathbf{q} \sum_\mathbf{k} \frac{f_{k-q} - f_k}{\hbar(\omega + i\delta + \epsilon_{k-q} - \epsilon_k)} \;. \tag{8.25}$$

We see that $n_\mathbf{q}$ can be cancelled from both sides. Introducing the first-order approximation $P^1(q,\omega)$ to the *polarization function* $P(q,\omega)$ as

$$\frac{4\pi e^2}{\epsilon_0 q^2 L^3}\; P^1(q,\omega) = V_q \sum_\mathbf{k} \frac{f_{k-q} - f_k}{\hbar(\omega + i\delta + \epsilon_{k-q} - \epsilon_k)} \tag{8.26}$$

we can write the solution of Eq. (8.25) as

$$V_q\; P^1(q,\omega) = 1 \tag{8.27}$$

which determines the eigenfrequencies $\omega = \omega_q$

$$V_q \sum_\mathbf{k} \frac{f_{k-q} - f_k}{\hbar(\omega_q + i\delta + \epsilon_{k-q} - \epsilon_k)} = 1 \;. \tag{8.28}$$

To analyze the solutions of Eq. (8.28), let us first discuss the long-wavelength limit for a three dimensional plasma. Long wavelength means $\lambda \to \infty$, and hence $q \propto 1/\lambda \to 0$. We expand the RHS of Eq. (8.28) in terms of q and drop higher-order corrections to get

$$E_{k-q} - E_k = \frac{\hbar^2}{2m}(k^2 - 2\mathbf{k}\cdot\mathbf{q} + q^2) - \frac{\hbar^2 k^2}{2m} \cong -\frac{\hbar^2 \mathbf{k}\cdot\mathbf{q}}{m} \tag{8.29}$$

and

$$f_{k-q} - f_k = f_k - \mathbf{q}\cdot\nabla_{\mathbf{k}} f_k + \cdots - f_k \cong -\mathbf{q}\cdot\nabla_{\mathbf{k}} f_k . \tag{8.30}$$

Inserting these expansions into Eq. (8.28) yields

$$1 \cong -V_q \sum_{\mathbf{k},i} \frac{q_i \dfrac{\partial f}{\partial k_i}}{\hbar\omega_0 - \hbar^2\mathbf{k}\cdot\mathbf{q}/m}$$

$$\cong -\frac{V_q}{\hbar\omega_0} \sum_{\mathbf{k},i} q_i \frac{\partial f}{\partial k_i}\left[1 + \frac{\hbar\mathbf{k}\cdot\mathbf{q}}{m\omega_0}\right], \tag{8.31}$$

where we have set $\omega_{q\to 0} = \omega_0$. The term proportional to $\Sigma\, \partial f/\partial k$ vanishes because it yields the distribution function for $k \to \infty$. So we are left with

$$1 = -\frac{V_q}{\hbar\omega_0} \sum_{\mathbf{k},i} q_i \frac{\partial f}{\partial k_i} \frac{\hbar\mathbf{k}\cdot\mathbf{q}}{m\omega_0} . \tag{8.32}$$

Partial integration gives

$$1 = \frac{V_q q^2}{m\omega_0^2} \sum_{\mathbf{k}} f_k = \frac{V_q q^2 N}{m\omega_0^2} = \frac{4\pi e^2}{\epsilon_0 q^2 L^3} \frac{q^2 N}{m\omega_0^2} , \tag{8.33}$$

or

$$\omega_0^2 = \omega_{pl}^2 = \frac{4\pi e^2 n}{m} . \tag{8.34}$$

Equation (8.34) shows that in the long-wavelength limit, $q \to 0$, $\omega_{q\to 0} = \omega_{pl}$, i.e., we recover the classical result for the eigenfrequency of the electron plasma.

Next we discuss the solution of Eq. (8.28) for general wavelengths. First we write the LHS of Eq. (8.28) in the form

$$\sum_{\mathbf{k}} \frac{f_{k-q}-f_k}{\hbar(\omega_q+i\delta+\epsilon_{k-q}-\epsilon_k)} = \sum_{\mathbf{k}} \frac{f_k}{\hbar}\left[\frac{1}{\omega_q+i\delta+\epsilon_k-\epsilon_{k+q}} - \frac{1}{\omega_q+i\delta+\epsilon_{k-q}-\epsilon_k}\right]. \tag{8.35}$$

This expression shows that the poles occur at

$$\omega_q = \epsilon_{k+q} - \epsilon_k = \frac{\hbar kq}{m}\cos(\theta) + \frac{\hbar q^2}{2m} \tag{8.36}$$

and

$$\omega_q = \epsilon_k - \epsilon_{k-q} = \frac{\hbar kq}{m}\cos(\theta) - \frac{\hbar q^2}{2m}, \tag{8.37}$$

where θ is the angle between **k** and **q**. As schematically shown in Fig. 8.1, we can find the solutions of Eq. (8.28) as the intersections of the LHS of Eq. (8.28) with the straight line "1", which is the RHS of Eq. (8.28). From Fig. 8.1 we see that these intersection points are close to the poles of the LHS. For illustration we discuss in the following the low-temperature situation. Here we know that the extrema of the allowed k values are $k' = \pm k_F$. Considering only $\omega_q > 0$, we obtain from Eq. (8.36)

$$\omega_q^{max} = \frac{\hbar k_F q}{m} + \frac{\hbar q^2}{2m}, \tag{8.38}$$

for $\cos(\theta)=1$ and

$$\omega_q^{min} = -\frac{\hbar k_F q}{m} + \frac{\hbar q^2}{2m}, \tag{8.39}$$

for $\cos(\theta)=-1$. From Eq. (8.37) we get no solution for $\cos(\theta)=-1$ and for $\cos(\theta)=1$ we obtain

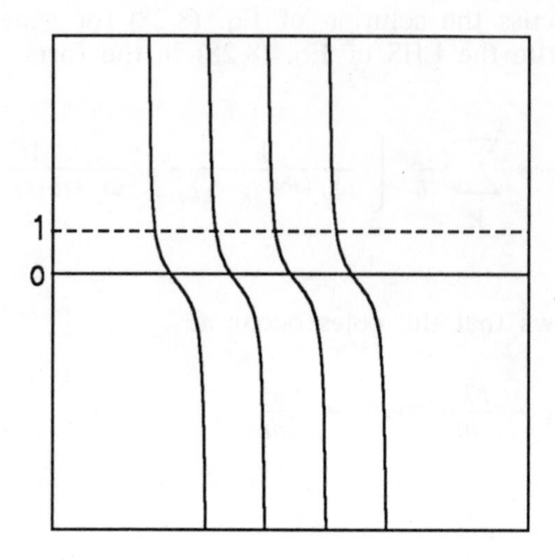

Fig. 8.1: The full lines are a schematic drawing of the LHS of Eq. (8.28) and the dashed line "1" is the RHS of Eq. (8.28).

$$\omega_q^{ext} = \frac{\hbar k_F q}{m} - \frac{\hbar q^2}{2m} . \tag{8.40}$$

As shown in Fig. 8.2, Eqs. (8.38) and (8.39) define two parabolas that are displaced from the origin by $\pm\ k_F$. The region between these parabolas for $\omega_q > 0$ is the region where we find the poles. Physically these solutions represent the transition of an electron from **k** to **k**±**q**, i.e., these are *pair excitations*. They are called pair excitations because the pair of states **k** and **k**±**q** is involved in the transition. The region between the parabolas is therefore called the *continuum of electron-pair excitations*. These pair excitations are not to be confused with electron-hole pairs, which we discuss in later chapters of this book. Note that the pair excitations need an empty final state to occur, and at low temperatures typically involve scattering from slightly below the Fermi surface to slightly above. The lack of empty final states for scattering with small momentum transfer prevents conduction in an insulator, although there is no lack of electrons. When a plasma mode hits the continuum of pair

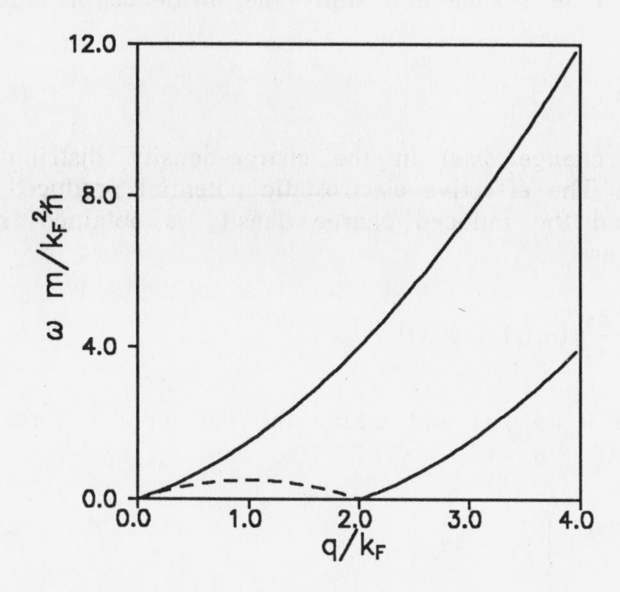

Fig. 8.2: The full lines show the boundary of the continuum of pair excitations at $T = 0\ K$, according to Eqs. (8.38) and (8.39), respectively. The dashed line is the result of Eq. (8.40).

excitations, it gets damped heavily (Landau damping), causing the collective plasmon excitation to decay into pair excitations. At finite temperatures the boundaries of the pair-excitation spectrum are not sharp, but qualitatively the picture remains similar to the $T = 0$ result.

8-2. Plasma Screening

One of the most important effects of the many-body interactions in an electron plasma is the phenomenon of plasma screening. Before we go into the quantum theory of this effect, we briefly review the classical theory.

8-2.1 Classical Theory

We assume a *3d* system and study the influence of a local test charge

$$n_t(\mathbf{r}) = e\,\delta(\mathbf{r}) \tag{8.41}$$

which induces a change $\delta n(\mathbf{r})$ in the charge-density distribution within the plasma. The effective electrostatic potential produced by the test charge and the induced charge density is obtained from Poisson's equation as

$$\nabla^2\phi_{eff}(\mathbf{r}) = -\frac{4\pi}{\epsilon_0}\,[n_t(\mathbf{r}) + \delta n(\mathbf{r})] \ . \tag{8.42}$$

Introducing $V_{eff}(\mathbf{r}) = e\phi_{eff}(\mathbf{r})$ and taking the Fourier transform of Eq. (8.42) yields

$$q^2 V_{eff}(q) = \frac{4\pi e^2}{\epsilon_0}\left[\frac{1}{L^3} + \delta n_q\right] , \tag{8.43}$$

where δn_q is the Fourier transform of the induced charge density

$$\delta n_q = \int \frac{d^3r}{L^3}\,\delta n(\mathbf{r})\,e^{i\mathbf{q}\cdot\mathbf{r}}$$

has been used. Assuming thermal equilibrium in the plasma, the electron density is given by the Fermi distribution function

$$n_q = \frac{1}{L^3}\,\frac{1}{e^{\beta(E_q + V_{eff}(q) - \mu)} + 1} \ . \tag{8.44}$$

Assuming that the change induced by the potential $V_{eff}(q)$ is sufficiently small, we can write

$$\delta n_q \cong -\frac{\partial n}{\partial \mu}\,V_{eff}(q) \ . \tag{8.45}$$

Inserting Eq. (8.45) into Eq. (8.43) yields

$$\left[q^2 + \frac{4\pi e^2}{\epsilon_0}\frac{\partial n}{\partial \mu}\right] V_{eff}(q) = \frac{4\pi e^2}{\epsilon_0 L^3} \; . \tag{8.46}$$

If we now define the inverse *screening length*, i.e., the *screening wavenumber* as

$$\boxed{\kappa = \sqrt{\frac{4\pi e^2}{\epsilon_0}\frac{\partial n}{\partial \mu}}} \tag{8.47}$$

3d screening wavenumber

we find the effective or screened potential

$$\boxed{V_{eff}(q) \equiv V_s(q) = \frac{4\pi e^2}{\epsilon_0 L^3}\,\frac{1}{q^2 + \kappa^2}} \; . \tag{8.48}$$

3d statically screened Coulomb potential

Introducing the static dielectric function $\epsilon(q,0) \equiv \epsilon(q,\omega{=}0)$ through the relation

$$V_s(q) = \frac{V_q}{\epsilon(q,0)}, \tag{8.49}$$

we obtain

$$\boxed{\epsilon(q,0) = 1 + \frac{\kappa^2}{q^2}} \tag{8.50}$$

3d static dielectric function

We can see already from this classical result that the plasma screening removes the divergence at $q \to 0$ from the Coulomb potential. Taking the Fourier transformation of Eq. (8.48) yields, compare Eq. (7.13),

$$V_s(r) = \sum_{\mathbf{q}} \frac{4\pi e^2}{\epsilon_0 L^3 (q^2+\kappa^2)}\, e^{i\mathbf{q}\cdot\mathbf{r}} = \frac{e^2}{\epsilon_0 r}\, e^{-\kappa r} \; . \tag{8.51}$$

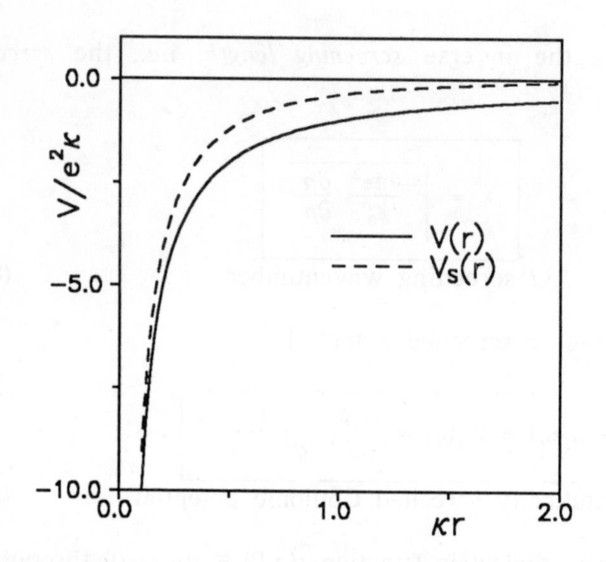

Fig. 8.3: Screened (dashed line) and unscreened (full line) Coulomb potential for a three-dimensional system.

This screened Coulomb potential is plotted in Fig. 8.3 together with the bare potential. The comparison shows that the long-ranged bare Coulomb potential is screened to a distance $1/\kappa$. The screened Coulomb potential, Eq. (8.51), is often called *Yukawa potential*.

8-2.2 Quantum Theory

In the remainder of this chapter we now discuss the quantum theory of plasma screening. For this purpose we start with the effective single-particle Hamiltonian

$$\mathcal{H} = \int d^3r \; \psi^\dagger(\mathbf{r}) \left[- \frac{\hbar^2 \nabla^2}{2m} \right] \psi(\mathbf{r}) + \int d^3r \; V_{eff}(r) \; \psi^\dagger(\mathbf{r}) \psi(\mathbf{r}) \; , \quad (8.52)$$

where

$$V_{eff}(r) = V(r) + V_s(r) \tag{8.53}$$

is the sum of the Coulomb potential $V(r)$ of a test charge and the induced potential $V_s(r)$ of the screening particles. The effective potential V_{eff} is different to V_{eff} derived from the classical theory, Eq. (8.48), and has to be determined self-consistently. The Fourier transform of Eq. (8.52) is

$$\mathscr{H} = \sum_{\mathbf{k}} E_k a_{\mathbf{k}}^\dagger a_{\mathbf{k}} + L^3 \sum_{\mathbf{p}} V_{eff}(p)\, n_{-\mathbf{p}} \,, \tag{8.54}$$

and the equation of motion for $a_{\mathbf{k-q}}^\dagger a_{\mathbf{k}}$ is

$$\frac{d}{dt}\, a_{\mathbf{k-q}}^\dagger a_{\mathbf{k}} = \frac{i}{\hbar}\,[\mathscr{H},\, a_{\mathbf{k-q}}^\dagger a_{\mathbf{k}}]$$

$$= i\,(\epsilon_{k-q} - \epsilon_k)\, a_{\mathbf{k-q}}^\dagger a_{\mathbf{k}} - \frac{i}{\hbar} \sum_{\mathbf{p}} V_{eff}(p)\, (a_{\mathbf{k-q}}^\dagger a_{\mathbf{k+p}} - a_{\mathbf{k-p-q}}^\dagger a_{\mathbf{k}}). \tag{8.55}$$

Using the random phase approximation in the last two terms and taking the expectation value yields

$$\frac{d}{dt}\, \langle a_{\mathbf{k-q}}^\dagger a_{\mathbf{k}} \rangle = i\,(\epsilon_{k-q} - \epsilon_k)\, \langle a_{\mathbf{k-q}}^\dagger a_{\mathbf{k}} \rangle - \frac{iV_{eff}(q)}{\hbar}\,(f_{k-q} - f_k)\,. \tag{8.56}$$

We assume that the test charge varies periodically in time as $\exp(-i(\omega+i\delta)t)$, where $\omega+i\delta$ establishes an adiabatic switch-on of the test charge potential. Making the ansatz that the driven density has the same time dependence

$$\langle a_{\mathbf{k-q}}^\dagger a_{\mathbf{k}} \rangle \propto e^{-i(\omega+i\delta)t} \,, \tag{8.57}$$

Eq. (8.56) yields

$$\hbar(\omega+i\delta+\epsilon_{k-q}-\epsilon_k)\, \langle a_{\mathbf{k-q}}^\dagger a_{\mathbf{k}} \rangle = V_{eff}(q)\,(f_{k-q} - f_k)\,, \tag{8.58}$$

and therefore

$$\langle a^{\dagger}_{\mathbf{k-q}} a_{\mathbf{k}} \rangle = V_{eff}(q) \frac{(f_{k-q} - f_k)}{\hbar(\omega + i\delta + \epsilon_{k-q} - \epsilon_k)} , \tag{8.59}$$

or, after division and summation over **k**,

$$\langle n_q \rangle = \frac{1}{L^3} V_{eff}(q) P^1(q,\omega), \tag{8.60}$$

where P^1 again is the polarization function defined in Eq. (8.26). The potential of the screening particles obeys Poisson's equation in the form

$$\nabla^2 V_s(r) = -\frac{4\pi e^2 n(r)}{\epsilon_0} , \tag{8.61}$$

or

$$V_s(q) = \frac{4\pi e^2}{\epsilon_0 q^2} n(q,\omega) = \frac{4\pi e^2}{\epsilon_0 q^2 L^3} V_{eff}(q) P^1(q,\omega) , \tag{8.62}$$

Inserting this result into the Fourier-transform of Eq. (8.53) yields

$$V_{eff}(q) = V_q \left[1 + V_{eff}(q) P^1(q,\omega) \right] \tag{8.63}$$

or

$$V_{eff}(q) = \frac{V_q}{1 - V_q P^1(q,\omega)} = \frac{V_q}{\epsilon(q,\omega)} \equiv V_s(q,\omega) \tag{8.64}$$

where $V_s(q,\omega)$ is the dynamically screened Coulomb potential and $\epsilon(q,\omega)$ is given by

$$\epsilon(q,\omega) = 1 - V_q P^1(q,\omega) . \tag{8.65}$$

Inserting Eq. (8.26) into Eq. (8.65), we finally obtain

$$\boxed{\epsilon(q,\omega) = 1 - V_q \sum_{\mathbf{k}} \frac{f_{k-q} - f_k}{\hbar(\omega + i\delta + \epsilon_{k-q} - \epsilon_k)}} \tag{8.66}$$

Lindhard formula
for the longitudinal dielectric function

The Lindhard formula describes a complex retarded dielectric function, i.e., the poles are in the lower complex frequency plane, and it includes spatial dispersion (q dependence) and temporal dispersion (ω dependence). Eq. (8.66) is valid both in 3 and 2 dimensions. In the derivation we sometimes used the *3d* expressions, but that could have been avoided without changing the final result.

The longitudinal plasma eigenmodes are obtained from

$$\boxed{\epsilon(q,\omega) = 0 \text{ or } 1 = V_q \, P^1(q,\omega)}$$

longitudinal eigenmodes (8.67)

This equation is identical to the plasma eigenmode equation (8.27). Hence, our discussion of plasma screening of the Coulomb potential and of the collective plasma oscillations obtained from $\epsilon(q, \omega) = 0$, shows that screening and plasmons are intimately related phenomena.

8-3. Analysis of the Lindhard Formula

To appreciate the Lindhard (or RPA) result, we discuss some important limiting cases in *3d* and *2d* systems.

A) Three Dimensions

In the long-wavelength limit, $q \to 0$, we repeat the steps described by Eqs. (8.29) - (8.34) to obtain

$$\epsilon(0,\omega) = 1 - \frac{\omega_{pl}^2}{\omega^2}, \tag{8.68}$$

the classical (or Drude) dielectric function, which is the same as the result obtained for the oscillator model in Chap. 1.

In the static limit, $\omega \to 0$, Eq. (8.66) yields

$$\epsilon(q,0) = 1 - V_q \sum_{\mathbf{k}} \frac{f_{k-q} - f_k}{E_{k-q} - E_k}. \tag{8.69}$$

Using again the expansions (8.29) and (8.30), where we now write

$$\sum_i q_i \frac{\partial f_k}{\partial k_i} = - \sum_i q_i \frac{\partial f_k}{\partial \mu} \frac{\partial \epsilon_k}{\partial k_i} = - \sum_i q_i k_i \frac{\hbar^2}{m} \frac{\partial f_k}{\partial \mu} , \tag{8.70}$$

we find

$$\begin{aligned} \epsilon(q,0) &= 1 + \frac{4\pi e^2}{\epsilon_0 q^2} \frac{\partial}{\partial \mu} \frac{1}{L^3} \sum_{\mathbf{k}} f_k \\ &= 1 + \frac{4\pi e^2}{\epsilon_0 q^2} \frac{\partial n}{\partial \mu} = 1 + \frac{\kappa^2}{q^2} \end{aligned} \tag{8.71}$$

which is identical to our classical result, Eq. (8.50).

The inverse screening length given by Eq. (8.47) can be calculated explicitly for the two limiting cases of i) a degenerate electron gas where the Fermi function is the unit-step function (Thomas-Fermi theory of screening) and ii) for the non-degenerate case where the distribution function is approximated as Boltzmann distribution (Debye-Hückel theory of screening). Using Eqs. (6.15) and (6.16) for the degenerate electron gas one can write Eq. (8.71) as

$$\kappa^2 = \frac{4\pi e^2}{\epsilon_0} \frac{\partial n}{\partial k_F} \frac{\partial k_F}{\partial \mu} \tag{8.72}$$

with the result

$$\boxed{\kappa = \sqrt{\frac{6\pi e^2 n}{\epsilon_0 E_F}}}$$

3d Thomas-Fermi screening wavenumber (8.73)

The chemical potential for a nondegenerate Boltzmann distribution is given in Eq. (6.27) and

$$\frac{\partial \mu}{\partial n} = \frac{1}{\beta n} . \tag{8.74}$$

Inserting this result into Eq. (8.47) we obtain

$$\boxed{\kappa = \sqrt{\frac{4\pi e^2 n\beta}{\epsilon_0}}}$$

3d Debye-Hückel screening wavenumber (8.75)

B) Two Dimensions

To investigate the long-wavelength limit of the Lindhard formula for a two-dimensional system, we again insert the expansions (8.29) - (8.31) into Eq. (8.66) to obtain

$$\epsilon(q\to 0, \omega) - 1 = V_q \frac{L^2}{m\omega^2}\, 2 \int \frac{d^2k}{(2\pi)^2} \sum_{i,j} q_i q_j k_j \left.\frac{\partial f_k}{\partial k_i}\right\}, \qquad (8.76)$$

where the factor 2 comes from the spin summation implicitly included in $\Sigma_{\mathbf{k}}$. Partial integration on the RHS of Eq. (8.76) yields with

$$2 \int \frac{d^2k}{(2\pi)^2}\, k_j \frac{\partial f_k}{\partial k_i} = -\, 2 \int \frac{d^2k}{(2\pi)^2}\, f_k \frac{\partial k_i}{\partial k_j} = -\, n\, \delta_{ij}$$

the Drude result

$$\epsilon(q\to 0, \omega) = 1 - \frac{\omega_{pl}^2(q)}{\omega^2}\,, \qquad (8.77)$$

with the *2d* plasma frequency

$$\boxed{\omega_{pl}(q) = \sqrt{\frac{2\pi e^2 nq}{\epsilon_0 m}}}\,,$$

2d plasma frequency (8.78)

where n is the *2d* particle density N/L^2.

To study the static limit, ω=0, of the Lindhard formula, we use the *2d* Coulomb potential, Eq. (7.36) and obtain

$$\epsilon(q,0) = 1 + V_q \frac{\partial n}{\partial \mu}$$

or

$$\boxed{\epsilon(q,0) = 1 + \frac{\kappa}{q}}$$

2d static dielectric function (8.79)

where the inverse screening length in quasi-two dimensions is given as

$$\kappa = \frac{2\pi e^2}{\epsilon_0} \frac{\partial n}{\partial \mu} \tag{8.80}$$

Hence, the statically screened *2d* Coulomb potential is

$$V_s(q) = \frac{2\pi e^2}{\epsilon_0 L^2} \frac{1}{q + \kappa} . \tag{8.81}$$

For the chemical potential of the two-dimensional Fermi gas we have the explicit result given in Eq. (6.41) yielding

$$\frac{\partial \mu}{\partial n} = \frac{\hbar^2 \pi}{m} \frac{1}{1 - e^{-(\hbar^2 \beta \pi n/m)}} \tag{8.82}$$

and thus

$$\boxed{\kappa = \frac{2me^2}{\epsilon_0 \hbar^2} (1 - e^{-\hbar^2 \beta \pi n/m})}$$

2d screening wavenumber (8.83)

This expression is correct for all densities and temperatures. It is interesting to note that the screening wavenumber in *2d* becomes n-independent for low temperatures and high densities, whereas the corresponding *3d* result always remains n-dependent.

8-4. Plasmon-pole approximation

Eqs. (8.68) and (8.77) show that in the long-wavelength limit, both in *3d* and *2d*, the inverse dielectric function

$$\frac{1}{\epsilon(q\to 0,\omega)} = \frac{\omega^2}{(\omega + i\delta)^2 - \omega_{pl}^2} = 1 + \frac{\omega_{pl}^2}{(\omega + i\delta)^2 - \omega_{pl}^2} \tag{8.84}$$

has just one pole. We use this observation to construct an approximation for the full dielectric function $\epsilon(q,\omega)$, which tries to replace the continuum of poles contained in the Lindhard formula by one effective plasmon pole at ω_q

$$\boxed{\frac{1}{\epsilon(q,\omega)} = 1 + \frac{\omega_{pl}^2}{(\omega + i\delta)^2 - \omega_q^2}}$$

plasmon-pole approximation (8.85)

The effective plasmon frequency ω_q is chosen to fulfill certain sum rules (Mahan, 1981) which can be derived from the Kramers-Kronig relation, Eq. (1.23),

$$\epsilon'(q,\omega) - 1 = \frac{2}{\pi} P\int_0^\infty d\omega' \frac{\omega'\epsilon''(q,\omega')}{\omega'^2 - \omega^2} \tag{8.86}$$

We make use of the static long-wavelength limit, which for *3d* is given by Eq. (8.50),

$$\epsilon'(q,0) - 1 = \frac{\kappa^2}{q^2} = \frac{2}{\pi} \lim_{q\to 0} P\int_0^\infty d\omega' \frac{\epsilon''(q,\omega')}{\omega'} . \tag{8.87}$$

This is the so-called conductivity sum rule. According to Eq. (8.85), we have

$$\epsilon(q,\omega) = 1 + \frac{\omega^2 - \omega_q^2}{(\omega+i\delta-\Omega_q)(\omega+i\delta+\Omega_q)} \tag{8.88}$$

where

$$\Omega_q^2 = \omega_q^2 - \omega_{pl}^2 \,. \tag{8.89}$$

Using the Dirac identity in Eq. (8.88) and inserting the result into the RHS of Eq. (8.87), we find

$$\frac{\kappa^2}{q^2} = \lim_{q\to 0} \frac{\omega_{pl}^2}{\Omega_q^2} \,, \tag{8.90}$$

which yields

$$\lim_{q\to 0} \omega_q^2 = \omega_{pl}^2 \left[1 + \frac{q^2}{\kappa^2} \right] . \tag{8.91}$$

Following Lundquist (1967), we therefore choose the form

$$\boxed{\omega_q^2 = \omega_{pl}^2 \left[1 + \frac{q^2}{\kappa^2} \right] + \nu_q^2}$$

3d effective plasmon frequency (8.92)

The last term ν_q^2 in this equation is added in order to simulate the contribution of the pair continuum. Usually one chooses

$$\nu_q^2 = C\, q^4 \tag{8.93}$$

where C is a numerical constant. Practical applications show that it is often sufficient to use the much simpler plasmon-pole approximation instead of the full RPA dielectric function to obtain good qualitative results for the effects of screening.

Similarly, one gets for two-dimensional systems (see problem 8.2)

$$\boxed{\omega_q^2 = \omega_{pl}^2(q) \left[1 + \frac{q}{\kappa} \right] + \nu_q^2}$$

2d effective plasmon frequency (8.94)

REFERENCES

G.D. Mahan, *Many Particle Physics*, Plenum Press, New York (1981)

B.I. Lundquist, Phys. Konden. Mat. **6**, 193 and 206 (1967)

D. Pines and P. Nozieres, *The Theory of Quantum Liquids*, Benjamin, Reading, Mass. (1966)

For the modifications of the plasmon-pole approximation in an electron-hole plasma see, e.g.,

R. Zimmermann, *Many-Particle Theory of Highly Excited Semiconductors*, Teubner, Berlin (1988)

H. Haug and S. Schmitt-Rink, Prog. Quantum Electron. **9**, 3 (1984).

PROBLEMS

Problem 8.1: Use the quasi-*2d* Coulomb potential

$$V_k = \frac{2\pi e^2}{L^2 \epsilon_0 k}$$

and apply the classical theory outlined in Chap. 1 to verify that the *2d* plasma frequency is given by Eq. (8.78).

Problem 8.2: Derive the effective plasmon frequency, Eq. (8.94), of the *2d* plasmon-pole approximation.

Chapter 9
RETARDED GREEN'S FUNCTION FOR ELECTRONS

In the treatment of the classical oscillator model in Chap. 1, we have seen how a retarded Green's function determines the response of the oscillator to a driving field. In this chapter we will now discuss how a retarded Green's function can also be introduced for quantum-mechanical many-body systems, such as the electron gas of Chap. 7. The retarded Green's function contains information about the spectral properties of the system, i.e., about the changes of the single-particle energies which occur due to the interactions with other particles or fields. Furthermore, retarded Green's functions of many-body systems determine the linear response to external fields in the same way as for the classical oscillator.

9-1. Definitions

We define the retarded Green's function for electrons as

$$G^r_{ss'}(\mathbf{r}t,\mathbf{r}'t') = -\frac{i}{\hbar}\,\theta(t-t')\,\langle[\psi_s(\mathbf{r},t)\,,\,\psi^\dagger_{s'}(\mathbf{r}',t')]_+\rangle \tag{9.1}$$

The average $\langle....\rangle$ stands for $tr\rho$ .... , where the statistical operator ρ is taken at some fixed initial time, e.g., t=-∞. Due to the step function $\theta(t-t')$, G^r is retarded with $G^r(t,t') = 0$ for $t' > t$.

The following manipulations show that, as its classical counterpart, also the quantum mechanical Green's function G^r obeys a differential equation with an inhomogeneity given by delta functions in space and time. The equation of motion for the retarded Green's function is

$$i\hbar \frac{\partial}{\partial t} G^r_{ss}(\mathbf{r}t,\mathbf{r}'t')$$

$$= \frac{\partial\theta(t-t')}{\partial t} \langle[\psi_s(\mathbf{r},t) \ , \ \psi^\dagger_s(\mathbf{r}',t')]_+\rangle - \frac{i}{\hbar}\theta(t-t') \left\langle \left[i\hbar \frac{\partial\psi_s(\mathbf{r},t)}{\partial t} \ , \ \psi^\dagger_{s'}(\mathbf{r}',t') \right]_+ \right\rangle$$

$$= \delta(t-t')\langle[\psi_s(\mathbf{r},t) \ , \ \psi^\dagger_s(\mathbf{r}',t)]_+\rangle$$

$$- \frac{i}{\hbar}\theta(t-t') \langle[\ [\psi_s(\mathbf{r},t) \ , \ \mathscr{H}(t)] \ , \ \psi^\dagger_{s'}(\mathbf{r}',t')]_+\rangle, \tag{9.2}$$

where the Heisenberg equation $i\hbar \ \partial\psi/\partial t = [\psi,\mathscr{H}]$ has been used. For most practical calculations it is sufficient to consider only Green's functions which are diagonal in the spin index since the Coulomb interaction is spin independent. The equal time anti-commutator in the first term of (9.2) is

$$[\psi_s(\mathbf{r},t) \ , \ \psi^\dagger_s(\mathbf{r}'t)]_+ = \delta(\mathbf{r}-\mathbf{r}') \tag{9.3}$$

so that indeed a differential equation for the Green's function results which has a singular inhomogeneity in space and time

$$i\hbar \frac{\partial}{\partial t} G^r_{ss}(\mathbf{r}t,\mathbf{r}'t') =$$

$$\delta(t-t')\delta(\mathbf{r}-\mathbf{r}')) - \frac{i}{\hbar}\theta(t-t')\langle[\ [\psi_s(\mathbf{r},t) \ , \ \mathscr{H}(t)] \ , \ \psi^\dagger_{s'}(\mathbf{r}',t')]_+\rangle. \tag{9.4}$$

Let us first consider the ideal Fermi gas with the Hamiltonian (6.4). Evaluating the commutator in the second term on the RHS of Eq. (9.4) we obtain the equation of motion for the retarded Green's function, also called *Dyson equation*, for this simple case as

$$\left[i\hbar \frac{\partial}{\partial t} + \frac{\hbar^2\nabla^2}{2m} \right] G^r_{0,ss}(\mathbf{r}t,\mathbf{r}'t') = \delta(t-t')\delta(\mathbf{r}-\mathbf{r}') \ . \tag{9.5}$$

Here we have introduced the subscript 0 to indicate hat we are dealing with the non-interacting situation. Eq. (9.5) shows that G^r_0 depends only on the relative coordinates $\boldsymbol{\rho}$=$\mathbf{r}-\mathbf{r}'$ and τ=t-t'. Taking the Fourier transform of Eq. (9.5) with respect $\boldsymbol{\rho}$ and τ and introducing the notation

$$G^r_{0,ss}(\boldsymbol{\rho},\tau) = \sum_{\mathbf{k}} \int \frac{d\omega}{2\pi} e^{i(\mathbf{k}\boldsymbol{\rho} - \omega\tau)} G^r_{0,ss}(\mathbf{k},\omega) , \tag{9.6}$$

the Dyson equation for the ideal Fermi gas becomes

$$\left[\hbar\omega - \frac{\hbar^2 k^2}{2m}\right] G^r_{0,ss}(\mathbf{k},\omega) = 1 \tag{9.7}$$

or

$$G^r_{0,ss}(\mathbf{k},\omega) = \frac{1}{\hbar} \frac{1}{\omega - \epsilon_k + i\delta} , \tag{9.8}$$

with $\hbar\epsilon_k = \hbar^2k^2/2m$. In Eq. (9.8) we have included the infinitesimal imaginary part $+i\delta$ which shifts the poles to the lower half of the complex frequency plane. As discussed in Chap. 1, this shift guarantees the correct retardation, i.e., $G^r(\tau<0)=0$. It is seen from Eq. (9.8) that the poles of the retarded Green's function give the single-particle energies ϵ_k, or actually single-particle frequencies, but, as usual in the literature, we will not stress this trivial difference of a factor $\hbar$.

9-2. Interacting Electron Gas

Using the electron gas Hamiltonian (7.1), we obtain for the commutator in Eq. (9.4)

$$\begin{aligned}[\psi_s(\mathbf{r},t) , \mathscr{H}(t)] = &- \frac{\hbar^2\nabla^2}{2m} \psi_s(\mathbf{r},t) \\ &+ \sum_{s'} \int d^3r'' V(\mathbf{r}-\mathbf{r}'') \psi^\dagger_{s'}(\mathbf{r}'',t)\psi_{s'}(\mathbf{r}'',t)\psi_s(\mathbf{r},t) \\ &- \int d^3r'' V(\mathbf{r}-\mathbf{r}'') n(\mathbf{r}'')\psi_s(\mathbf{r},t) ,\end{aligned} \tag{9.9}$$

where $n(\mathbf{r})$ is the background charge distribution. Inserting this result into Eq. (9.4), we see that the Coulomb interaction gives rise to two additional terms in the Dyson equation, which are

$$- \frac{i}{\hbar}\, \theta(t-t') \int d^3r''\ V(\mathbf{r}-r'')\ \{\ \langle[\psi^{\dagger}_{s'}(r''\mathrm{t})\psi_{s'}(r''\mathrm{t})\psi_s(r,t)\ ,\ \psi^{\dagger}_s(r',\mathrm{t}')]_+\rangle$$

$$- \langle[\psi_s(\mathbf{r},t)n(\mathbf{r}'')\ ,\ \psi^{\dagger}_s(\mathbf{r}',t')]_+\rangle\ \}. \tag{9.10}$$

Hence, as a consequence of the interaction, also higher order Green's functions with more that two electron operators appear. The simplest possible approximation is to factorize the expectation value of products of four operators in the first term in (9.10) into products of expectation values of two electron operators as indicated by the arrows

$$- \frac{i}{\hbar}\, \theta(t-t')\ \langle[\psi^{\dagger}_{s'}(\mathbf{r}''t)\psi_{s'}(\mathbf{r}''t)\psi_s(\mathbf{r},t), \psi^{\dagger}_s(\mathbf{r}',t')]_+\rangle$$

$$= \langle\psi^{\dagger}_{s'}(\mathbf{r}''t)\psi_{s'}(\mathbf{r}''t)\rangle\ G^r_{ss}(\mathbf{r}t,\mathbf{r}'t') - \langle\psi^{\dagger}_{s'}(\mathbf{r}''t)\psi_s(\mathbf{r},t)\rangle\ G^r_{s's}(\mathbf{r}''t,\mathbf{r}'t'). \tag{9.11}$$

The minus sign in the second term arises because we had to commute two electron annihilation operators before taking the expectation values. Note, that the second term, which is the exchange term, contributes only if $s=s'$, because two states with opposite spins are not correlated. Inserting Eq. (9.11) into Eq. (9.10) yields

$$\int d^3r''V(\mathbf{r}-r'')\{\ (\ \langle\psi^{\dagger}_{s'}(r''\mathrm{t})\psi_{s'}(r''\mathrm{t})\rangle - n(r'')\ )\ G^r_{ss}(r\mathrm{t},r'\mathrm{t}')$$

$$- \delta_{ss'}\ \langle\psi^{\dagger}_s(\mathbf{r}''t)\psi_s(\mathbf{r},t)\rangle\ G^r_{s's}(\mathbf{r}''t,\mathbf{r}'t')\}. \tag{9.12}$$

The first term, also called the *Hartree term*, contributes only if charge neutrality is locally disturbed, i.e.,

$$\langle\psi^{\dagger}_{s'}(\mathbf{r}''t)\psi_{s'}(\mathbf{r}''t)\rangle \neq n(\mathbf{r}'')\ . \tag{9.13}$$

However, the second term, also called *Fock* or *exchange term*, always yields an energy reduction due to the exchange-hole around each Fermion, as discussed in Chap. 7.

Generally, all those terms in the Dyson equation which appear as consequence of the interaction, can be written in the form

$$\left[i\hbar\frac{\partial}{\partial t} + \frac{\hbar^2\nabla^2}{2m}\right]G^r_{ss}(\mathbf{r}t,\mathbf{r}'t') =$$

$$\delta(t-t')\delta(\mathbf{r}-\mathbf{r}')) + \int d^3r''dt'' \ \hbar\Sigma^r_{ss}(\mathbf{r}t,\mathbf{r}''t'')G^r_{ss}(\mathbf{r}''t'',\mathbf{r}'t'), \tag{9.14}$$

where we have introduced Σ^r as the *retarded self-energy*. In order to reproduce the results of our factorization approximation, we have to choose the Hartree-Fock self-energy as

$$\hbar\Sigma^r_{ss}(\mathbf{r}t,\mathbf{r}''t'') =$$

$$\delta(t-t'')\left\{\delta(\mathbf{r}-r'')\int d^3\mathbf{r}' \ V(\mathbf{r}-\mathbf{r}')\left[\sum_{s'}\langle\psi^\dagger_{s'}(\mathbf{r}'t)\psi_{s'}(\mathbf{r}'t)\rangle - n(\mathbf{r}')\right]\right.$$

$$\left. - \ V(\mathbf{r}-\mathbf{r}'')\langle\psi^\dagger_s(\mathbf{r}''t)\psi_s(\mathbf{r},t)\rangle\right\}, \tag{9.15}$$

where the second term is the *exchange self-energy*. The Hartree-Fock self-energy approximation is instantaneous, i.e., not retarded.

For homogeneous stationary systems G^r and Σ^r depend only on the relative coordinates $\mathbf{r}-\mathbf{r}'=\boldsymbol{\rho}$ and $t-t'=\tau$, and not on the center-of-mass coordinates $(\mathbf{r}+\mathbf{r}')/2$ and $(t+t')/2$. Taking the Fourier transform with respect to $\boldsymbol{\rho}$ and τ, as defined in Eq. (9.6), the Dyson equation (9.14) takes the form

$$\left[\hbar\omega - \frac{\hbar^2k^2}{2m}\right]G^r_{ss}(\mathbf{k},\omega) = 1 + \hbar\Sigma^r_{ss}(\mathbf{k},\omega)G^r_{ss}(\mathbf{k},\omega)$$

or

$$\boxed{G^r_{ss}(\mathbf{k},\omega) = \frac{1}{\hbar}\ \frac{1}{\omega + i\delta - \epsilon_k - \Sigma^r_{ss}(\mathbf{k},\omega)}}$$

retarded electron Green's function (9.16)

Eq. (9.16) shows that the free-particle energy ϵ_k is replaced by ϵ_k +

$\Sigma^r(\mathbf{k},\omega)$ as a consequence of the interactions, thus explaining the origin of the name *self-energy*. Usually, the self-energy is a complex frequency-dependent function. In the static limit (ω=0), the real part of Σ^r describes a shift of the single-particle energies due to the many-body interactions, and the imaginary part describes the corresponding damping (inverse damping time).

As illustrative example we now evaluate the self-energy, Eq. (9.15), for spatially and temporally homogeneous systems. As mentioned earlier, the Hartree term vanishes in this case as a consequence of charge neutrality. For the exchange term we get

$$\hbar\Sigma^r_{exc,ss}(\mathbf{k},\omega) = -\sum_{\mathbf{q}} V_{|\mathbf{k}-\mathbf{q}|}\, n_{\mathbf{q}} \,, \tag{9.17}$$

where $\psi(\mathbf{r},t) = \sum_{\mathbf{k}} e^{i\mathbf{kr}} a_{\mathbf{k}}(t)$ (with $L^3 \equiv 1$ for simplicity) and

$$\langle\psi^\dagger(\mathbf{r}'',t)\psi(\mathbf{r},t)\rangle = \sum_{\mathbf{k},k'} \langle a^\dagger_{\mathbf{k}'}(t) a_{\mathbf{k}}(t)\rangle\; e^{i(\mathbf{kr}-\mathbf{k}'\mathbf{r}'')}$$

$$= \sum_{\mathbf{k}} n_{\mathbf{k}}\, e^{i\mathbf{k}(\mathbf{r}-\mathbf{r}'')} \tag{9.18}$$

have been used. For a thermal electron distribution $n_{\mathbf{k}}$ is given by the Fermi distribution f_k, and the exchange self-energy is

$$\hbar\Sigma^r_{exc,ss}(\mathbf{k},\omega) = -\sum_{\mathbf{q}} V_{|\mathbf{k}-\mathbf{q}|}\, f_q \,. \tag{9.19}$$

Hence, as a consequence of the exchange interaction, all single-particle energies ϵ_k are replaced by the renormalized energies

$$e_k = \epsilon_k + \Sigma^r_{exc}(k) \,. \tag{9.20}$$

In order to estimate the numerical value of the energy shift for the case of a degenerate electron plasma, we approximate the relevant momentum transfer by the Fermi wavenumber

$$|\mathbf{k}-\mathbf{q}| \cong k_F . \tag{9.21}$$

Then Eq. (9.19) yields

$$\begin{aligned}\hbar\Sigma^r_{exc,ss} &\cong - \sum_{\mathbf{q}} V_{k_F} f_q \\ &= - \frac{1}{2} V_{k_F} n ,\end{aligned} \tag{9.22}$$

where the factor 1/2 appears since we do not sum over the spin. Using the expression for the Fermi wavenumber, Eq. (6.18), and for the three-dimensional Coulomb potential, Eq. (7.13), we see that the exchange energy varies with the plasma density as

$$\hbar\Sigma^r_{x,ss} \cong - \frac{2}{3^{2/3}\pi^{1/3}} \frac{e^2}{\epsilon_0} n^{1/3} . \tag{9.23}$$

Eq. (9.23) shows that the energy reduction per particle increases with increasing plasma density n. This increase is proportional to $n^{1/3}$, i.e., proportional to the inverse mean inter-particle distance. Basically the same result has already been obtained in Chap. 7, Eq. (7.26), where we analyzed the total exchange energy of the system. All these results are a consequence of the net energy gain (reduced Coulomb repulsion) resulting form the Pauli exclusion principle causing particles with equal spin to avoid each other. Qualitatively, we can estimate this energy gain as the Coulomb energy $e^2/\epsilon_0 r$ evaluated for $r=n^{-1/3}$.

For a quasi-two dimensional system, we have to evaluate Eq. (9.22) with V_q and k_F given by Eqs. (7.36) and (7.46), respectively. The resulting exchange self-energy varies as $-n^{1/2}$, again in agreement with the corresponding result in Eq. (7.37). In two dimensions the energy gain due to the exchange hole is $e^2/\epsilon_0 n^{-1/2}$, where $n^{-1/2}$ measures the distance between particles.

9-3. Screened Hartree-Fock Approximation

Semiconductors under optical excitation or with an injected current are ideally suited to study the density-dependent renormalizations of the single-particle energies, because one can vary the plasma density in these systems over many orders of magnitude. To obtain a realistic estimate for the single-particle energy renormalization, we now make use of the plasmon-pole approximation. Basically, we extend the Hartree-Fock approximation of the previous section by replacing the bare Coulomb potential V_q with the screened one, $V_s(q,\omega) = V_q/\epsilon(q,\omega)$. However, we will see that we have to include at the same time a correlation self-enegy, the so-called Coulomb-hole self-energy, which describes the energy reduction due to the depletion shell around a given charged particle.

As a consequence of the frequency dependence of $V_s(q,\omega)$, the resulting retarded self-energy is also frequency-dependent and cannot simply be interpreted as single-particle energy shift. In order to avoid this complication we restrict our treatment to the static approximation

$$V_s(q) = \frac{V_q}{\epsilon(q,0)} \ . \tag{9.24}$$

Replacing V_q by $V_s(q)$ in the exchange self-energy, Eq. (9.19), we obtain

$$\boxed{\hbar\Sigma^r_{exc,ss}(k) = -\sum_{\mathbf{k}'} V_s(|\mathbf{k}-\mathbf{k}'|)\, f_{k'} \equiv \hbar\Sigma^r_{SX}(k)}$$

screened exchange self-energy (9.25)

In Sec. 9-2 we have discussed that the momentum transfer in a degenerate plasma is of the order of k_F, allowing us to approximate Eq. (9.25) as

$$\hbar\Sigma^r_{SX}(k{\cong}0) \cong -\frac{1}{2}\, n\, V_s(k_F) \tag{9.26}$$

This expression can be evaluated using the plasmon-pole approximation. We obtain for the three-dimensional system

$$\hbar\Sigma_{SX}^{r} \cong -\frac{1}{2}\, n\, \frac{4\pi e^2}{\epsilon_0 k_F^2}\left[1 - \frac{\omega_{pl}^2}{\omega_{k_F}^2}\right] \tag{9.27}$$

where ω_q^2 and ω_{pl}^2 are given by Eqs. (8.92) and (8.34), respectively. In the two-dimensional case we get

$$\hbar\Sigma_{SX}^{r} \cong -\frac{1}{2}\, n\, \frac{2\pi e^2}{\epsilon_0 k_F}\left[1 - \frac{\omega_{pl}^2(k_F)}{\omega_{k_F}^2}\right] \tag{9.28}$$

where ω_q^2 and ω_{pl}^2 are defined in Eqs. (8.94) and (8.78), respectively.

A comparison of these results with the experiment and with a detailed derivation shows that the screened exchange self-energy underestimates the true single-particle energy shifts. This is not astonishing since, as we have mentioned before, there has to be an additional contribution due to the Coulomb correlation effects not included in the Hartree-Fock theory. The corresponding term which appears in addition to the screened exchange self-energy, is called the *Coulomb hole self-energy*, Σ_{CH}. The origin of this name is due to the strong reduction of the pair correlation function $R_{ss'}(|\mathbf{r}-\mathbf{r}'|=0)$ also for electrons with different spin, $s\neq s'$, when the proper Coulomb correlations are included in the many-electron wavefunction. In analogy to the "exchange hole" discussed in Chap. 7, this correlation effect is referred to as "Coulomb hole". Hence, the Coulomb-hole self-energy describes the reduction of the total energy due to the fact that the electrons avoid each other because of their mutual Coulomb repulsion. The renormalized single-particle energy is instead of Eq. (9.20)

$$e_k = \epsilon_k + \Sigma_{SX}(k) + \Sigma_{CH}\ . \tag{9.20a}$$

The derivation of the Coulomb-hole contribution to the self-energy is given in Chaps. 10, 12, and the Appendix, where we also discuss the corresponding change of the effective semiconductor band-gap. Here we only want to mention that this contribution can be calculated as the change of the self-interaction of a particle with and without the presence of a plasma, i.e.,

$$\hbar\Sigma_{CH} = \frac{1}{2} \lim_{r\to 0} \left[V_s(r) - V(r) \right] . \tag{9.29}$$

Ignoring for the moment the term $\propto q^4$ in the effective plasmon frequency ω_q of the single-plasmon-pole approximation, one obtains for $V_s(r)$ the Yukawa potential, Eq. (8.51), both in *2d* and in *3d*. Inserting Eq. (8.51) into Eq. (9.29) we obtain the Coulomb-hole self-energy as

$$\Sigma_{CH} = - \frac{e^2}{2\epsilon_0} \kappa \tag{9.30}$$

where we have included static screening through ϵ_0 and κ is the screening wavenumber in *2d* or *3d*, respectively.

For the three-dimensional system it is possible to improve Eq. (8.51) by analytically evaluating $V_s(r)$ using the full static plasmon-pole approximation,

$$V_s(r) = \frac{L^3}{(2\pi)^3} \int d^3q \; V_s(q) \; e^{-i\mathbf{q}\cdot\mathbf{r}} , \tag{9.31}$$

where

$$V_s(q) = \frac{4\pi e^2}{\epsilon_0 q^2} \left(1 + \frac{1}{1 + \frac{q^2}{\kappa^2} + \left(\frac{\nu_q}{\omega_{pl}} \right)^2} \right) . \tag{9.32}$$

The resulting expression is

$$\begin{aligned} V_s(r) &= \frac{e^2}{\epsilon_0 r} e^{-ra_+} \left[\cos(ra_-) + (4/u^2 - 1)^{-1/2} \sin(ra_-) \right] && \text{for } u \le 2 \\ &= \frac{e^2}{2\epsilon_0 r} \left[b_- e^{-rc_+} + b_+ e^{-rc_-} \right] && \text{for } u > 2 , \end{aligned} \tag{9.33}$$

where

$$u = \frac{\omega_{pl}}{\kappa^2 C^{1/2}} \,,$$

$$a_\pm = \frac{\kappa u}{2}\sqrt{\frac{2}{u} \pm 1} \,,$$

$$b_\pm = 1 \pm \left[1 - \frac{4}{u^2}\right]^{-1/2} \,,$$

$$c_\pm = \frac{\kappa u}{\sqrt{2}}\left[1 \pm \left(1 - \frac{4}{u^2}\right)^{1/2}\right]^{1/2} \,.$$

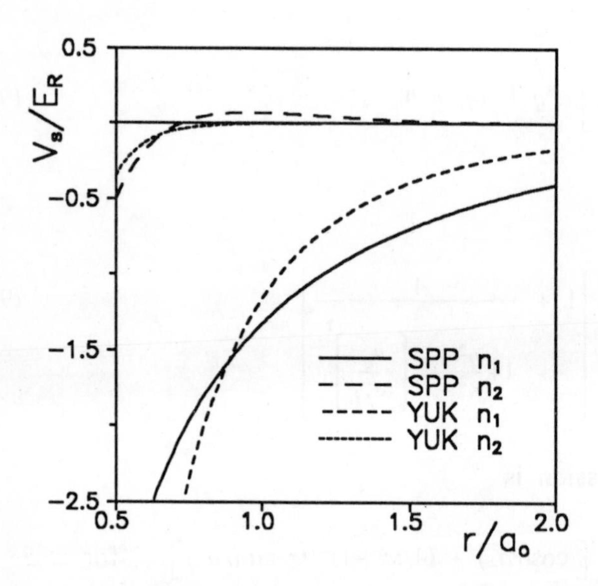

Fig. 9.1: Comparison of the screened Coulomb potential $V_s(r)$ using the single-plasmon pole approximation (SPP), Eq. (9.33) and the Yukawa potential (YUK), Eq. (8.51), for the plasma densities $n_1 = 10^{-3}\ a_0^{-3}$ and $n_2 = a_0^{-3}$, respectively. Here, a_0 is the *3d* Bohr radius, Eq. (7.40), and $E_R = E_0$ is given by Eq. (9.35).

The screened Coulomb interaction potential according to Eq. (9.33) is plotted in Fig. 9.1 for low and high plasma density and compared to the Yukawa potential, Eq. (8.51), for the same parameter values. As most prominent feature we notice the more effective screening with increasing plasma density causing the strong reduction of the effective Coulomb potential. Whereas the Yukawa potential is always negative, indicating electron-electron repulsion, the plasmon-pole approximation yields an effective Coulomb potential, which becomes attractive at large distances and sufficiently high plasma densities.

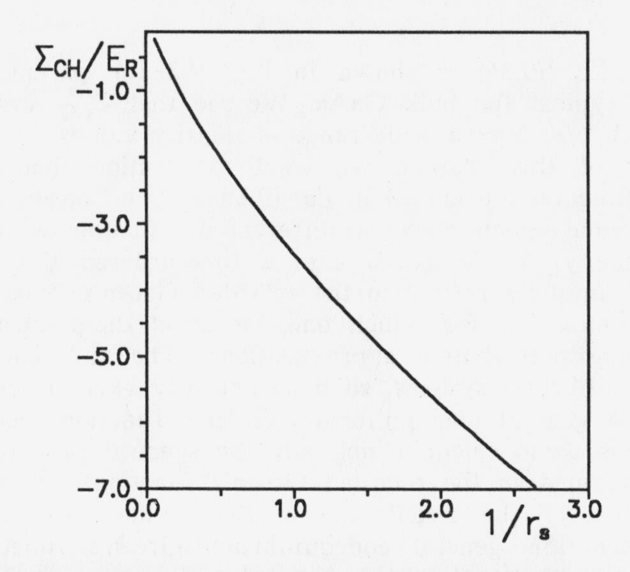

Fig. 9.2: Coulomb hole self-energy, Eq. (9.34), as function of particles density expressed in terms of $1/r_s$, Eq. (7.39). The used parameter values are a_0=1.096·10^{-6}cm, T=10K, E_0=5.32meV, m_e/m=1.12, m_h/m=9.1, where m is the reduced electron-hole mass.

Evaluating the Coulomb hole self-energy, Eq. (9.29), for the potential given by Eq. (9.33) we obtain

$$\Sigma_{CH} = -\, a_0 \kappa E_0 \left[\frac{2}{1-4/u^2} \right]^{1/2} (\, b_+ - b_-) \qquad \text{for } u \geq 2$$

$$= -\, a_0 \kappa E_0 u \left[(1 + 2/u)^{1/2} - (1 + 2/u)^{-1/2} \right] \qquad \text{for } u < 2 \tag{9.34}$$

where

$$E_0 = \frac{e^2}{2 a_0 \epsilon_0} = E_R . \tag{9.35}$$

An evaluation of Eq. (9.34) is shown in Fig. 9.2 for parameter values which are typical for bulk GaAs. We see that Σ_{CH} varies almost linearly with $1/r_s$ over a wide range of density values.

At the end of this chapter, we want to mention that the retarded Green's function introduced in Eq. (9.1) is by no means the only Green's function which obeys a differential equation with a singular inhomogeneity. There exists, e.g., a time-ordered Green's function which is uniquely related to the retarded Green's function in equilibrium systems, but for which one has strict diagrammatic rules for developing perturbative approximations. The situation is different in nonequilibrium systems, such as optically excited semiconductors. Here, a general nonequilibrium Green's function theory exists, which allows us to calculate not only the spectral properties of the system (contained in the retarded Green's function) but also the kinetic evolution of the distribution of the excitations in the system. Furthermore, this general nonequilibrium Green's function theory provides also strict diagrammatic rules for the retarded Green's function. An introduction to the theory of nonequilibrium Green's functions is given in the Appendix of this book.

REFERENCES

S. Doniach and E.H. Sondheimer, *Green's Functions for Solid State Physicists*, Benjamin, Reading, Mass. (1974)

A.L. Fetter and J.D. Walecka, *Quantum Theory of Many-Particle Physics*, McGraw-Hill, New York (1971)

G.D. Mahan, *Many Particle Physics*, Plenum Press, New York (1981)

D. Pines and P. Nozieres, *The Theory of Quantum Liquids*, Benjamin, Reading, Mass. (1966)

PROBLEMS

Problem 9.1: Follow the steps discussed in this chapter to prove Eq. (9.15).

Problem 9.2: Evaluate Eq. (9.15) for a quasi-*2d* system. Use the approximations discussed for the *3d* case to obtain an estimate for the exchange self-energy.

Chapter 10
INTERBAND KINETICS AND EXCITONS

In the preceding chapters we have discussed the quantum statistics and the many-body Coulomb effects in an electron plasma with positive background (jellium). The obtained results are clearly relevant for a metal, since it has a partially filled conduction band. Moreover these results are also important for a dielectric medium such as a semiconductor, since they describe the carrier-carrier interactions within the same band, i.e., the *intraband interactions*. However, as already discussed in Chap. 4, the optical properties of semiconductors are mostly related to *interband transitions* between the valence and conduction bands.

As it turns out, one can separate the many-body treatment of the electron-hole system in excited semiconductors quite naturally into the determination of *spectral properties* and *kinetic properties*. As spectral we denote energy shifts and the broadening due to interactions, i.e., the renormalizations of the states due to the many-body interactions. The kinetics, on the other hand, deals with the development of the particle distributions in the renormalized states. The optical properties are mainly linked with the *interband kinetics*, whereas the transport properties are connected with the *intraband kinetics* of electrons and/or holes, depending on the kind of free carriers in the semiconductor.

The analysis of the semiconductor interband transitions for variable excitation conditions is presented in Chap. 13. In the present chapter we consider only the low-excitation regime, where an extremely small density of electrons and holes exist. We therefore concentrate on the analysis of one-electron-hole pair effects.

10-1. The Interband Polarization

In Chap. 4 we have already discussed the optical susceptibility due to free-carrier transitions in semiconductors. In the present chapter we now generalize this treatment to include also the Coulomb effects. Again, we compute the optical susceptibility from which we then get the absorption coefficient and the refractive index. .

First we analyze the macroscopic interband polarization induced by the coherent monochromatic classical light field $\mathbf{E}(\mathbf{r},t)$. The polarization $\mathbf{P}(t)$ is defined as the expectation value of the electric dipole $e\mathbf{r}$ as

$$\mathbf{P}(t) = \int d^3r \; \mathbf{P}(\mathbf{r},t) = \sum_s \int d^3r \; \langle \hat{\psi}_s^\dagger(\mathbf{r},t) \; e\mathbf{r} \; \hat{\psi}_s(\mathbf{r},t) \rangle$$

$$= \sum_s \int d^3r \; \mathrm{tr} \; [\rho_0 \; \hat{\psi}_s^\dagger(\mathbf{r},t) \; e\mathbf{r} \; \hat{\psi}_s(\mathbf{r},t)] \; . \tag{10.1}$$

In Eq. (10.1) the average is taken with the equilibrium statistical operator ρ_0 describing the system at the initial time before the field was switched on. If one introduces the reduced one-particle density matrix $\rho(\mathbf{r},\mathbf{r}',t)$ as the single-time correlation function

$$\rho_{ss'}(\mathbf{r},\mathbf{r}',t) = \langle \hat{\psi}_s^\dagger(\mathbf{r},t) \; \hat{\psi}_{s'}(\mathbf{r}',t) \rangle \; , \tag{10.2}$$

one can rewrite Eq. (10.1) as

$$\mathbf{P}(t) = \sum_s \int d^3r \; \rho_{ss'}(\mathbf{r},\mathbf{r}',t) \Big|_{r=r',\,s=s'} e\mathbf{r} \quad . \tag{10.3}$$

These general definitions now have to be adapted to a semiconductor by expanding the electron field operators into an appropriate basis.

A) Spatially Homogeneous Systems

In the case of a spatially homogeneous system one is dealing with completely delocalized electrons. Then the Bloch wavefunctions $\psi_{\lambda k}(\mathbf{r})$, Eq. (3.26), are the appropriate set for expanding the field operators

$$\hat{\psi}_s(\mathbf{r},t) = \sum_{\lambda,\mathbf{k}} a_{\lambda,\mathbf{k},s}(t)\ \psi_{\lambda,\mathbf{k}}(\mathbf{r})\ . \tag{10.4}$$

Inserting this expansion into Eq. (10.1) yields

$$\mathbf{P}(t) = \sum_{s,\lambda,\lambda',\mathbf{k},\mathbf{k}'} \langle a^\dagger_{\lambda,\mathbf{k},s}\ a_{\lambda',\mathbf{k}',s}\rangle \int d^3r\ \psi^*_{\lambda,\mathbf{k}}(\mathbf{r})\ e\mathbf{r}\ \psi_{\lambda',\mathbf{k}'}(\mathbf{r})\ . \tag{10.5}$$

As in Chap. 3, we now split the integral over the total crystal volume $\mathcal{V} = \Omega N$ into an integral over the volume of a unit cell Ω and a sum over all N unit cells. Introducing

$$\mathbf{r} = \mathbf{R}_n + \mathbf{r}_n\ , \tag{10.6}$$

where $\mathbf{R}_n$ labels the unit cells and $\mathbf{r}_n$ varies within the cell, we can rewrite the integral in Eq. (10.5) as

$$\int d^3r\ \psi^*_{\lambda,\mathbf{k}}(\mathbf{r})\ e\mathbf{r}\ \psi_{\lambda',\mathbf{k}'}(\mathbf{r}) = \sum_{n=1}^{N} \frac{e^{i(\mathbf{k}-\mathbf{k}')\mathbf{R}_n}}{N}$$

$$\times \int \frac{d^3r_n}{\Omega}\ u^*_{\lambda,\mathbf{k}}(\mathbf{r}_n)\ e[\mathbf{R}_n+\mathbf{r}_n]\ u_{\lambda',\mathbf{k}'}(\mathbf{r}_n)\ , \tag{10.7}$$

where Eq. (3.26) has been used, and the weak variation of the plane-wave factors within the unit cells has been disregarded, as usual. From Chap. 3 we know that the Bloch functions $u_{\lambda,\mathbf{k}}$ are orthogonal for different $(\lambda,\mathbf{k})$ and normalized within an elementary cell, Eq. (3.32). Therefore, the term $\propto \mathbf{R}_n$ in Eq. (10.7) is zero, except when $\lambda=\lambda'$, i.e., for *intraband* transitions. Here we are not interested in intraband transitions and consider only interband transitions, i.e., $\lambda \neq \lambda'$. For these interband transitions we obtain from the integral in Eq. (10.7)

$$\delta_{\mathbf{k},\mathbf{k}'} \int \frac{d^3r}{\Omega}\, u^*_{\lambda,\mathbf{k}}(r)\; e\mathbf{r}\; u_{\lambda',\mathbf{k}}(r) = \delta_{\mathbf{k},\mathbf{k}'}\, \mathbf{d}_{\lambda\lambda',k} \;, \tag{10.8}$$

where $\lambda \neq \lambda'$, and $d_{\lambda\lambda',k}$ is the interband dipole matrix element. Inserting Eqs. (10.7) and (10.8) into Eq. (10.5), we obtain the polarization in a spatially homogeneous system as

$$\begin{aligned} \mathbf{P}(t) &= \sum_{s,\mathbf{k},\lambda,\lambda'} \langle a^\dagger_{\lambda,\mathbf{k},s}(t)\; a_{\lambda',\mathbf{k},s}(t)\rangle\; \mathbf{d}_{\lambda\lambda',k} \\ &= \sum_{\mathbf{k},s,\lambda,\lambda'} P_{\lambda\lambda'}(\mathbf{k},s,t)\; \mathbf{d}_{\lambda\lambda',k} \;, \end{aligned} \tag{10.9}$$

where we have introduced the pair function

$$P_{\lambda\lambda'}(\mathbf{k},s,t) = \langle a^\dagger_{\lambda,\mathbf{k},s}(t)\; a_{\lambda',\mathbf{k},s}(t)\rangle . \tag{10.10}$$

B) Spatially Inhomogeneous Systems

If we want to treat systems with a finite geometry, such as semiconductor microcrystallites, or other systems which exhibit spatial variations, we have to use localized electron wavefunctions as the basis set for the expansion in Eq. (10.4). As one example, we may use the so-called *Wannier functions* $w_\lambda(\mathbf{r}-\mathbf{r}_i)$ describing an electron localized around the i-th atom. The electron field operator is then

$$\hat{\psi}_s(\mathbf{r},t) = \sum_{\lambda,i} \hat{\psi}_{\lambda,s}(\mathbf{r}_i,t)\; w_\lambda(\mathbf{r}-\mathbf{r}_i) \;, \tag{10.11}$$

where $\hat{\psi}_\lambda(\mathbf{r}_i,t)$ annihilates an electron at the i-th atom in the state λ. With this expansion the polarization of Eq. (10.1) becomes

$$\mathbf{P}(t) = \sum_{s,\lambda,\lambda',i,j} \langle \hat{\psi}^{\dagger}_{\lambda,s}(\mathbf{r}_i,t)\, \hat{\psi}_{\lambda',s}(\mathbf{r}_j,t)\rangle \int d^3r\; w^*_{\lambda}(\mathbf{r}-\mathbf{r}_i)\; e\mathbf{r}\; w_{\lambda'}(\mathbf{r}-\mathbf{r}_j)$$

$$\cong \sum_{s,\lambda,\lambda\neq\lambda',i} \langle \hat{\psi}^{\dagger}_{\lambda,s}(\mathbf{r}_i,t)\, \hat{\psi}_{\lambda',s}(\mathbf{r}_i,t)\rangle\; \mathbf{d}_{\lambda\lambda'}(\mathbf{r}_i)\; \mathcal{V}, \tag{10.12}$$

where we have defined

$$\mathbf{d}_{\lambda\lambda'}(\mathbf{r}_i) = \int \frac{d^3r}{L^3}\; e\mathbf{r}\; w^*_{\lambda}(\mathbf{r}-\mathbf{r}_i)\; w_{\lambda'}(\mathbf{r}-\mathbf{r}_i). \tag{10.13}$$

Again, the diagonal element $d_{\lambda\lambda}$ vanishes due to the symmetry of the Wannier functions, and we have neglected the overlap of wave-functions localized at different atoms, i.e.,

$$\int d^3r\; \mathbf{r}\; w^*_{\lambda}(\mathbf{r}-\mathbf{r}_i)\; w_{\lambda}(\mathbf{r}-\mathbf{r}_j) \cong 0 \quad \text{for } i\neq j\ . \tag{10.14}$$

On a macroscopic scale the summation over the lattice cells i can be treated as an integral

$$\sum_i \rightarrow \int \frac{d^3r}{L^3}\ , \tag{10.15}$$

so that

$$\mathbf{P}(\mathbf{r},t) = \sum_{s,\lambda,\lambda'\neq\lambda} \langle \hat{\psi}^{\dagger}_{\lambda,s}(\mathbf{r},t)\, \hat{\psi}_{\lambda',s}(\mathbf{r},t)\rangle\; \mathbf{d}_{\lambda\lambda'}(\mathbf{r})$$

$$\equiv \sum_{\lambda,\lambda'\neq\lambda} P_{\lambda\lambda'}(\mathbf{r},s,t)\; \mathbf{d}_{\lambda\lambda'}(\mathbf{r})\ . \tag{10.16}$$

As in Chap. 4 we now restrict the treatment to the optical transitions between valence and conduction band of the semiconductor (two-band approximation). Furthermore, we suppress the spin index s from now on since the following calculations will not depend on s. The spin summation in the definition of the polarization (10.1) will lead to an extra prefactor of 2 in the final result. Choosing $\lambda=v$ and $\lambda'=c$, the pair function defined in Eq. (10.10) becomes

$$P_{vc}(\mathbf{k},t) = \langle a^{\dagger}_{v,\mathbf{k}}(t)\, a_{c,\mathbf{k}}(t)\rangle \tag{10.17}$$

and the corresponding real-space function, introduced in Eq. (10.16), is

$$P_{vc}(\mathbf{r},t) = \langle\hat{\psi}^{\dagger}_{v}(\mathbf{r},t)\hat{\psi}_{c}(\mathbf{r},t)\rangle\ . \tag{10.18}$$

These quantities can be regarded as different representations of the off-diagonal elements of the reduced density matrix [compare Eqs. (4.12) - (4.18)]. In an equilibrium system without permanent dipole moment, these interband density-matrix elements would always vanish. However, the presence of the light field induces optical transitions between the bands. Therefore, the interband polarization function P_{vc} in such an externally driven system is finite.

C) Hamiltonian Interband Polarization Dynamics

In second quantization we write the interaction Hamiltonian, Eq. (2.4), between the electric field and the semiconductor electrons as

$$\mathscr{H}_I = \int d^3r\ \hat{\psi}^{\dagger}(\mathbf{r})\ [-e\mathbf{r}]\cdot\mathcal{E}(\mathbf{r},t)\ \hat{\psi}(\mathbf{r})\ . \tag{10.19}$$

Assuming that the electric field is a monochromatic coherent plane wave

$$\mathcal{E}(\mathbf{r},t) = \frac{\mathcal{E}_0}{2}\,[\ e^{i(\mathbf{q}\mathbf{r}-\omega t)} + \text{c.c.}\ ] \tag{10.20}$$

and using the expansion (10.4) for spatially homogeneous systems, we obtain

$$\mathscr{H}_I \cong -\sum_{\mathbf{k}} \left[a^\dagger_{c,\mathbf{k}} a_{v,\mathbf{k}} d_{cv,\mathbf{k}} \frac{\mathscr{E}_0}{2} e^{-i\omega t} + \text{h.c.} \right], \tag{10.21}$$

where we have taken the limit $q \cong 0$ (dipole approximation) and defined d as the projection of the dipole $\mathbf{d}_{cv,k}$ in the field direction $\boldsymbol{\mathscr{E}}_0/\mathscr{E}_0$. For spatially inhomogeneous systems we use the expansion (10.11) to get

$$\mathscr{H}_I \cong -\int d^3r \left[\hat{\psi}^\dagger_c(\mathbf{r})\, \hat{\psi}_v(\mathbf{r})\, d_{cv} \frac{\mathscr{E}_0}{2} e^{i(\mathbf{qr} - \omega t)} + \text{h.c} \right]. \tag{10.22}$$

In Eqs. (10.21) and (10.22) we considered only the resonant terms, i.e., we again made the rotating wave approximation, see Eq. (4.10). Both forms of the interaction Hamiltonian show how the applied field causes transitions of electrons between valence and conduction band.

Besides the interaction with the external field we also have to consider the kinetic and Coulomb contributions from the electrons. These effects are described by the Hamiltonian (7.17). For our present purposes, we have to extend the treatment of Chap. 7 by including also the band index $\lambda = c,v$. For the two-band model in effective mass approximation the electron Hamiltonian (7.17) then becomes

$$\mathscr{H}_{el} = \sum_{\lambda,\mathbf{k}} E_{\lambda,k} a^\dagger_{\lambda,\mathbf{k}} a_{\lambda,\mathbf{k}} + \frac{1}{2} \sum_{\substack{\mathbf{k},\mathbf{k}' \\ q \neq 0 \\ \lambda,\lambda'}} V_q\, a^\dagger_{\lambda,\mathbf{k}+\mathbf{q}} a^\dagger_{\lambda',\mathbf{k}'-\mathbf{q}} a_{\lambda',\mathbf{k}'} a_{\lambda,\mathbf{k}} . \tag{10.23}$$

This Hamiltonian is obtained from Eq. (7.17) by including the summation over the band indices $\lambda,\lambda' = c,v$. Furthermore, we have omitted all those Coulomb terms which do not conserve the number of electrons in each band. Such terms are neglected because they would describe interband scattering, i.e., promotion of an electron from valence to conduction band or vice versa due to the Coulomb interaction, which is energetically very unfavorable (see problem 10.1). As in Chap. 4 we define the conduction and valence band energies as

$$E_{c,k} = \hbar\epsilon_{c,k} = E_g + \hbar^2k^2/2m_c\ , \tag{10.24}$$

and

$$E_{v,k} = \hbar\epsilon_{v,k} = \hbar^2k^2/2m_v\ . \tag{10.25}$$

The electronic Hamiltonian in real-space representation can be written as

$$\mathcal{H}_{e\ell} = \sum_{\lambda} \int d^3r\ \hat{\psi}^{\dagger}_{\lambda}(\mathbf{r})\ \mathcal{H}^0_{\lambda}\ \hat{\psi}_{\lambda}(\mathbf{r})$$

$$+ \frac{1}{2} \sum_{\lambda,\lambda'} \int d^3r\ d^3r'\hat{\psi}^{\dagger}_{\lambda}(\mathbf{r})\ \hat{\psi}^{\dagger}_{\lambda'}(\mathbf{r}')\ V(|\,\mathbf{r}-\mathbf{r}'|)\ \hat{\psi}_{\lambda'}(\mathbf{r}')\ \hat{\psi}_{\lambda}(\mathbf{r})\ , \tag{10.26}$$

where

$$\mathcal{H}^0_c = -\ (\hbar^2\nabla^2_c)/(2m_c) + E_g \tag{10.27}$$

and

$$\mathcal{H}^0_v = -\ (\hbar^2\nabla^2_v)/(2m_v). \tag{10.28}$$

As usual, we have assumed that the Coulomb potential varies slowly over one lattice unit cell.

The full Hamiltonian of the electrons in valence and conduction band interacting with the light field is now given by

$$\mathcal{H} = \mathcal{H}_I + \mathcal{H}_{e\ell} \tag{10.29}$$

To derive the equation of motion of the polarization we use the Heisenberg equation for the individual operators. An elementary but lengthy calculation yields for the isotropic, homogeneous case

$$\hbar\left[i\frac{d}{dt} - (\epsilon_{c,k} - \epsilon_{v,k})\right] P_{vc}(\mathbf{k}) = (n_{c,k} - n_{v,k})\ d_{cv,k}\ e^{-i\omega t}\ \mathcal{E}_0/2$$

$$+ \sum_{\mathbf{k}',q\neq 0} V_q\ [\ \langle a^{\dagger}_{c,\mathbf{k}'+\mathbf{q}}a^{\dagger}_{v,\mathbf{k}-\mathbf{q}}a_{c,\mathbf{k}'}a_{c,\mathbf{k}}\rangle + \langle a^{\dagger}_{v,\mathbf{k}'+\mathbf{q}}a^{\dagger}_{v,\mathbf{k}-\mathbf{q}}a_{v,\mathbf{k}'}a_{c,\mathbf{k}}\rangle$$

$$+ \langle a^{\dagger}_{v,\mathbf{k}'}a^{\dagger}_{c,\mathbf{k}'-\mathbf{q}}a_{c,\mathbf{k}'}a_{c,\mathbf{k}-\mathbf{q}}\rangle + \langle a^{\dagger}_{v,\mathbf{k}'}a^{\dagger}_{v,\mathbf{k}'-\mathbf{q}}a_{v,\mathbf{k}'}a_{c,\mathbf{k}-\mathbf{q}}\rangle\ ], \tag{10.30}$$

where, as usual,

$$n_{\lambda,k} = \langle a^{\dagger}_{\lambda,k} a_{\lambda,k} \rangle . \tag{10.31}$$

The expectation values of the four operator terms in Eq. (10.30) are now split in the spirit of a generalized Hartree-Fock approximation into products of densities and interband polarizations. For example, we approximate

$$\langle a^{\dagger}_{c,\mathbf{k'+q}} a^{\dagger}_{v,\mathbf{k-q}} a_{c,\mathbf{k'}} a_{c,\mathbf{k}} \rangle \cong P_{vc}(\mathbf{k})\; n_{c,\mathbf{k}}\; \delta_{\mathbf{k-q},\mathbf{k'}} \tag{10.32}$$

and similarly also the other terms. As result we obtain

$$\hbar \left[i \frac{d}{dt} - (e_{c,k} - e_{v,k}) \right] P_{vc}(\mathbf{k},t) =$$

$$(n_{c,k} - n_{v,k}) \left[d_{cv,k} \frac{\mathcal{E}_0}{2}\, e^{-i\omega t} + \sum_{q \neq k} V_{|\mathbf{k-q}|}\; P_{vc}(\mathbf{q},t) \right] , \tag{10.33}$$

where we have introduced the renormalized frequencies

$$e_{\lambda,k} = \epsilon_{\lambda,k} + \Sigma_{exc,\lambda}(k) \tag{10.34}$$

with the exchange self-energy [compare Eq. (9.17)]

$$\hbar \Sigma_{exc,\lambda}(k) = - \sum_{q \neq k} V_{|\mathbf{k-q}|}\; n_{\lambda,q} \; . \tag{10.35}$$

For the case of non-interacting particles, $V_q \equiv 0$, Eq. (10.33) becomes

$$\hbar \left[i \frac{d}{dt} - (\epsilon_{c,k} - \epsilon_{v,k}) \right] P^{0}_{vc}(\mathbf{k},t) = (n_{c,k} - n_{v,k})\; d_{cv,k} \frac{\mathcal{E}_0}{2}\, e^{-i\omega t} \tag{10.36}$$

With the ansatz

$$P^0_{vc}(\mathbf{k},t) = P^0_{vc}(\mathbf{k})\, e^{-i(\omega+i\delta)t} \tag{10.37}$$

we obtain

$$P^0_{vc}(\mathbf{k}) = \frac{(n_{c,k} - n_{v,k})\, d_{cv,k}}{\hbar[\omega + i\delta - (\epsilon_{ck} - \epsilon_{vk})]} \frac{\mathcal{E}_0}{2}, \tag{10.38}$$

which is just our old free-particle result, compare Eq. (4.21).

10-2. Wannier Equation

In general, Eq. (10.33) is an inhomogeneous integro-differential equation. In order to solve this equation we have to know the carrier densities $n_{\lambda k}$. The case of finite carrier densities will be addressed in later chapters (e.g., Chap. 12). Here we concentrate on the linear optical properties, i.e., the situation of an unexcited crystal where

$$n_{c,k} \equiv 0 \text{ and } n_{v,k} \equiv 1. \tag{10.39}$$

Inserting (10.39) and

$$P_{vc}(\mathbf{k},t) = P_{vc}(\mathbf{k})\, e^{-i(\omega+i\delta)t} \tag{10.40}$$

into Eq. (10.33) yields

$$\left[\hbar(\omega+i\delta) - E_g + \frac{\hbar^2 k^2}{2m_r}\right] P_{vc}(\mathbf{k}) = - \left[d_{cv,k}\frac{\mathcal{E}_0}{2} + \sum_{q \neq k} V_{|\mathbf{k}-\mathbf{q}|}\, P_{vc}(\mathbf{q},t)\right] \tag{10.41}$$

where

$$\frac{1}{m_r} = \frac{1}{m_c} - \frac{1}{m_v} \tag{10.42}$$

is the inverse reduced mass. As a reminder we note again at this point that $m_v < 0$, compare Chaps. 3 and 4.

For the solution it turns out to be advantageous to transform Eq. (10.41) into real-space. Multiplying Eq. (10.41) from the left by

$$\frac{L^3}{(2\pi)^3} \int d^3k \; ...$$

and using the Fourier transformation in the form

$$f(\mathbf{r}) = \frac{L^3}{(2\pi)^3} \int d^3q \; f(\mathbf{q}) \; e^{-i\mathbf{q}\cdot\mathbf{r}}$$

$$f(\mathbf{q}) = \frac{1}{L^3} \int d^3r \; f(\mathbf{r}) \; e^{i\mathbf{q}\cdot\mathbf{r}} \tag{10.43}$$

we obtain

$$\left[\hbar(\omega+i\delta) - E_g + \frac{\hbar^2\nabla_{\mathbf{r}}^2}{2m_r} + V(r)\right] P_{vc}(\mathbf{r}) = - d_{cv} \frac{\mathcal{E}_0}{2} \delta(\mathbf{r})L^3 \; , \tag{10.44}$$

where we have ignored the k-dependence of the interband dipole matrix element. Note, however, that this approximation holds only for allowed transitions in the band-edge region. For optically forbidden transitions the matrix element strongly depends on k, and even for allowed transitions one has to take the wavevector dependence of $d_{cv,k}$ into account to study the optical properties higher in the band.

One way to solve the inhomogeneous Eq. (10.44) is to expand P_{vc} into the solution of the corresponding homogeneous equation

$$-\left[\frac{\hbar^2\nabla_{\mathbf{r}}^2}{2m_r} + V(r)\right] \psi(\mathbf{r}) = E_\nu \psi(\mathbf{r}) \; .$$

Wannier equation (10.45)

Eq. (10.45) has exactly the form of a two-particle Schrödinger equation for the relative motion of an electron and a hole interacting via the attractive Coulomb potential $V(r)$. This equation is known as the

Wannier equation. Obviously one has a one-to-one correspondence with the hydrogen atom, if one replaces the proton by the valence-band hole. Remember, however, that the pair equation (10.36) was derived under the assumption that the Coulomb potential varies little within one unit cell. This assumption is valid only if the resulting electron-hole-pair Bohr radius a_0 which determines the extension of the ground state wave function is considerably larger than a lattice constant.

To solve the Wannier equation, we proceed analogously to the solution of the hydrogen problem which is discussed in many quantum mechanics textbooks, such as Landau and Lifshitz (1958) or Schiff (1968). Introducing the scaled radius

$$\rho = r\alpha \tag{10.46}$$

Eq. (10.45) then becomes

$$\left[-\nabla_\rho^2 - \frac{\lambda}{\rho}\right] \psi(\rho) = \frac{2m_r E_\nu}{\hbar^2\alpha^2} \psi(\rho) , \tag{10.47}$$

where

$$\lambda = e^2 2m_r/(\epsilon_0 \hbar^2 \alpha). \tag{10.48}$$

As in the hydrogen-atom case the energy E_ν is negative for bound states ($E_{bound} < E_g$) and positive for the ionization continuum. We define

$$\alpha^2 = -\frac{8m_r E_\nu}{\hbar^2} \tag{10.49}$$

to rewrite Eq. (10.47) as

$$\left[-\nabla_\rho^2 - \frac{\lambda}{\rho}\right] \psi(\rho) = -\frac{1}{4}\psi(\rho) , \tag{10.50}$$

and Eq. (10.48) becomes

$$\lambda = \frac{e^2}{\hbar\epsilon_0} \sqrt{-\frac{m_r}{2E_\nu}} . \tag{10.51}$$

With this choice of α^2 the parameter λ will be real for bound states

Laplace operator in spherical/polar coordinates as

$$\nabla_\rho^2 = \frac{1}{\rho^2}\frac{\partial}{\partial\rho}\rho^2\frac{\partial}{\partial\rho} - \frac{\mathscr{L}^2}{\rho^2} \quad \text{in } 3d$$

$$\nabla_\rho^2 = \frac{1}{\rho}\frac{\partial}{\partial\rho}\rho\frac{\partial}{\partial\rho} - \frac{\mathscr{L}_z^2}{\rho^2} \quad \text{in } 2d, \tag{10.52}$$

where $\mathscr{L}$ and $\mathscr{L}_z$ are the operators of the total angular momentum and its z component,

$$\mathscr{L}^2 = -\left[\frac{1}{\sin^2\theta}\frac{\partial^2}{\partial\rho^2} + \frac{1}{\sin\theta}\frac{\partial}{\partial\theta}\sin\theta\frac{\partial}{\partial\theta}\right] \tag{10.53}$$

and

$$\mathscr{L}_z^2 = -\frac{\partial^2}{\partial\phi^2}\,. \tag{10.54}$$

These operators obey the following eigenvalue equations

$$\mathscr{L}^2 Y_{\ell,m}(\theta,\phi) = \ell(\ell+1)Y_{\ell,m}(\theta,\phi) \quad \text{with } |m| \le \ell \tag{10.55}$$

and

$$\mathscr{L}_z \frac{1}{\sqrt{2\pi}}e^{im\phi} = m\frac{1}{\sqrt{2\pi}}e^{im\phi}\,, \tag{10.56}$$

where the functions $Y_{\ell,m}(\theta,\phi)$ are the spherical harmonics. With the ansatz

$$\psi(\boldsymbol{\rho}) = f_\ell(\rho)Y_{\ell,m} \quad \text{in } 3d \tag{10.57}$$

$$\psi(\boldsymbol{\rho}) = f_m(\rho)\frac{1}{\sqrt{2\pi}}e^{im\phi} \quad \text{in } 2d \tag{10.58}$$

we find the equation for the radial part $f(\rho)$ of the wave function as

$$\left[\frac{1}{\rho^2}\frac{\partial}{\partial\rho}\rho^2\frac{\partial}{\partial\rho} + \frac{\lambda}{\rho} - \frac{1}{4} - \frac{\ell(\ell+1)}{\rho^2}\right]f_\ell(\rho) = 0 \quad \text{in } 3d$$

$$\left[\frac{1}{\rho}\frac{\partial}{\partial\rho}\,\rho\,\frac{\partial}{\partial\rho} + \frac{\lambda}{\rho} - \frac{1}{4} - \frac{m^2}{\rho^2}\right] f_m(\rho) = 0 \quad \text{in } 2d. \tag{10.59}$$

10-3. Excitons

Now we discuss the bound-state solutions which are commonly referred to as *Wannier excitons*. In order to solve Eq. (10.59) for this case, we first determine the asymptotic form of the wavefunctions for large radii. For $\rho \to \infty$ the leading terms in Eq. (10.59) are

$$\left[\frac{d^2}{d\rho^2} - \frac{1}{4}\right] f_\infty(\rho) = 0 \tag{10.60}$$

and the convergent solution is

$$f_\infty(\rho) = e^{-\rho/2} . \tag{10.61}$$

Writing $f(\rho) = f_0(\rho) f_\infty(\rho)$ and studying the asymptotic behavior for small ρ suggests that $f_0(\rho)$ for $\rho \to 0$ should vary like ρ^ℓ (*3d*) or $\rho^{|m|}$ (*2d*), respectively. Thus we make the ansatz for the total wavefunctions

$$f_\ell(\rho) = \rho^\ell e^{-\frac{\rho}{2}} R(\rho) \text{ in } 3d$$

$$f_m(\rho) = \rho^{|m|} e^{-\frac{\rho}{2}} R(\rho) \text{ in } 2d. \tag{10.62}$$

Inserting (10.62) into Eq. (10.59) yields

$$\rho\,\frac{\partial^2 R}{\partial\rho^2} + (2(\ell+1)-\rho)\,\frac{\partial R}{\partial\rho} + (\lambda-\ell-1)R = 0 \quad \text{in } 3d$$

$$\rho\,\frac{\partial^2 R}{\partial\rho^2} + (2|m|+1-\rho)\,\frac{\partial R}{\partial\rho} + \left[\lambda-|m|-\frac{1}{2}\right]R = 0 \quad \text{in } 2d. \tag{10.63}$$

Both equations are of the form

$$\rho \frac{\partial^2 R}{\partial \rho^2} + (p+1-\rho) \frac{\partial R}{\partial \rho} + qR = 0 \tag{10.64}$$

with

$$p = 2\ell + 1 \ , \ q = \lambda - \ell - 1 \quad \text{in } 3d,$$

$$p = 2|m| \ , \quad q = \lambda - |m| - \frac{1}{2} \text{ in } 2d. \tag{10.65}$$

The solution of Eq. (10.64) can be obtained by a power expansion

$$R(\rho) = \sum_{\nu=0} \beta_\nu \rho^\nu \ . \tag{10.66}$$

Inserting (10.66) into (10.64) and comparing the coefficients of the different powers of ρ yields the recurrence relation

$$\beta_{\nu+1} = \beta_\nu \frac{\nu-q}{(\nu+1)(\nu+p+1)} \ . \tag{10.67}$$

In order to get a normalizable result, the series must terminate for $\nu = \nu_{max}$, so that all $\beta_{\nu \geq \nu_{max}} = 0$. Thus $\nu_{max} - q = 0$, or, using (10.65) for the $3d$ case

$$\nu_{max} + \ell + 1 = \lambda \equiv n \tag{10.68a}$$

where the main quantum number n can assume the values $n = 1$, $2, \cdots$ for $\ell = \nu_{max} = 0$; $\ell = 1$ and $\nu_{max} = 0$, or $\ell = 0$, $\nu_{max} = 1$, etc. Correspondingly, we have in $2d$

$$\nu_{max} + |m| + \frac{1}{2} = \lambda \equiv n + \frac{1}{2} \ . \tag{10.68b}$$

Here the allowed values of the main quantum number are $n = 0, 1, 2, \cdots$. The bound-state energies follow from Eq. (10.51) as

$$E_n = - E_0 \frac{1}{n^2} \quad \text{with } n = 1, 2, \ldots$$

3d exciton binding energy (10.69*a*)

$$E_n = - E_0 \frac{1}{(n+1/2)^2} \quad \text{with } n = 0, 1, \ldots$$

2d exciton binding energy (10.69*b*)

where

$$E_0 = \frac{\hbar^2}{2m_r a_0^2} = \frac{e^2}{2\epsilon_0 a_0} = \frac{e^4 m_r}{2\epsilon_0^2 \hbar^2} \tag{10.70}$$

is the exciton Rydberg energy and

$$a_0 = \frac{\hbar^2 \epsilon_0}{e^2 m_r} \tag{10.71}$$

is the exciton Bohr radius which we introduced already in Eq. (7.40) as characteristic length scale.

The binding energy of the exciton ground state is E_0 in *3d* and $4E_0$ in *2d*, respectively. The larger binding energy in *2d* can be understood by considering quantum well structures with decreasing width. The wavefunction tries to conserve its spherical symmetry as much as possible (the admixture of *p*-wave functions is energetically not favorable) and decreases thus the Bohr radius also in the direction perpendicular to the quantum well below the *3d* limit. In fact the exciton radius is obtained from the exponential term

$$e^{-\rho/2} = e^{-\alpha r/2}$$

with $\alpha = 2/a_0 n$ in *3d* and $\alpha = 2/a_0(n+1/2)$ in *2d*. In the ground state the *3d* exciton radius is thus simply the exciton Bohr radius a_0, but it is only $a_0/2$ in *2d*.

Eq. (10.70) shows that the exciton Rydberg energy E_0 is inversely proportional to the square of the background dielectric constant. The background dielectric constant describes the screening of the Coulomb interaction in an unexcited crystal. This screening requires virtual interband transitions which become more and more unlikely for materials with larger bandgap energy E_g. As a result the exciton Rydberg is large for wide-gap materials such as I-VII

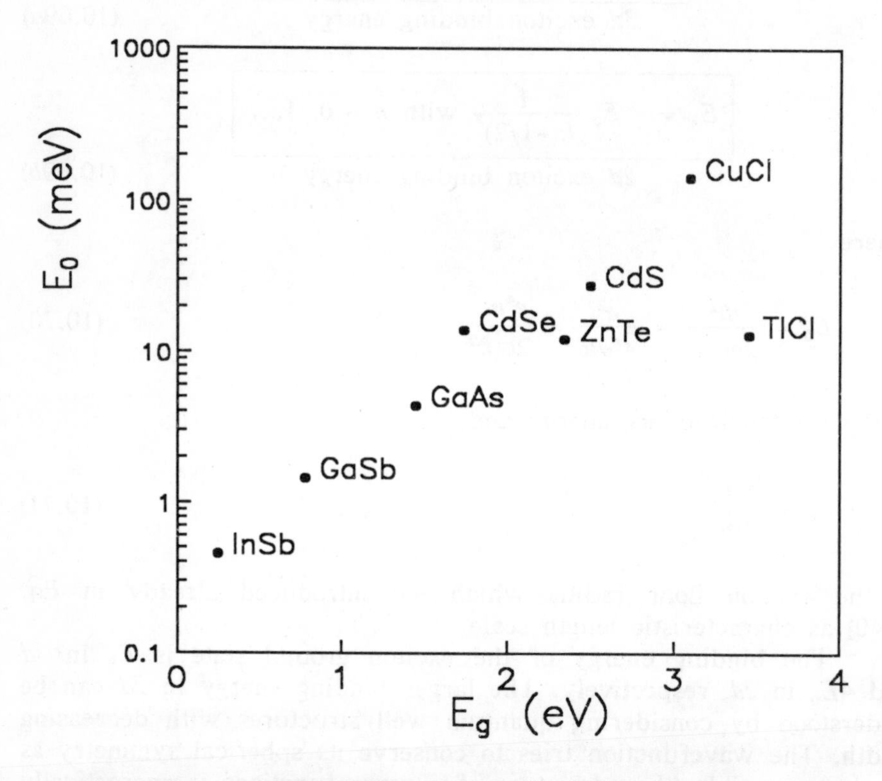

Fig. 10.1: Experimental values for the exciton binding energy E_0 versus bandgap energy.

compound semiconductors, where one element is from the first group of the periodic table and the other one from the seventh. As is shown in Fig. 10.1 for some direct-gap material, E_0 decreases when one goes through the series of I-VII, II-VI, and III-V compound semiconductors.

From Eqs. (10.62) and (10.66) we can now obtain the complete radial exciton wavefunctions. These functions still have to be normalized such that

$$\int_0^\infty dr\ r^{d-1}\ |f(r)|^2 = 1\ . \tag{10.72}$$

The resulting first few normalized functions are in *3d*

ν_{max}	n	ℓ	$f_{n,\ell}(\rho) = C\ \rho^\ell e^{-\frac{\rho}{2}} \sum_\nu \beta_\nu\ \rho^\nu$	E_n
0	1	0	$f_{1,0}(r) = a_0^{-3/2} 2\ e^{-r/a_0}$	$E_1 = -\ E_0$
1	2	0	$f_{2,0}(r) = (2a_0)^{-3/2}\ (2 - r/a_0)\ e^{-r/2a_0}$	$E_2 = -\ \frac{E_0}{4}$
0	2	1	$f_{2,1}(r) = (2a_0)^{-3/2}\ \frac{r}{\sqrt{3}a_0}\ e^{-r/2a_0}$	$E_2 = -\ \frac{E_0}{4}$

3d exciton wavefunctions (10.73*a*)

The first few normalized eigenfunctions in *2d* are

ν_{max}	n	m	$f_{n,m}(\rho) = C\ \rho^{\lvert m\rvert} e^{-\frac{\rho}{2}} \sum_\nu \beta_\nu\ \rho^\nu$	E_n
0	0	0	$f_{0,0}(r) = \frac{1}{a_0}\ 4e^{-2r/a_0}$	$E_{n=0} = -\ 4E_0$
1	1	0	$f_{1,0}(r) = \frac{4}{a_0 3\sqrt{3}} \left[1 - \frac{4r}{3a_0}\right] e^{-2r/3a_0}$	$E_1 = -\ \frac{9E_0}{4}$
0	1	±1	$f_{1,\pm1}(r) = \frac{16}{a_0 9\sqrt{6}}\frac{r}{a_0}\ e^{-2r/3a_0}$	$E_1 = -\ \frac{9E_0}{4}$

2d exciton wavefunctions (10.73*b*)

The solution of the differential equation (10.64) with integer p and

q can be written in terms of the associate Laguerre polynomials[1]

$$L_q^p(\rho) = \sum_{\nu=0}^{q-p} (-1)^{\nu+p} \frac{(q!)^2 \, \rho^\nu}{(q-p-\nu)!(p+\nu)!\nu!} \, . \tag{10.74}$$

Thus the normalized exciton wavefunctions can be expressed in general by these orthogonal Laguerre polynomials as

$$\psi_{n,\ell,m}(\mathbf{r}) = -\sqrt{\left(\frac{2}{na_0}\right)^3 \frac{(n-\ell-1)!}{2n[(n+\ell)!]^3}} \; \rho^\ell e^{-\frac{\rho}{2}} L_{n+\ell}^{2\ell+1}(\rho) \, Y_{\ell,m}(\theta,\phi)$$

$$\text{with } \rho = \frac{2r}{na_0}$$

3d exciton wavefunction (10.75*a*)

and

$$\psi_{n,m}(\mathbf{r}) = \sqrt{\frac{1}{\pi a_0^2(n+1/2)^3} \frac{(n-|m|)!}{[(n+|m|)!]^3}} \; \rho^{|m|} e^{-\frac{\rho}{2}} \; L_{n+|m|}^{2|m|}(\rho) \; e^{im\phi}$$

$$\text{with } \rho = \frac{2r}{(n + 1/2)\, a_0}$$

2d exciton wavefunction (10.75*b*)

[1] This definition of the Laguerre polynomials is taken from Landau and Lifshitz (1958) and differs from some other textbooks.

10-4. The Ionization Continuum

For the continuous spectrum of the ionized states with $E_\nu \geq 0$ we put

$$E_\nu \equiv E_k = \frac{\hbar^2 k^2}{2m_r} \tag{10.76}$$

so that Eqs. (10.49) and (10.51) yield

$$\alpha = 2ik \ , \quad \lambda = -\, i\frac{e^2 m_r}{\epsilon_0 \hbar^2 k} = -\, \frac{i}{a_0 k} \tag{10.77}$$

Following the same argumentation as in the case of the bound-state solution we obtain from an asymptotic analysis for small and large ρ the prefactors $\rho^\ell \exp(-\rho/2)$ and $\rho^{|m|}\exp(-\rho/2)$ for *3d* and *2d*, respectively. Making an ansatz as in Eq. (10.52) we then obtain the equations

$$\rho\,\frac{\partial^2 R}{\partial \rho^2} + (2(\ell+1)-\rho)\,\frac{\partial R}{\partial \rho} - (i|\lambda|+\ell+1)R = 0 \ \text{in}\ 3d$$

$$\rho\,\frac{\partial^2 R}{\partial \rho^2} + (2|m|+1-\rho)\,\frac{\partial R}{\partial \rho} - \left[i|\lambda|+|m|+\frac{1}{2}\right]R = 0 \ \text{in}\ 2d. \tag{10.78}$$

These equations are solved by the confluent hypergeometric functions

$$F(\ell+1+i|\lambda|;\ 2(\ell+1);\ \rho)$$

$$F\left[|m|+\frac{1}{2}+i|\lambda|;\ 2|m|+1;\ \rho\right]. \tag{10.79}$$

where $F(a;b;z)$ is defined as

$$F(a;b;z) = 1 + \frac{a}{b\cdot 1}z + \frac{a(a+1)}{b(b+1)\cdot 1\cdot 2}z^2 + \cdots \tag{10.80}$$

The normalization of the continuum states has to be chosen such that it connects continuously with the normalization of the higher bound states, as discussed for the exciton problem by Elliot (1963) for *3d* and by Shinada and Sugano (1966) for *2d*. Therefore we normalize the wavefunctions in a sphere/circle of radius $\mathcal{R}$, where eventually $\mathcal{R} \to \infty$. The respective normalization integrals in *3d* and *2d* are

$$|A_{3d}| \int_0^{\mathcal{R}} dr\; r^2\, (2\mathrm{kr})^{2\ell}\, |F(\ell+1+i|\lambda|;\; 2(\ell+1);\; 2\mathrm{ikr})|^2 = 1$$

$$|A_{2d}| \int_0^{\mathcal{R}} dr\; r\, (2\mathrm{kr})^{2|m|}\, |F(|m|+1/2+i|\lambda|;\; 2|m|+1;\; 2\mathrm{ikr})|^2 = 1\,.$$

(10.81)

Since these integrals do not converge for $\mathcal{R} \to \infty$, we can use the asymptotic expressions for $z \to \infty$ for the confluent hypergeometric functions

$$F(a;b;z) = \frac{\Gamma(b) e^{i\pi a} z^{-a}}{\Gamma(b-a)}\left[1+\mathcal{O}\left(\frac{1}{|z|}\right)\right] + \frac{\Gamma(b) e^{z} z^{a-b}}{\Gamma(a)}\left[1+\mathcal{O}\left(\frac{1}{|z|}\right)\right], \quad (10.82)$$

where $\Gamma(n) = (n-1)!$, n integer, is the gamma function. Using (10.82) in (10.81) we compute the normalization factors A_{3d} and A_{2d} and finally obtain the normalized wave functions as

$$\psi_{k,\ell,m}(\mathbf{r}) = \frac{(\mathrm{i}2\mathrm{kr})^{\ell}}{(2\ell+1)!}\, e^{\frac{\pi|\lambda|}{2}} \sqrt{\frac{2\pi k^2}{\mathcal{R}|\lambda|\, \sinh(\pi|\lambda|)} \prod_{j=0}^{\ell} (j^2+|\lambda|^2)}$$

$$\times\, e^{-ikr}\, F(\ell+1+i|\lambda|;\; 2\ell+2;\; 2\mathrm{ikr})\, Y_{\ell,m}(\theta,\phi) \qquad (10.83a)$$

in *3d* and

$$\psi_{k,m}(\mathbf{r}) = \frac{(i2kr)^{|m|}}{(2|m|)!}\sqrt{\frac{\pi k}{\mathcal{R}(1/4+|\lambda|^2)\cosh(\pi|\lambda|)}\prod_{j=0}^{|m|}\left[\left(j-\frac{1}{2}\right)^2+|\lambda|^2\right]}$$

$$\times\ e^{\frac{\pi|\lambda|}{2}}\, e^{-ikr} F\left(|m|+\frac{1}{2}+i|\lambda|;\ 2|m|+1;\ 2ikr\right)\frac{e^{im\phi}}{\sqrt{2\pi}} \tag{10.83b}$$

in $2d$. For later reference it is important to note that the allowed k-values are defined by

$$k\mathcal{R} = \pi n \text{ and } \Delta k = \frac{\pi}{\mathcal{R}}$$

therefore we have in this case

$$\sum_k = \frac{\mathcal{R}}{\pi}\sum_k \Delta k \rightarrow \frac{\mathcal{R}}{\pi}\int dk.$$

10-5. Optical Spectra

With the knowledge of the exciton and continuum wave functions and the energy eigenvalues we can now solve the inhomogeneous equation (10.41) to obtain the interband polarization and thus calculate the optical spectrum of a semiconductor in the band edge region. We limit our treatment to the discussion of optically allowed transitions in direct-gap semiconductors, because these semiconductors are of main interest with respect to their use for electro-optical devices. Optical transitions across an indirect gap (where the extrema of the valence and conduction band are at different wavevectors) need the simultaneous participation of a photon and a phonon in order satisfy total momentum conservation. Here the phonon provides the necessary wavevector for the transition. Such two-quantum processes have a much smaller transition probability than the direct transitions.

To solve Eq. (10.41) we expand the polarization into the solutions of the Wannier equation

$$P_{vc}(\mathbf{r}) = \sum_{\nu} b_\nu \; \psi_\nu(\mathbf{r}) \; . \tag{10.84}$$

Inserting Eq. (10.84) into (10.44), multiplying by $\psi^*_\mu(\mathbf{r})$ and integrating over $\mathbf{r}$ we find

$$\sum_{\nu} b_\nu [\hbar(\omega+i\delta)-E_g-E_\nu] \int d^3r \; \psi^*_\mu(\mathbf{r})\psi_\nu(\mathbf{r}) = - \, d_{cv} \, \frac{\mathcal{E}_0}{2} \, L^3 \psi^*_\mu(\mathbf{r}{=}0) \; , \tag{10.85}$$

or

$$b_\mu = - \frac{d_{cv} L^3 \; \psi^*_\mu(\mathbf{r}{=}0)}{\hbar(\omega_0+i\delta) \; - \; E_g \; - \; E_\mu} \, \frac{\mathcal{E}_0}{2} . \tag{10.86}$$

From Eq. (10.84), we see that

$$P_{vc}(\mathbf{r}) = - \sum_{\nu} \frac{\mathcal{E}_0}{2} \, \frac{d_{cv} \;\, L^3 \; \psi^*_\nu(\mathbf{r}{=}0)}{\hbar(\omega_0+i\delta) \; - \; E_g \; - \; E_\nu} \; \psi_\nu(\mathbf{r}) \tag{10.87}$$

and therefore

$$P_{vc}(\mathbf{k}) = - \sum_{\nu} \frac{\mathcal{E}_0}{2} \, \frac{d_{cv} \psi^*_\nu(\mathbf{r}{=}0)}{\hbar(\omega_0+i\delta) \; - \; E_g \; - \; E_\nu} \int d^3r \; \psi_\nu(\mathbf{r}) \; e^{i\mathbf{k}\cdot\mathbf{r}} \; . \tag{10.88}$$

Inserting Eqs. (10.88) and (10.40) into Eq. (10.9) we get

$$\begin{aligned} P(t) &= - \, 2 \sum_{\mathbf{k}} d_{vc} P_{vc}(\mathbf{k}) \; e^{-i(\omega+i\delta)t} \; + \text{c.c.} \\ &= 2 \sum_{\nu} \frac{L^3 \; |d_{cv}|^2 \; |\psi_\nu(\mathbf{r}{=}0)|^2}{\hbar(\omega_0+i\delta) \; - \; E_g \; - \; E_\nu} \, \frac{\mathcal{E}_0}{2} e^{-i(\omega+i\delta)t} \; + \text{c.c.} \; , \end{aligned} \tag{10.89}$$

where the factor 2 results from the spin summation, Eq. (10.1). As in Chap. 4, Eq. (4.26), we obtain the optical susceptibility $\chi(\omega)$ from

$$\frac{P(t)}{L^3} = \chi(\omega) \frac{\mathcal{E}_0}{2} e^{-i(\omega+i\delta)t} \tag{10.90}$$

with the result

$$\boxed{\chi(\omega) = -2 |d_{cv}|^2 \sum_{\mu} \frac{|\psi_\mu(\mathbf{r}=0)|^2}{\hbar(\omega+i\delta) - E_g - E_\mu}} \; .$$

electron-hole-pair susceptibility (10.91)

Eq. (10.91) shows that the optical susceptibility is the sum over all states μ, where the oscillator strength of each transition is determined by the probability to find the conduction-band electron and the valence-band hole at the origin, i.e., within the same lattice unit cell. The wavefunctions which are finite in the origin are those with ℓ=0 and m=0 in *3d* and those with m=0 in *2d*. Using the wavefunctions (10.75), (10.83) and

$$Y_{0,0} = \frac{1}{\sqrt{4\pi}} ,$$

the optical susceptibility of a three-dimensional semiconductor becomes

$$\chi(\omega) = - \frac{2|d_{cv}|^2}{\pi E_0 a_0^3} \left\{ \sum_n \frac{1}{n^3} \frac{E_0}{\hbar(\omega+i\delta) - E_g - E_n} + \frac{1}{2} \int dx \frac{x e^{\pi/x}}{\sinh(\pi/x)} \frac{E_0}{\hbar(\omega+i\delta) - E_g - E_0 x^2} \right\} . \tag{10.92}$$

Inserting (10.92) into Eq. (1.53) we obtain the band-edge absorption spectrum as

$$\alpha(\omega) = \alpha_0^{3d} \frac{\hbar\omega}{E_0} \left[\sum_{n=1}^{\infty} \frac{4\pi}{n^3} \delta(\Delta + 1/n^2) + \Theta(\Delta) \frac{\pi\, e^{\pi/\sqrt{\Delta}}}{\sinh(\pi/\sqrt{\Delta})} \right]$$

3d Elliott formula (10.93)

where

$$\Delta = (\hbar\omega - E_g)/E_0 \tag{10.94}$$

and α_0^d has been defined in Eq. (4.47).

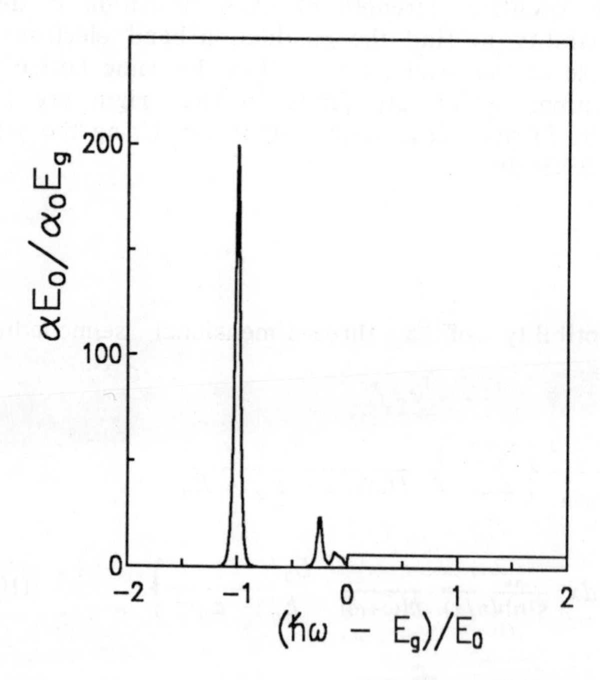

Fig. 10.2: Computed *3d* band-edge absorption spectrum, Eq. (10.93). The δ-functions for the bound states are replaced by $\delta_\Gamma(x)=1/[\pi\Gamma\cosh(x/\Gamma)]$ with $\hbar\Gamma=0.02E_0$. $\alpha_0=\alpha_0^{3d}$ is defined in Eq. (4.47).

Eq. (10.93) is often called the Elliot formula. The *3d* exciton absorption spectrum consist of a series of sharp lines with a rapidly decreasing oscillator strength $\propto n^{-3}$ and a continuum absorption due to the ionized states. A comparison of the continuum part α_{cont} of (10.93) with Eq. (4.46) describing the free-carrier absorption spectrum α_{free}, shows that one can write

$$\alpha_{cont} = \alpha_{free}\ C(\omega) \tag{10.95}$$

where

$$C(\omega) = \frac{\frac{\pi}{\sqrt{\Delta}}\ e^{\pi/\sqrt{\Delta}}}{\sinh(\pi/\sqrt{\Delta})} \tag{10.96}$$

is the so-called Sommerfeld or Coulomb enhancement factor. For $\Delta \to 0$, $C(\omega) \to 2\pi/\sqrt{\Delta}$ so that the continuum absorption assumes a constant value at the band gap, in striking difference to the square-root law of the free-carrier absorption. This shows that the attractive Coulomb interaction not only creates the bound states but has also a pronounced influence on the ionization continuum. If one takes into account a realistic broadening of the single particle-energy eigenstates, e.g., caused by scattering of electron-hole pairs with phonons, only few bound states can be spectrally resolved. The energetically higher bound states merge continuously with the absorption of the ionized states. Fig. 10.2 shows a spectrum which has been calculated using the broadening $\hbar\Gamma = 0.02\ E_0$. The dominant feature is the *1s*-exciton absorption peak. The *2s*-exciton is also resolved, but its height is only 1/8-th of the *1s*-resonance. The higher exciton states appear only as a small peak just below the bandgap and the continuum absorption is almost constant in the shown spectral region. Such spectra are indeed observed at very low temperatures in extremely good-quality semiconductors.

For the two-dimensional limit we obtain the optical susceptibility as

$$\chi(\omega) = -\frac{|d_{cv}|^2}{\pi a_0^2 E_0}\left[\sum_{n=0}^{\infty} \frac{2}{(n+1/2)^3}\ \frac{E_0}{\hbar(\omega+i\delta) - E_g - E_n} + \int dx\ \frac{x\ e^{\pi/x}}{\cosh(\pi/x)}\ \frac{E_0}{\hbar(\omega+i\delta) - E_g - E_k}\right] \tag{10.97}$$

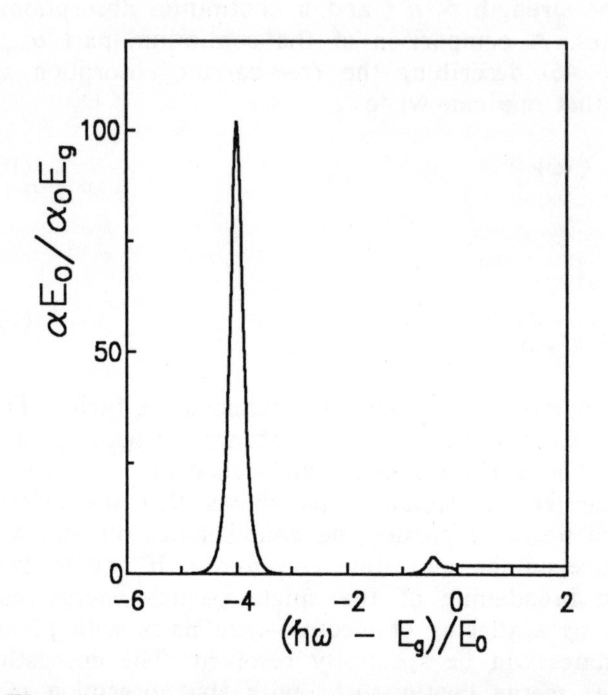

Fig. 10.3: Computed *2d* band-edge absorption spectrum, Eq. (10.98). The broadening has been choosen as $\hbar\Gamma$=0.1E_0 and α_0=α_0^{2d} is defined in Eq. (4.47). For smaller broadenings only the *1s*-resonance would be visible in the spectrum.

and the resulting absorption spectrum is

$$\alpha(\omega) = \alpha_0^{2d} \frac{\hbar\omega}{E_0} \left[\sum_{n=0}^{\infty} \frac{4}{(n+1/2)^3} \, \delta\left[\Delta + \frac{1}{(n+1/2)^2}\right] + \Theta(\Delta) \frac{e^{\pi/\sqrt{\Delta}}}{\cosh(\pi/\sqrt{\Delta})} \right]$$

2d Elliott formula (10.98)

The *2d* Coulomb enhancement factor

$$C(\omega) = \frac{e^{\pi/\sqrt{\Delta}}}{\cosh(\pi/\sqrt{\Delta})} \tag{10.99}$$

approaches 2 for $\Delta \to 0$. The absorption at the band edge is thus twice the free-carrier continuum absorption. For a finite damping the absorption of the ionized states and the absorption in the higher bound states again join continuously. Fig. 10.3 shows the computed absorption spectrum using Eq. (10.98) for a broadening $\hbar\Gamma = 0.1\ E_0$. In comparison to the *3d*-case the *2d 1s*-exciton is spectrally far better resolved as a consequence of the four times larger binding energy in *2d*.

REFERENCES

R.J. Elliot, in *Polarons and Excitons*, eds. C.G. Kuper and G.D. Whitefield, Oliver and Boyd, (1963), *p.* 269

L.D. Landau and E.M. Lifshitz, *Quantum Mechanics*, Pergamon, New York (1958)

L.I. Schiff, *Quantum Mechanics*, 3rd *ed*., McGraw Hill, New York (1968)

M. Shinada and S. Sugano, J. Phys. Soc. Japan **21**, 1936 (1966).

For further discussion of the theory of excitons see:

Polarons and Excitons, eds. C.G. Kuper and G.D. Whitefield, Oliver and Boyd, (1963)

R.S. Knox, *Theory of Excitons*, Academic, New York (1963).

PROBLEMS

Problem 10.1: Derive the Hamiltonian (10.23) from Eq. (7.17) by including the band index λ with all the operators and by summing over $\lambda=c,v$. Discuss the terms which have been omitted in (10.23).

Problem 10.2: Use the Heisenberg equation with the Hamiltonian (10.23) to compute the equation of motion (10.30) for the interband polarization. Make the Hartree Fock approximation in the four-operator terms to obtain Eq. (10.33).

Problem 10.3: Derive Eq. (10.44) using the real-space representation of Sec. 10-1.B.

Problem 10.4: Verify that Eqs. (10.93) and (10.98) correctly reduce to the respective free-particle result without Coulomb interaction. Hint: Formally one can let the Bohr radius $a_0 \to \infty$ to obtain the free-particle limit.

Chapter 11
POLARITONS

In direct-gap semiconductors excitons and photons are strongly coupled. To account for this coupling, it is often useful to introduce new quasi-particles, the *polaritons*, which combine exciton and photon properties. This polariton concept has been very helpful for explaining optical measurements in *3d*-semiconductors with a large direct gap at low excitation densities.

11-1. Dielectric Theory of Polaritons

In Chap. 10 we derived the Wannier equation, Eq. (10.45), for the relative motion of an electron-hole pair. If we include also the center-of-mass motion, this equation becomes

$$-\left[\frac{\hbar^2 \nabla^2_{\mathbf{R}}}{2M} + \frac{\hbar^2 \nabla^2_{\mathbf{r}}}{2m_r} + V(r) \right] \psi(\mathbf{R},\mathbf{r}) = E_{tot}\ \psi(\mathbf{R},\mathbf{r})\ , \tag{11.1}$$

where $M = m_e + m_h$. As in the case of the hydrogen atom, the center-of-mass motion is described by a plane wave

$$\psi(\mathbf{R},\mathbf{r}) = \frac{e^{i\mathbf{k}\cdot\mathbf{R}}}{L^{3/2}}\ \psi(\mathbf{r})\ . \tag{11.2}$$

Correspondingly, the total energy eigenvalue of Wannier excitons is

$$E_{tot} = E_g + E_n + \frac{\hbar^2 K^2}{2M}\ . \tag{11.3}$$

As in the case of an hydrogen atom, the center-of-mass wavefunction of the exciton is a plane wave $\propto \exp(i\mathbf{k}\cdot\mathbf{R})$.

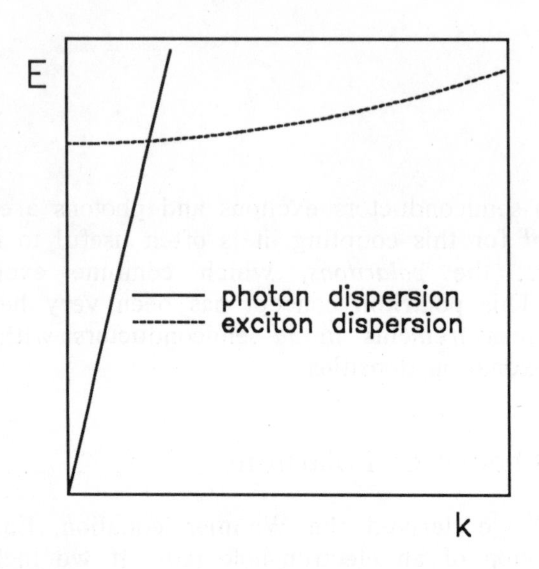

Fig. 11.1: Schematic drawing of photon and *1s*-exciton dispersion.

In Fig. 11.1 we plot the dispersion of the *1s*-exciton together with the dispersion of the light

$$\hbar\omega_k = \frac{\hbar ck}{n_0} , \tag{11.4}$$

where n_0 is the background refractive index of the medium. The figure shows that both dispersions intersect, i.e., the dispersions are degenerate at the intersection point. The exciton-photon interaction removes that degeneracy introducing a modified joint dispersion for both particles. The quasi-particles associated with this new dispersion are the exciton-polaritons. In general, polaritons are also formed by transverse optical phonons and the light, called phonon-polaritons. Since we do not discuss phonon-polaritons in this book, we refer to the exciton-polaritons simply as polaritons.

The optical dielectric function for the excitons is

$$\epsilon(\omega) = \epsilon_0(1 + 4\pi\chi(\omega))$$

$$= \epsilon_0 \left[1 - 8\pi \; |d_{cv}|^2 \sum_n \frac{|\psi_n(\mathbf{r}=0)|^2}{\hbar(\omega + i\delta) - E_g - E_n} \right], \tag{11.5}$$

where $\epsilon_0 = n_0^2$ and Eq. (10.91) has been used. In this form only resonant terms are contained and the momentum of the emitted or absorbed photon has been neglected. In order to find the propagating electromagnetic modes in a dielectric medium, we have to insert $\epsilon(\omega)$ into the wave equation (1.43) for the transverse electric field of a light beam. For a plane wave we get

$$\left[- k^2 + \frac{\omega^2}{c^2}\epsilon(\omega) \right] \mathcal{E}_\omega \; e^{i(\mathbf{k}\cdot\mathbf{r} - \omega t)} = 0. \tag{11.6}$$

The transverse eigenmodes of the medium are obtained from the requirement that $\mathcal{E}_\omega \neq 0$, so that

$$\boxed{c^2 k^2 = \omega^2 \epsilon(\omega)} \tag{11.7}$$

transverse eigenmodes

A) Polaritons without spatial dispersion and damping

Let us analyze Eq. (11.7) considering only the contribution of the lowest exciton level,

$$\epsilon(\omega) \cong \epsilon_0 \left[1 - 8\pi \; |d_{cv}|^2 \; \frac{|\psi_1(\mathbf{r}=0)|^2}{\hbar(\omega + i\delta) - E_g + E_0} \right], \tag{11.8}$$

or

$$\epsilon(\omega) = \epsilon_0 \left[1 - \frac{\Delta}{\omega - \omega_0 + i\delta} \right], \tag{11.9}$$

where

$$\hbar\Delta = 8\pi \; |d_{cv}|^2 \; |\psi_1(\mathbf{r}=0)|^2 \quad \text{and} \quad \hbar\omega_0 = E_g - E_0 \; . \tag{11.10}$$

In order to satisfy Eq. (11.7), the wavenumber k has to be chosen complex

$$k = k' + ik'' . \tag{11.11}$$

Then we can write the real and imaginary part of Eq. (11.7) as

$$\frac{\omega^2\epsilon_0}{c^2}\left[1-\frac{\Delta}{\omega-\omega_0}\right] = k'^2 - k''^2 \tag{11.12}$$

and

$$\pi\delta(\omega-\omega_0)\,\frac{\omega_0^{\;2}\epsilon_0}{c^2} = 2k'k'' , \tag{11.13}$$

respectively, where the Dirac identity, Eq. (1.17), has been used. The imaginary part of the wavenumber is proportional to $\delta(\omega-\omega_0)$ describing absorption at the exciton resonance. Outside the resonance $(\omega\neq\omega_0)$ we obtain the undamped polariton modes

$$\omega\sqrt{\frac{\omega-\omega_0-\Delta}{\omega-\omega_0}} = \frac{ck'}{\sqrt{\epsilon_0}} . \tag{11.14}$$

The resulting dispersion is plotted in Fig. 11.2. From Eq. (11.14) one obtain for low frequencies, $\omega<<\omega_0$, a photon-like dispersion

$$\omega \cong \frac{ck'}{\sqrt{\epsilon_0(1+\Delta/\omega_0)}} \tag{11.15}$$

with a light velocity slightly smaller than c/n_0. For $\omega \to \omega_0$ the wavenumber diverges, $k' \to \infty$. No solution of Eq. (11.14) exists for frequencies between ω_0 and $\omega_0 + \Delta$. In other words, there is a *stop band* between ω_0 and $\omega_0 + \Delta$ separating the lower and upper polariton branch. At $\omega_0 + \Delta$, k' is zero and for $\omega>>\omega_0$ one again obtains a photon-like dispersion with the light velocity c/n_0.

The longitudinal eigenmodes of a dielectric medium are determined by

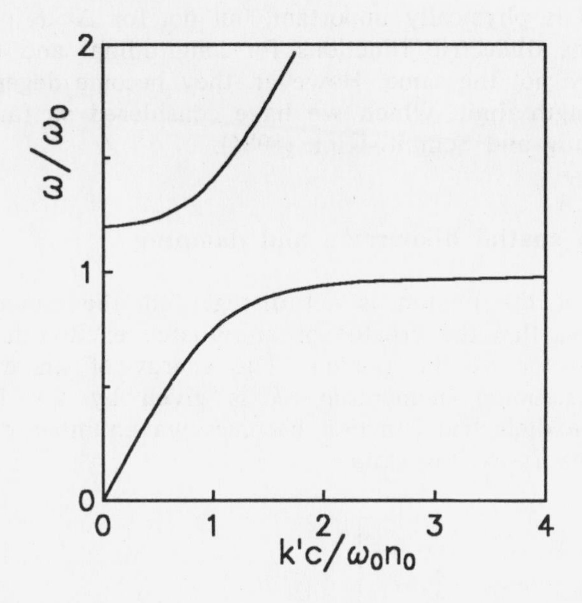

Fig. 11.2: Polariton dispersion without spatial dispersion for Δ/ω_0=0.2.

$$\boxed{\epsilon(\omega) = 0} \quad ,$$

longitudinal eigenmodes (11.16)

which yields the eigenfrequency of the longitudinal exciton $\omega_\ell = \omega_0 + \Delta$, while the transverse exciton frequency $\omega_t = \omega_0$. Therefore Δ is called the longitudinal-transverse (LT) splitting. For the longitudinal excitons the polarization is parallel to the wavevector $\mathbf{k}$. Eq. (11.10) shows that the longitudinal-transverse splitting is proportional to the so-called optical matrix element

$$|d_{cv}|^2 \ |\psi_1(\mathbf{r}=0)|^2 \propto \frac{|d_{cv}|^2}{a_0^d} , \tag{11.17}$$

where Eq. (10.75) has been used. The region in which no transverse waves propagate is thus particularly large for crystals with small excitons Bohr radii, i.e., in wide-gap semiconductors such as CuCl.

Whether the LT-splitting can be observed experimentally depends on the ratio of Δ to the exciton damping δ, which is finite in real crystals, e.g., as a consequence of the exciton-phonon interaction. For $\Delta \gg \delta$ the stop band is physically important, but not for $\Delta \ll \delta$.

In general the dielectric functions for longitudinal and transverse excitations are not the same. However, they become degenerate in the long-wavelength limit, which we have considered so far. For more details see Haug and Schmitt-Rink (1984).

B) Polaritons with spatial dispersion and damping.

If the momentum of the photon is not disregarded, the momentum conservation requires that the created or annihilated exciton has the same finite wavevector as the photon. The energy of an exciton with a finite translational momentum $\hbar k$ is given by Eq. (11.3). Then the transverse dielectric function becomes wavenumber dependent yielding for the *1s*-exciton state

$$\epsilon(\mathbf{k},\omega) = \epsilon_0 \left[1 - \frac{\Delta}{\omega - \omega_0 - \frac{\hbar k^2}{2M} + i\gamma} \right] . \tag{11.18}$$

Here we treat the damping γ as a generally non-vanishing constant. However, we note that a realistic description of an exciton absorption line often requires a frequency dependent damping $\gamma(\omega)$. A constant γ results in a Lorentzian absorption line, but in reality one observes at elevated temperatures nearly universally an exponential decrease of the exciton absorption $\alpha(\omega)$ for frequencies below the exciton resonance, i.e.,

$$\boxed{\alpha(\omega) = \alpha_0 \, e^{-(\omega_0 - \omega)/\sigma} \quad \text{for } \omega < \omega_0} \tag{11.19}$$

Urbach rule

The derivation of the Urbach rule needs a damping $\gamma(\omega)$ which decreases with increasing detuning $\omega_0 - \omega$. The physical origin of the dynamical or frequency-dependent damping is the following. The absorption of a photon with insufficient energy $\hbar\omega < \hbar\omega_0$ requires the scattering of the virtually created exciton with energy $\hbar\omega$ into a state $E_k = \hbar\omega_0 + \hbar^2 k^2/2M$ under the absorption or scattering of an already present excitation in the crystal. In polar semiconductors the relevant excitation will be a longitudinal optical (LO) phonon. Now it is evident that $\gamma(\omega)$ decreases rapidly with decreasing frequency because the probability to absorb n LO phonons decreases rapidly with increasing n. From a microscopic point of view the damping is the imaginary part of the exciton self-energy $\Sigma(\mathbf{k},\omega)$, which in general is both frequency and momentum dependent.

Now let us return to the simple form of Eq. (11.18) to discuss the transverse eigenmodes. Because a momentum-dependent dielectric function means a nonlocal response in real space, one speaks in this case also of a dielectric function with spatial dispersion. Inserting Eq. (11.18) into the eigenmode equation (11.7) yields

$$\frac{c^2k^2}{\epsilon_0} = \omega^2\left(1 - \frac{\Delta}{\omega-\omega_0-\frac{\hbar k^2}{2M}+i\gamma}\right) \tag{11.20}$$

The solution of this equation for the real and imaginary part of the wavenumber is shown in Fig. 11.3. With spatial dispersion and finite damping one finds for all frequencies two branches, ω_1 and ω_2. At high frequencies, $\omega/\omega_0 > 1$, Fig. 11.3 again shows a photon-like and an exciton-like branch. In the range of the LT-split $\omega_2(k')$ has some structure which results in a negative group velocity (negative slope of $\omega_2(k')$). One sees however that in this range the damping of this mode increases strongly. In a region with damping the group velocity looses its meaning. Instead the Poynting vector has to be calculated, from which one gets an energy velocity. At still lower frequency the branch ω_2 continues as a Tscherenkov mode with $c_T > c/n_0$. But again this has no physical significance due to the very large damping. The mode ω_1 is damped in the range of the LT-splitting and becomes photon-like with little damping below the exciton resonance, $\omega/\omega_0=1$.

If one considers the influence of the interface between the crystal and, e.g., vacuum, the presence of two propagating modes require an additional boundary condition (ABC). Often Pekar's additional boundary condition is chosen which states that the normal component of the polarization has to be zero at the interface. The need of the additional boundary condition stems from the fact that one has used a three dimensional spatial Fourier transform, which is not allowed for a crystal with a surface. To avoid this problem, one has to calculate the two-point polarization function $P_{cv}(\mathbf{r},\mathbf{r}',t)$ for a spatially inhomogeneous system.

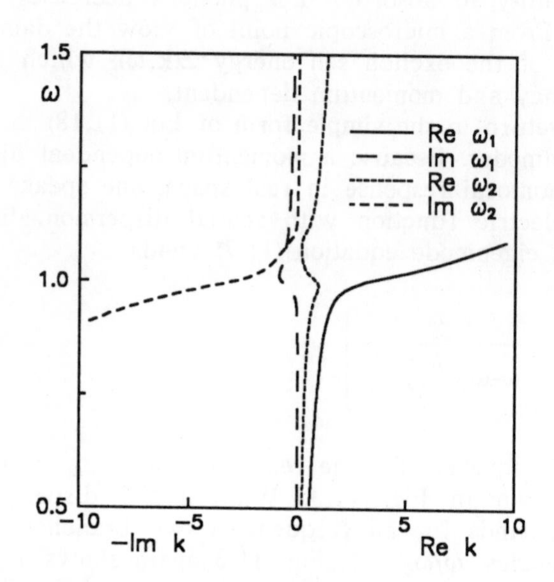

Fig. 11.3: Polariton dispersion with spatial dispersion and damping for Δ/ω_0=0.1, γ/ω_0=0.01, $\hbar\omega_0\epsilon_0/2Mc^2 = 0.001$. The frequency ω is in units of ω_0 and k is in units of $c/\omega_0 n_0$, respectively.

11-2. Hamiltonian Theory of Polaritons

Polaritons can also be discussed directly as mixed exciton-photon excitations in a semiconductor. For this end we will use exciton creation and destruction operators, which, as we will show, obey approximately Bose commutation relations. For the subsequent calculations it is convenient to introduce a hole operator

$$\beta^{\dagger}_{-\mathbf{k},-s} \equiv a_{v,\mathbf{k},s} \tag{11.21}$$

indicating that the annihilation of an electron with $(\mathbf{k},s)$ in the valence band corresponds to the creation of a hole with the opposite momentum and spin. To keep the notation symmetrical, we define

$$\alpha^{\dagger}_{\mathbf{k},s} \equiv a^{\dagger}_{c,\mathbf{k},s} \tag{11.22}$$

where the electron operators α, $\alpha^{\dagger}$ operate exclusively in the conduction band and the hole operators β, $\beta^{\dagger}$ operate in the valence band, respectively.

Suppressing the spin index, and using the electron-hole-pair operator

$$\alpha^{\dagger}_{\mathbf{k}}\beta^{\dagger}_{-\mathbf{k}'} , \tag{11.23}$$

the operator describing generation of an exciton in the state ν with total momentum $\mathbf{K}$ can be written as

$$B^{\dagger}_{\nu,\mathbf{K}} = \sum_{\mathbf{k},\mathbf{k}'} \delta(\mathbf{K}-(\mathbf{k}-\mathbf{k}'))\ \psi_{\nu}\left[\frac{\mathbf{k}+\mathbf{k}'}{2}\right]\ \alpha^{\dagger}_{\mathbf{k}}\beta^{\dagger}_{-\mathbf{k}'}$$

$$= \sum_{\mathbf{k}} \psi_{\nu}(\mathbf{k}-\mathbf{K}/2)\ \alpha^{\dagger}_{\mathbf{k}}\beta^{\dagger}_{\mathbf{K}-\mathbf{k}} \ . \tag{11.24}$$

The derivation of this relation is best obtained in the Dirac representation, where

$$B^{\dagger}_{\nu,\mathbf{K}} = |\nu\mathbf{K}\rangle\langle 0| \ . \tag{11.25}$$

Multiplying Eq. (11.25) with the completeness relation in terms of the electron-hole pairs

$$\sum_{\mathbf{k},\mathbf{k}'} |\mathbf{k},-\mathbf{k}'\rangle\langle\mathbf{k},-\mathbf{k}'| = 1 \tag{11.26}$$

one gets

$$B^{\dagger}_{\nu,\mathbf{K}} = \sum_{\mathbf{k},\mathbf{k}'} |\mathbf{k},-\mathbf{k}'\rangle\langle\mathbf{k},-\mathbf{k}'|\nu\mathbf{K}\rangle\langle 0|$$

$$= \sum_{\mathbf{k},\mathbf{k}'} \langle \mathbf{k},-\mathbf{k}' | \nu\mathbf{K}\rangle \; |\mathbf{k},-\mathbf{k}'\rangle\langle 0|$$

$$= \sum_{\mathbf{k},\mathbf{k}'} \langle \mathbf{k},-\mathbf{k}' | \nu\mathbf{K}\rangle \; \alpha^{\dagger}_{\mathbf{k}} \; \beta^{\dagger}_{-\mathbf{k}'} \; . \tag{11.27}$$

Furthermore,

$$\langle \mathbf{k},-\mathbf{k}' | \nu\mathbf{K}\rangle = \int d^3r \; d^3r' \; \langle \mathbf{k},-\mathbf{k}' | \mathbf{r},\mathbf{r}'\rangle\langle \mathbf{r},\mathbf{r}' | \nu\mathbf{K}\rangle$$

$$= \int d^3r \; d^3r' \; e^{-i\mathbf{k}\cdot\mathbf{r}} e^{i\mathbf{k}'\cdot\mathbf{r}'} e^{i\mathbf{K}\cdot(\mathbf{r}+\mathbf{r}')/2} \; \psi_\nu(\mathbf{r}-\mathbf{r}')$$

$$= \delta(\mathbf{K}-(\mathbf{k}-\mathbf{k}')) \; \psi_\nu \left[\frac{\mathbf{k}+\mathbf{k}'}{2}\right] \tag{11.28}$$

where $\psi_\nu(\mathbf{k})$ is the Fourier transform of the exciton wavefunction for the relative motion. Explicitly, the Fourier transforms for the ground-state wavefunctions are

$$\psi_0(k) = 8\sqrt{\pi\alpha_0^3} \frac{1}{(1+(ka_0)^2)^2} \; ; \qquad \text{in } 3d \tag{11.29}$$

and

$$\psi_0(k) = \sqrt{2\pi} \; \alpha_0^2 \; \frac{1}{(1+(ka_0/2)^2)^{3/2}} \; . \qquad \text{in } 2d \tag{11.30}$$

We see that $\psi_0(k)$ is roughly constant for $0 < k < 1/a_0$ or $2/a_0$ and falls off rapidly for large k values.

The exciton is a bound states of two Fermions and can thus be expected to be approximately a Boson. The commutator of exciton operators is

$$[B_{0,0}, B_{0,0}^{\dagger}] = \sum_{\mathbf{k},\mathbf{k}'} \psi_0(k)\, \psi_0(k') \left[\beta_{-\mathbf{k}}\alpha_{\mathbf{k}} \,,\, \alpha_{\mathbf{k}'}^{\dagger}\beta_{-\mathbf{k}'}^{\dagger} \right]$$

$$= \sum_{\mathbf{k}} |\psi_0(k)|^2 \left(1 - \alpha_{\mathbf{k}}^{\dagger}\alpha_{\mathbf{k}} - \beta_{-\mathbf{k}}^{\dagger}\beta_{-\mathbf{k}} \right) \tag{11.31}$$

so that

$$\langle [B_{0,0}, B_{0,0}^{\dagger}] \rangle = 1 - O(na_0^d) \,. \tag{11.32}$$

The deviation from the Boson commutator is proportional to the mean number of electrons and holes contained in an excitonic volume. This shows that excitons approximately become Bosons in the low-density limit, i.e., in weakly excited crystals.

The Hamiltonian for free excitons can be written as

$$\mathcal{H}_0 = \sum_{\mathbf{k}} \hbar e_{\nu k}\; B_{\nu,\mathbf{k}}^{\dagger}\; B_{\nu,\mathbf{k}} \,. \tag{11.33}$$

In order to express the interaction Hamiltonian $\mathcal{H}_I$ of the electron system with the light field in terms of exciton operators, we multiply Eq. (11.24) with $\psi_\nu^*(\boldsymbol{\kappa})$ and sum over ν to get

$$\sum_{\nu} \psi_\nu^*(\boldsymbol{\kappa})\; B_{\nu,\mathbf{K}}^{\dagger} = \sum_{\nu} \psi_\nu^*(\boldsymbol{\kappa})\; \psi_\nu(\mathbf{k}-\mathbf{K}/2)\; \alpha_{\mathbf{k}}^{\dagger}\; \beta_{\mathbf{K}-\mathbf{k}}^{\dagger}$$

$$= \sum_{\nu} \delta_{\boldsymbol{\kappa},\mathbf{k}-\mathbf{K}/2}\; \alpha_{\mathbf{k}}^{\dagger}\; \beta_{\mathbf{K}-\mathbf{k}}^{\dagger} = \alpha_{\frac{1}{2}\mathbf{K}+\boldsymbol{\kappa}}^{\dagger}\; \beta_{\frac{1}{2}\mathbf{K}-\boldsymbol{\kappa}}^{\dagger} \,. \tag{11.34}$$

Taking the finite wavenumber q of the light explicitly into account, $\mathcal{H}_I$ can then be written as

$$\mathcal{H}_I = -\sum_{\mathbf{k},\mathbf{q}} d_{cv} \left\{ \alpha^\dagger_{\frac{1}{2}\mathbf{q}+\boldsymbol{\kappa}} \beta^\dagger_{\frac{1}{2}\mathbf{q}-\boldsymbol{\kappa}} \frac{\mathcal{E}(\mathbf{q})}{2} e^{-i\omega_q t} + \text{h.c.} \right\}$$

$$= -\sum_{\mathbf{k},\mathbf{q},\nu} d_{cv} \left\{ \psi_\nu(k) B^\dagger_{\nu,\mathbf{q}} \frac{E(\mathbf{q})}{2} e^{-i\omega_q t} + \text{h.c.} \right\}. \qquad (11.35)$$

Using

$$\mathcal{E}(\mathbf{q},t) = -\frac{1}{c}\frac{\partial}{\partial t} A(\mathbf{q},t) \qquad (10.36)$$

and Eq. (5.66) we obtain

$$\mathcal{E}(\mathbf{q},t) = \mathcal{E}(\mathbf{q})\, e^{-i\omega_q t} = i\sqrt{2\pi\hbar\omega_q}\,(\, c_\mathbf{q} - \text{h.c.}\,) \qquad (11.37)$$

where we denote the photon annihilation operator by $c_\mathbf{q}$ and

$$\omega_q = \frac{cq}{\sqrt{\epsilon_0}}\,. \qquad (11.38)$$

The exciton-photon interaction Hamiltonian in the resonant approximation is thus

$$\mathcal{H}_I = -i\hbar \sum_{\mathbf{q},\nu} g_{\nu\mathbf{q}}\,(\, B^\dagger_{\nu\mathbf{q}}\, c_\mathbf{q} - \text{h.c.}\,) \qquad (11.39)$$

with the optical matrix element

$$\hbar g_{\nu\mathbf{q}} = d_{cv}\, \psi_\nu(\mathbf{r}{=}0)\, \sqrt{\pi\hbar\omega_q/2}\,. \qquad (11.40)$$

The total exciton-photon Hamiltonian is now

$$\mathcal{H} = \hbar\sum_{\mathbf{q}} \left\{ \sum_\nu e_{\nu q} B^\dagger_{\nu\mathbf{q}} B_{\nu\mathbf{q}} + \omega_q c^\dagger_\mathbf{q} c_\mathbf{q} - i\sum_\nu g_{\nu\mathbf{q}}\,(B^\dagger_{\nu\mathbf{q}} c_\mathbf{q} - \text{h.c.}) \right\} \qquad (11.41)$$

In the following we consider only the lowest exciton level. The bilinear Hamiltonian (11.41) can be diagonalized by introducing *polariton operators* $p_{\mathbf{q}}$ as linear combination of exciton and photon operators

$$p_{\mathbf{q}} = u_{\mathbf{q}} B_{\mathbf{q}} + v_{\mathbf{q}} c_{\mathbf{q}} \; . \tag{11.42}$$

The polariton operators have to obey Bose commutation relations

$$[p_{\mathbf{q}} \, , \, p_{\mathbf{q}}^{\dagger}] = |u_{\mathbf{q}}|^2 + |v_{\mathbf{q}}|^2 = 1 \; . \tag{11.43}$$

Now we choose the unknown coefficients $u_{\mathbf{q}}$ and $v_{\mathbf{q}}$ so that the Hamiltonian (11.41) becomes diagonal in the polariton operators

$$\mathscr{H} = \hbar \sum_{\mathbf{q}} \Omega_q \, p_{\mathbf{q}}^{\dagger} \, p_{\mathbf{q}} \; . \tag{11.44}$$

It turns out that the transformation coefficients $u_{\mathbf{q}}$ and $v_{\mathbf{q}}$ and the polariton spectrum $\Omega_{\mathbf{q}}$ can best be found by evaluating the commutator $[p,H]/\hbar$ once directly using the polariton Hamiltonian (11.44) and once using Eq. (11.42) together with the exciton-photon Hamiltonian. We obtain

$$\begin{aligned} \Omega_{\mathbf{q}} p_{\mathbf{q}} &= \Omega_{\mathbf{q}} (u_{\mathbf{q}} B_{\mathbf{q}} + v_{\mathbf{q}} c_{\mathbf{q}}) \\ &= u_{\mathbf{q}} \, (e_{\mathbf{q}} B_{\mathbf{q}} + i g_{\mathbf{q}} c_{\mathbf{q}}) + v_{\mathbf{q}} (\omega_{\mathbf{q}} c_{\mathbf{q}} - i g_{\mathbf{q}} B_{\mathbf{q}}) \; . \end{aligned} \tag{11.45}$$

Comparing the coefficients of $B_{\mathbf{q}}$ and $c_{\mathbf{q}}$ we find

$$0 = (\Omega_{\mathbf{q}} - e_{\mathbf{q}}) \, u_{\mathbf{q}} + i g_{\mathbf{q}} v_{\mathbf{q}}$$

and

$$0 = - \, i g_{\mathbf{q}} u_{\mathbf{q}} + (\Omega_{\mathbf{q}} - \omega_{\mathbf{q}}) \, v_{\mathbf{q}} \; , \tag{11.46}$$

respectively. The determinant of the coefficients of $u_{\mathbf{q}}$ and $v_{\mathbf{q}}$ has to vanish, i.e,

$$(\Omega_{\mathbf{q}} - e_{\mathbf{q}}) \, (\Omega_{\mathbf{q}} - \omega_{\mathbf{q}}) - g_{\mathbf{q}}^2 = 0,$$

or

$$\boxed{\Omega_{\mathbf{q},1,2} = \frac{1}{2}(e_\mathbf{q} + \omega_\mathbf{q}) \pm \frac{1}{2}\sqrt{(e_\mathbf{q}+\omega_\mathbf{q})^2 + 4g_\mathbf{q}^2}} \; .$$

polariton spectrum (11.47)

We see immediately that $\Omega_{1,2}$ are the frequencies of the upper and lower polariton branches which we found already in Sec. 11.1. For $q \to 0$ the upper branch approaches $e_0 + g^2/e_0$ showing that the LT splitting in this formulation is given as g^2/e_0.

Using Eqs. (11.43) and (11.46) we find for the upper polariton branch

$$u_{\mathbf{q},1} = \sqrt{\frac{\Omega_{\mathbf{q},1} - \omega_\mathbf{q}}{2\Omega_{\mathbf{q},1} - e_\mathbf{q} - \omega_\mathbf{q}}} \quad \text{and} \quad v_{\mathbf{q},1} = i\sqrt{\frac{\Omega_{\mathbf{q},1} - e_\mathbf{q}}{2\Omega_{\mathbf{q},1} - e_\mathbf{q} - \omega_\mathbf{q}}} \tag{11.48}$$

and for the lower polariton branch with $\Omega_2 < e, \omega$ the coefficients

$$u_{\mathbf{q},2} = \sqrt{\frac{\Omega_{\mathbf{q},2} - \omega_\mathbf{q}}{2\Omega_{\mathbf{q},2} - e_\mathbf{q} - \omega_\mathbf{q}}} \quad \text{and} \quad v_{\mathbf{q},2} = -i\sqrt{\frac{\Omega_{\mathbf{q},2} - e_\mathbf{q}}{2\Omega_{\mathbf{q},2} - e_\mathbf{q} - \omega_\mathbf{q}}} \tag{11.49}$$

Fig. (11.4) shows the wavenumber dependence of $|u_1|^2$ and $|v_1|^2$ for the upper polariton branches according to Eq. (11.48), for simplicity without spatial dispersion, i.e., with $e_q = e_0$. The upper branch polariton is for small q-values an exciton-like ($|u_1|^2 \cong 1$) and changes around the exciton resonance successively into a photon-like excitation ($|v_1|^2 \cong 1$). The lower branch polariton shows the reverse properties. After this transformation, the resulting polariton Hamiltonian is

$$\mathcal{H} = \hbar \sum_{\mathbf{q};i=1,2} \Omega_{i,q}\, p_{i,\mathbf{q}}^\dagger\, p_{i,\mathbf{q}} \; . \tag{11.50}$$

All formulas derived in this chapter hold naturally only if the nonresonant interaction terms are small. The full interaction Hamiltonian with nonresonant terms has the form

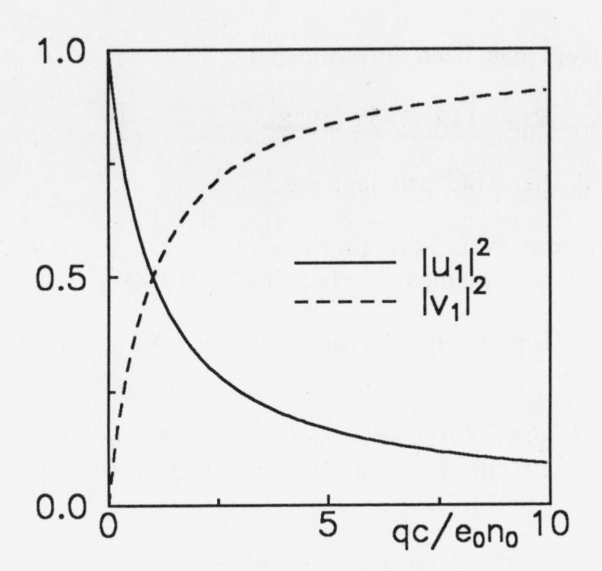

Fig. 11.4: Dispersion of $|u_i|^2$ and $|v_i|^2$ for the upper polariton branch.

$$\mathscr{H}_I = i \sum g_q \,(B_{\mathbf{q}} + B^{\dagger}_{-\mathbf{q}})(c_{-\mathbf{q}} - c^{\dagger}_{\mathbf{q}}) \ . \tag{11.51}$$

The total Hamiltonian can still be diagonalized by a transformation which mixes $B_{\mathbf{q}}$, $c_{\mathbf{q}}$ and $B^{\dagger}_{-\mathbf{q}}$, $c^{\dagger}_{-\mathbf{q}}$ operators linearly

$$p_{i\mathbf{q}} = u_{i1\mathbf{q}}\, B_{\mathbf{q}} + u_{i2\mathbf{q}}\, c_{\mathbf{q}} + u_{i3\mathbf{q}}\, B^{\dagger}_{-\mathbf{q}} + u_{i4\mathbf{q}}\, c^{\dagger}_{-\mathbf{q}} \ . \tag{11.52}$$

This slightly more general transformation is called the Hopfield (1958) polariton transformation.

If we compare the dielectric formalism used at the beginning of this chapter with the diagonalization procedure of this section, one sees that an advantage of the first approach is that it allows quite naturally to include the damping of the modes.

REFERENCES

The polariton concept has been introduced by:

J.J. Hopfield, Phys. Rev. **112**, 1555 (1958).

For a discussion of the ABC problem see:

V.M. Agranovich and V.L. Ginzburg, *Crystal Optics with Spatial Dispersion and Excitons*, Springer Verlag, Berlin (1984)

K. Cho, *Excitons*, Topics in Current Physics, Vol. 14, Springer Verlag, Berlin (1979)

A. Stahl and I. Balslev, *Electrodynamics of the Semiconductor Band Edge*, Springer Tracts in Modern Physics 110, Springer Verlag, Berlin (1987).

For further reading on the dielectric formalism see:

H. Haug and S. Schmitt-Rink, Prog. Quant. Electron. **9**, 3 (1984).

PROBLEMS

Problem 11.1: Generalize Eq. (11.20) to include the energetically higher (n = 2, 3, ..) exciton levels.

Problem 11.2: Compute the Fourier transform of the *1s*-exciton wavefunction in *2d* and *3d*.

Problem 11.3: Compute the polariton transformation coefficients $u_{\mathbf{q}}$ and $v_{\mathbf{q}}$ and the polariton spectrum $\Omega_{\mathbf{q}}$ by evaluating the commutator

$$[p, H]$$

once using the polariton Hamiltonian, Eq. (11.44), and once using Eq. (11.42) together with the exciton-photon Hamiltonian.

Problem 11.4: Show that the operators for the lower and upper polariton branches commute, e.g.,

$$[p_1, p_2^\dagger] = 0.$$

Chapter 12
SEMICONDUCTOR BLOCH EQUATIONS

In the previous two chapters we analyzed the interband transitions in semiconductors for very low excitation conditions, where the Coulomb attraction between the conduction-band electron and the valence-band hole turned out to be very important, but no interactions between different electron-hole pairs had to be considered. These interactions become increasingly important for higher electron-hole-pair densities. In this case we have to deal with finite numbers of electrons and holes in the conduction and valence band which are coupled dynamically to the interband polarization. In this chapter we derive the set of coupled equations which governs the medium dynamics in the band-gap region. We call these equations the *semiconductor Bloch equations* since they have a similar structure and play a similar role as the optical Bloch equations in the theory of atomic systems.

12-1. Hamiltonian Equations

As in Chap. 10 we start from the Hamiltonian

$$\mathcal{H} = \mathcal{H}_{e\ell} + \mathcal{H}_I \tag{12.1}$$

where $\mathcal{H}_{e\ell}$ is the Hamiltonian of the semiconductor electron-hole system in two-band approximation and $\mathcal{H}_I$ dscribes the interband dipole coupling to the light field. In detail we have

$$\mathcal{H}_{e\ell} = \sum_{\mathbf{k}} E_{c,k} a^\dagger_{c,\mathbf{k}} a_{c,\mathbf{k}} + \sum_{\mathbf{k}} E_{v,k} a^\dagger_{v,\mathbf{k}} a_{v,\mathbf{k}}$$

$$+ \frac{1}{2} \sum_{\mathbf{k},\mathbf{k}',q\neq 0} V_q \Big[a^\dagger_{c,\mathbf{k}+\mathbf{q}} a^\dagger_{c,\mathbf{k}'-\mathbf{q}} a_{c,\mathbf{k}'} a_{c,\mathbf{k}} + a^\dagger_{v,\mathbf{k}+\mathbf{q}} a^\dagger_{v,\mathbf{k}'-\mathbf{q}} a_{v,\mathbf{k}'} a_{v,\mathbf{k}}$$

$$+ 2a^\dagger_{c,\mathbf{k}+\mathbf{q}} a^\dagger_{v,\mathbf{k}'-\mathbf{q}} a_{v,\mathbf{k}'} a_{c,\mathbf{k}} \Big] , \tag{12.2}$$

and the interaction Hamiltonian with the light field is written as

$$\mathcal{H}_I = - \sum_{\mathbf{k}} \Big[d_{cv,\mathbf{k}} \frac{\mathcal{E}_0(t)}{2} a^\dagger_{c,\mathbf{k}} a_{v,\mathbf{k}} + \text{h.c.} \Big] , \tag{12.3}$$

compare Eqs. (10.22) and (10.23). For the subsequent calculations it is convenient to transform the total Hamiltonian into the electron-hole representation (see Chap. 11). Inserting the definitions (11.21) and (11.22) into the Hamiltonian (12.1) and restoring normal ordering of the electron and hole operators, we obtain

$$\mathcal{H} = \sum_{\mathbf{k}} [E_{e,k} \alpha^\dagger_{\mathbf{k}} \alpha_{\mathbf{k}} + E_{h,k} \beta^\dagger_{-\mathbf{k}} \beta_{-\mathbf{k}}]$$

$$+ \frac{1}{2} \sum_{\mathbf{k},\mathbf{k}',q\neq 0} V_q [\alpha^\dagger_{\mathbf{k}+\mathbf{q}} \alpha^\dagger_{\mathbf{k}'-\mathbf{q}} \alpha_{\mathbf{k}'} \alpha_{\mathbf{k}} + \beta^\dagger_{\mathbf{k}+\mathbf{q}} \beta^\dagger_{\mathbf{k}'-\mathbf{q}} \beta_{\mathbf{k}'} \beta_{\mathbf{k}} - 2\alpha^\dagger_{\mathbf{k}+\mathbf{q}} \beta^\dagger_{\mathbf{k}'-\mathbf{q}} \beta_{\mathbf{k}'} \alpha_{\mathbf{k}}]$$

$$- \sum_{\mathbf{k}} \Big[d_{cv,\mathbf{k}} \frac{\mathcal{E}_0(t)}{2} \alpha^\dagger_{\mathbf{k}} \beta^\dagger_{-\mathbf{k}} + \text{h.c.} \Big] , \tag{12.4}$$

where constant terms have been left out. The single particle energies in (12.4) are

$$E_{e,k} = E_{c,k} = \hbar\epsilon_{e,k}$$

$$E_{h,k} = - E_{v,k} + \sum_{q \neq 0} V_q = \hbar \epsilon_{k,k} \tag{12.5}$$

showing that the kinetic energy of the holes includes the Coulomb energy of the full valence band. The low-intensity interband transitions energy is therefore

$$\Delta E_k = E_{c,k} - E_{v,k} + \sum_{q \neq 0} V_q \ . \tag{12.6}$$

In extension of the analysis in Chap. 10, we now want to derive the coupled equations of motion for the following elements of the reduced density matrix

$$\begin{aligned} \langle \alpha^\dagger_{\mathbf{k}} \alpha_{\mathbf{k}} \rangle &= n_{e,\mathbf{k}}(t) \\ \langle \beta^\dagger_{-\mathbf{k}} \beta_{-\mathbf{k}} \rangle &= n_{h,\mathbf{k}}(t) \\ \langle \beta_{-\mathbf{k}} \alpha_{\mathbf{k}} \rangle &= P_{he}(\mathbf{k},t) \equiv P_{\mathbf{k}}(t) \ . \end{aligned} \tag{12.7}$$

We proceed as in Chap. 10, Eqs. (10.30) - (10.35), and compute the Hamiltonian equations of motion, but this time for all the expectations values (12.7). Straightforward operator algebra yields

$$\begin{aligned} \hbar \left[i \frac{d}{dt} - (\epsilon_{e,k} + \epsilon_{h,k}) \right] P_{\mathbf{k}} &= (n_{e,\mathbf{k}} + n_{h,\mathbf{k}} - 1)\, d_{vc,\mathbf{k}} \mathcal{E}_0(t)/2 \\ &+ \sum_{\mathbf{k}',q \neq 0} V_q \, [\, \langle \alpha^\dagger_{\mathbf{k}'+\mathbf{q}} \beta_{\mathbf{k}-\mathbf{q}} \alpha_{\mathbf{k}'} \alpha_{\mathbf{k}} \rangle + \langle \beta_{\mathbf{k}'+\mathbf{q}} \beta_{\mathbf{k}-\mathbf{q}} \beta^\dagger_{\mathbf{k}'} \alpha_{\mathbf{k}} \rangle \\ &+ \langle \beta_{\mathbf{k}'} \alpha^\dagger_{\mathbf{k}'-\mathbf{q}} \alpha_{\mathbf{k}'} \alpha_{\mathbf{k}-\mathbf{q}} \rangle + \langle \beta_{\mathbf{k}'} \beta_{\mathbf{k}'-\mathbf{q}} \beta^\dagger_{\mathbf{k}'} \alpha_{\mathbf{k}-\mathbf{q}} \rangle \,], \end{aligned} \tag{12.8}$$

$$\begin{aligned} \hbar \frac{\partial}{\partial t} n_{e,\mathbf{k}} &= -2 \, \mathrm{Im} \, [\, d_{cv,\mathbf{k}} \, \mathcal{E}_0(t)/2 \; P^*_{\mathbf{k}}] \\ &+ i \sum_{k',q \neq 0} V_q \, [\, \langle \alpha^\dagger_{\mathbf{k}} \alpha^\dagger_{\mathbf{k}'-\mathbf{q}} \alpha_{\mathbf{k}-\mathbf{q}} \alpha_{\mathbf{k}'} \rangle - \langle \alpha^\dagger_{\mathbf{k}+\mathbf{q}} \alpha^\dagger_{\mathbf{k}'-\mathbf{q}} \alpha_{\mathbf{k}} \alpha_{\mathbf{k}'} \rangle \\ &+ \langle \alpha^\dagger_{\mathbf{k}} \alpha_{\mathbf{k}-\mathbf{q}} \beta^\dagger_{\mathbf{k}'-\mathbf{q}} \beta_{\mathbf{k}'} \rangle - \langle \alpha^\dagger_{\mathbf{k}+\mathbf{q}} \alpha_{\mathbf{k}} \beta^\dagger_{\mathbf{k}'-\mathbf{q}} \beta_{\mathbf{k}'} \rangle \,] \ , \end{aligned} \tag{12.9}$$

and

$$\hbar \frac{\partial}{\partial t} n_{h,\mathbf{k}} = - 2 \,\mathrm{Im} \left[d_{cv,\mathbf{k}} \frac{\mathcal{E}_0(t)}{2} P^*_{\mathbf{k}}(t) \right]$$

$$+ i \sum_{k',q\neq 0} V_q \, [\langle \beta^\dagger_{-\mathbf{k}} \beta^\dagger_{\mathbf{k'}-\mathbf{q}} \beta_{-\mathbf{k}-\mathbf{q}} \beta_{\mathbf{k'}} \rangle - \langle \beta^\dagger_{-\mathbf{k}+\mathbf{q}} \beta^\dagger_{\mathbf{k'}-\mathbf{q}} \beta_{-\mathbf{k}} \beta_{\mathbf{k'}} \rangle$$

$$+ \langle \alpha^\dagger_{\mathbf{k'}+\mathbf{q}} \alpha_{\mathbf{k'}} \beta^\dagger_{-\mathbf{k}} \beta_{-\mathbf{k}+\mathbf{q}} \rangle - \langle \alpha^\dagger_{\mathbf{k'}+\mathbf{q}} \alpha_{\mathbf{k'}} \beta^\dagger_{-\mathbf{k}-\mathbf{q}} \beta_{-\mathbf{k}} \rangle \;]. \tag{12.10}$$

Now we split the four-operator terms into products of densities and interband polarizations. Without going into the technical details, we just mention here that one can derive systematic approximations to these correlation terms using projection operator techniques, see Lindberg and Koch (1988). This technique allows to separate the equations of motion into two parts

$$\frac{\partial}{\partial t} \langle A \rangle = \frac{\partial}{\partial t} \langle A \rangle_{HF} + \frac{\partial}{\partial t} \langle A \rangle \bigg|_{col} . \tag{12.11}$$

Here, the first term denotes the result obtained in time-dependent Hartree-Fock approximation and the second term stands for the corrections to Hartree-Fock (collision terms). Hartree-Fock approximation amounts to replacing the Hamiltonian, Eq. (12.3) by the effective Hartree-Fock Hamiltonian

$$\mathcal{H}_{eff} = \sum_{\mathbf{k}} (E_{e,k} \, \alpha^\dagger_{\mathbf{k}} \alpha_{\mathbf{k}} + E_{h,k} \, \beta^\dagger_{-\mathbf{k}} \beta_{-\mathbf{k}}) - \sum_{\mathbf{k},q\neq 0} V_q \Big[n_{e,\mathbf{k}-\mathbf{q}} \alpha^\dagger_{\mathbf{k}} \alpha_{\mathbf{k}}$$

$$+ n_{h,\mathbf{k}-\mathbf{q}} \beta^\dagger_{-\mathbf{k}} \beta_{-\mathbf{k}} + P^*_{\mathbf{k}-\mathbf{q}} \beta_{-\mathbf{k}} \alpha_{\mathbf{k}} + P_{\mathbf{k}-\mathbf{q}} \alpha^\dagger_{\mathbf{k}} \beta^\dagger_{-\mathbf{k}} \Big] . \tag{12.12}$$

In the equations of motion, Eqs. (12.8) - (12.10) this leads to a factorization of the four-operator expectation values into products of two-operator expectation values, such as in Eq. (10.32), or

$$\langle \alpha^\dagger_{\mathbf{k}} \alpha^\dagger_{\mathbf{k'}} \alpha_{\mathbf{p}} \alpha_{\mathbf{p'}} \rangle \cong (\delta_{\mathbf{k},\mathbf{p'}} \, \delta_{\mathbf{k'},\mathbf{p}} - \delta_{\mathbf{k},\mathbf{p}} \, \delta_{\mathbf{k'},\mathbf{p'}}) \, n_{e,\mathbf{k}} \, n_{e,\mathbf{k'}} . \tag{12.13}$$

As result of these manipulations we obtain the renormalized single-particle energies as

$$\hbar e_{i,k} = \hbar \epsilon_{i,k} + \Sigma_{exc,i}(k) \ , \ i= e, \ h \ , \tag{12.14}$$

compare Eq. (10.34), and the semiconductor Bloch equations in the form

$$\left[i \frac{\partial}{\partial t} - (e_{e,k}+e_{h,k}) \right] P_{\mathbf{k}} = \left. \frac{\partial P_{\mathbf{k}}}{\partial t} \right|_{col}$$

$$+ \ (n_{e,\mathbf{k}}+n_{h,\mathbf{k}} - 1) \left[d_{cv,\mathbf{k}} \frac{\mathcal{E}_0(t)}{2\hbar} + \sum_{q \neq k} \frac{V_{|\mathbf{k}-\mathbf{q}|}}{\hbar} P_{\mathbf{q}} \right] \tag{12.15}$$

$$\frac{\partial}{\partial t} n_{e,\mathbf{k}} = - \frac{2}{\hbar} \ \mathrm{Im} \left\{ \left[d_{cv,\mathbf{k}} \frac{\mathcal{E}_0(t)}{2} + \sum_{q \neq k} V_{|\mathbf{k}-\mathbf{q}|} P_{\mathbf{q}} \right] P^*_{\mathbf{k}} \right\}$$

$$+ \left. \frac{\partial n_{e,\mathbf{k}}}{\partial t} \right|_{col} \tag{12.16}$$

$$\frac{\partial}{\partial t} n_{h,\mathbf{k}} = - \frac{2}{\hbar} \ \mathrm{Im} \left\{ \left[d_{cv,\mathbf{k}} \frac{\mathcal{E}_0(t)}{2} + \sum_{q \neq k} V_{|\mathbf{k}-\mathbf{q}|} P_{\mathbf{q}} \right] P^*_{\mathbf{k}} \right\}$$

$$+ \left. \frac{\partial n_{h,\mathbf{k}}}{\partial t} \right|_{col} \tag{12.17}$$

semiconductor Bloch equations

The collision terms in Eqs. (12.15) - (12.17) describe, e.g., electron-electron (hole-hole) intraband scattering, elastic electron-hole interband scattering, etc. Since the explicit computation of these terms involves some lengthy operator manipulations we suppress all additional details here. For the purposes of this book it is sufficient to treat the scattering terms phenomenologically as decay and thermalization rates.

As limiting cases, the semiconductor Bloch equations reproduce both the optical Bloch equations for two-level systems (see, e.g., L. Allen and J. H. Eberly (1975); Sargent *et al.* (1974)) and the Wannier equation for electron-hole pairs. In Eq. (12.15) we recognize essentially the polarization equation (10.30). Compared to Eq. (10.30) we find in Eq. (12.15) the renormalized transition eneries, Eq. (9.19), and the damping due to collisions. The factor

$$[1 - n_{e,\mathbf{k}} - n_{h,\mathbf{k}}] = n_{v,\mathbf{k}} - n_{c,\mathbf{k}}$$

is again the population inversion of the state $\mathbf{k}$. Its effects on the optical absorption spectra are often denoted as *phase space filling*. The term

$$[1 - n_{e,\mathbf{k}} - n_{h,\mathbf{k}}]\, d_{cv,\mathbf{k}}\, \frac{\mathcal{E}_0(t)}{2}$$

on the RHS of Eq. (12.15) describes the coupling of the external field to the inversion and is well known from the optical Bloch equations for two-level systems. The resulting *band filling* nonlinearities in semiconductors have already been discussed in Chap. 4.

If one ignores all the Coulomb potential terms, $V(q) \to 0$, Eqs. (12.15) - (12.17) reduce to the usual optical Bloch equations. As shown in Chap. 10, the homogenous part of Eq. (12.15) becomes the generalized Wannier equation for electron-hole pairs with the effective Coulomb interaction being renormalized by the phase-space-filling factor. The term

$$-\,2\,\mathrm{Im}\left[\, d_{cv,\mathbf{k}}\, \frac{\mathcal{E}_0(t)}{2}\, P_{\mathbf{k}}^{*} \right]$$

in Eq. (12.16) describes the generation of electron-hole pairs by the absorption of light. The contribution proportional to the Coulomb potential may be regarded as a renormalization of the electromagnetic field inside the medium. As long as the scattering terms are ignored, the rate of change of the hole population, Eq. (12.17), is identical to the rate of change of the electron population, Eq. (12.16).

12-2. Low Excitation Coherent Regime

Until now, there exists no known analytical solution of the full semiconductor Bloch equations for arbitrary fields $\mathcal{E}(t)$. However, fully dynamic analytical solutions can be obtained if one ignores the exchange effects. This approximation has been used frequently to study the situation of not too strong ultrafast (femtosecond) excitation dynamics for the case when the density of generated electron-hole pairs is sufficiently low. Under this condition the coherent part of Eqs. (12.15) - (12.17) reduces to

$$\hbar \left[i \frac{\partial}{\partial t} - (\epsilon_{e,k} + \epsilon_{h,k}) \right] P_{\mathbf{k}} = (n_{e,\mathbf{k}} + n_{h,\mathbf{k}} - 1)\, d_{cv,\mathbf{k}} \frac{\mathcal{E}_0(t)}{2} - \sum_{q \neq k} V_{|\mathbf{k}-\mathbf{q}|} P_{\mathbf{q}} \,, \tag{12.18}$$

$$\hbar \frac{\partial}{\partial t} n_{e,\mathbf{k}} = i \left[d_{cv,\mathbf{k}} \frac{\mathcal{E}_0(t)}{2} P^*_{\mathbf{k}} - \text{c.c.} \right] , \tag{12.19}$$

and

$$n_{h,k} = n_{e,k} \,. \tag{12.20}$$

As initial conditions we assume

$$n_{h,\mathbf{k}} = n_{e,\mathbf{k}} = P_{cv,\mathbf{k}} = 0 \,.$$

As in Eqs. (10.43) and (10.44), we now transform the equations to real space with the result

$$\hbar i \frac{\partial}{\partial t} P(\mathbf{r}) = \mathcal{H}_{eh} P(\mathbf{r}) + d_{cv} \frac{\mathcal{E}_0(t)}{2} [2n(\mathbf{r}) - \delta(\mathbf{r})] \tag{12.21}$$

and

$$\hbar \frac{\partial}{\partial t} n(\mathbf{r}) = i \left[d_{cv} \frac{\mathcal{E}_0(t)}{2} P^*(-\mathbf{r}) - \text{c.c.} \right] , \tag{12.22}$$

where

$$\mathcal{H}_{eh} = E_g - \frac{\hbar^2\nabla^2}{2m_r} - V(r) \tag{12.23}$$

is the Wannier Hamiltonian, compare Eq. (10.45). Note, that under the present conditions the equations yield local charge neutrality,

$$n_e(\mathbf{r}) = n_h(-\mathbf{r}) \equiv n(\mathbf{r}) \ .$$

Now we multiply Eq. (12.21), its conjugate complex, and Eq. (12.22) by $\psi_\lambda^*(\mathbf{r})$, where $\psi_\lambda(\mathbf{r})$ is the eigenfunction of the Wannier equation (10.45). Then we integrate over $\mathbf{r}$ to obtain

$$\hbar i \frac{\partial}{\partial t} P_\lambda = \hbar\epsilon_\lambda P_\lambda + d_{cv} \frac{\mathcal{E}_0(t)}{2} [2n_\lambda - \psi_\lambda^*(x{=}0)] \ , \tag{12.24}$$

$$\hbar i \frac{\partial}{\partial t} \bar{P}_\lambda = -\hbar\epsilon_\lambda \bar{P}_\lambda - d_{cv}^* \frac{\mathcal{E}_0(t)}{2} [2n_\lambda - \psi_\lambda(x{=}0)] \ , \tag{12.25}$$

and

$$\hbar \frac{\partial}{\partial t} n_\lambda = i\, d_{cv} \frac{\mathcal{E}_0(t)}{2} \bar{P}_\lambda - i\, d_{cv}^* \frac{\mathcal{E}_0(t)}{2} P_\lambda \ , \tag{12.26}$$

where we have introduced the notation

$$\int d^3r\, \psi_\lambda^*(\mathbf{r})\, P(\mathbf{r}) = P_\lambda \ ,$$

$$\int d^3r\, \psi_\lambda^*(\mathbf{r})\, P^*(-\mathbf{r}) = \bar{P}_\lambda \ ,$$

$$\int d^3r\, \psi_\lambda^*(\mathbf{r})\, n(\mathbf{r}) = n_\lambda \ . \tag{12.27}$$

The set of Eqs. (12.24) - (12.26) is closed for each λ. The source terms in (12.24) and (12.25) are proportional to the electron-hole-pair wavefunction in the origin and therefore only the s-functions contribute. Hence, $\psi_\lambda(\mathbf{r})$ is a real function and depends only on $|\mathbf{r}|$.

Therefore

$$\bar{P}_\lambda = P^*_\lambda \quad \text{and} \quad n^*_\lambda = n_\lambda . \tag{12.28}$$

Introducing now

$$P_\lambda = \psi^*_\lambda(x{=}0)\, \tilde{P}_\lambda \quad \text{and} \quad n_\lambda = \psi^*_\lambda(x{=}0)\, \tilde{n}_\lambda , \tag{12.29}$$

we obtain the simplified equations of motion

$$\hbar i \frac{\partial}{\partial t} \tilde{P}_\lambda = \hbar \epsilon_\lambda \tilde{P}_\lambda + d_{cv} \frac{\mathcal{E}_0(t)}{2} (2\tilde{n}_\lambda - 1)$$

$$\hbar \frac{\partial}{\partial t} \tilde{n}_\lambda = i\, d_{cv} \frac{\mathcal{E}_0(t)}{2} \tilde{P}^*_\lambda - i\, d^*_{cv} \frac{\mathcal{E}_0(t)}{2} \tilde{P}_\lambda . \tag{12.30}$$

These equations are identical to the Bloch equations for the off-diagonal and diagonal elements of the density matrix of a two-level atom and one can use all the known solutions of these equations. From Eq. (10.1) we then get the optical polarization as

$$P = \sum_{\mathbf{k}} d_{cv}\, P_{cv}(\mathbf{k}) + \text{c.c.} = d_{cv}\, P(\mathbf{r}{=}0) + \text{c.c.}$$

$$= d_{cv} \sum_\lambda |\psi_\lambda(x{=}0)|^2\, \tilde{P}_\lambda + \text{c.c.} , \tag{12.31}$$

where the completeness of the functions $\psi_\lambda(x)$ has been used. Thus we see that within the present approximation scheme the optical response of the semiconductor can be computed as solution of an inhomogeneously broadened two-level system. The inhomogeneous broadening in semiconductors is an intrinsic consequence of the energy dispersion, $E(k)$. However, when summing over the energies, the density of states has to be weighted by the Coulomb enhancement factor.

12-3. High Excitation Quasi-Equilibrium Regime

In the following we use Eqs. (12.15) - (12.17) to study the absorption spectrum of a semiconductor under quasi-equilibrium steady-state excitation. For this purpose, we have to include screening, phase space filling and bandgap shifts since all these effects become important when the system is near the quasi-thermal equilibrium. Typically this situation is reached in resonant pump-probe experiments within a few hundred femtoseconds up to a few picoseconds. On this timescale, the carrier-carrier scattering caused the evolution of the originally nonthermal distributions of electrons and holes within their bands into quasi-Fermi distributions where the respective chemical potentials are defined within each band and the carriers are at an electronic temperature which is generally higher than the lattice temperature. The quasi-chemical potentials are determined by the numbers of excited electrons and holes and the plasma temperature is mainly determined through the electron-hole excess energy with respect to the band gap. The details of the electron-hole excitation process are unimportant since the rapid collision processes eliminated all memories of the excitation process. Therefore, the optical semiconductor nonlinearities in this regime do not depend on the exciting light field directly, but only on the number of the generated carriers. The semiconductor state is probed with a spectrally broad low intensity probe beam.

For the considered quasi-steady state high excitation situation it is important to include the effect of plasma screening of the Coulomb interaction between the charged particles. The screening may be described in a self-consistent way by replacing the unscreened potential V_q by the statically screened potential, $V_s(q)$. Because this replacement must be done in the original Hamiltonian (12.1), the single-particle energies are in this case

$$E_e^{sc}(k) = E_{c,k} = E_{e,k} \ , \tag{12.32}$$

and

$$E_h^{sc}(k) = -E_{v,k} + \sum_{q \neq 0} V_s(q) = E_{h,k} + \sum_{q \neq 0} [V_s(q) - V_q]$$

$$= E_{h,k} + \delta\epsilon_{Deb} \ . \tag{12.33}$$

where the last term, which is independent of the wave vector, is the Debye shift, or Coulomb-hole self energy of an electron in a plasma, see Eq. (9.29).

In order to compute the semiconductor susceptibility for a given distribution of electron-hole pairs, we solve the polarization equation in first order in the external probe field. Again, the scattering terms do not have to be included explicitly. These processes are responsible to establish the quasi-equilibrium situation in the first place, but when the system is in quasi-equilibrium, the scattering terms vanish. Under this condition the Fourier-transform of the equation of motion for the polarization, Eq. (12.15), is given by

$$\hbar i \frac{\partial}{\partial t} P(\mathbf{r}) = \left[\mathcal{H}_e(\mathbf{r}) + \mathcal{H}_h(\mathbf{r}) \right] P(\mathbf{r}) - \int \frac{d\mathbf{r}'}{L^3} V_s(\mathbf{r}') N(\mathbf{r}') P(\mathbf{r}-\mathbf{r}')$$

$$- \int d\mathbf{r}' \, V_s(\mathbf{r}') P(\mathbf{r}') \left[\delta(\mathbf{r}-\mathbf{r}') - \frac{N(\mathbf{r}-\mathbf{r}')}{L^3} \right] - d_{cv} \frac{\mathcal{E}_0(t)}{2} \left[\delta(\mathbf{r}) L^3 - N(\mathbf{r}) \right] \quad (12.34)$$

Here, $N(\mathbf{r}) = f_e(\mathbf{r}) + f_h(-\mathbf{r})$ with $f_i(\mathbf{r})$, i=e,h, denoting the quasi-thermal carrier (Fermi-Dirac) distribution. The two terms on the first line on the RHS of Eq. (12.34) are the nonlocal Hartree problem of electron and holes

$$\sum_{i=e,h} (\mathcal{H}^i_H \, y^i_\lambda)(\mathbf{r}) = \sum_{i=e,h} E^i_\lambda \, y^i_\lambda(\mathbf{r}) \,, \quad (12.35)$$

where $\mathcal{H}^i_H$ is defined as

$$(\mathcal{H}^i_H \, y^i_\lambda)(\mathbf{r}) = \mathcal{H}_i(\mathbf{r}) \, y^i_\lambda(\mathbf{r}) - \int \frac{d\mathbf{r}'}{L^3} V_s(\mathbf{r}') \, f_i(\mathbf{r}') \, y^i_\lambda(\mathbf{r}-r') \,. \quad (12.36)$$

In our approximation this problem is linear in the y's. However, since the energy arguments in the quasi-thermal distributions in the Hamiltonian (12.36) are given by the eigenenergies of (12.35), the eigenvalue problem is already nonlinear. Numerical studies have shown that a k-independent (rigid) band gap shift is a good approximation when dealing with the nonlinear optical spectra of laser-excited semiconductors. This approximation is equivalent to making the Hartree Hamiltonians local, i.e.,

$$\int d\mathbf{r}'\ V_s(\mathbf{r}')\ N(\mathbf{r}')\ P(\mathbf{r}'-\mathbf{r}) \cong \int d\mathbf{r}'\ V_s(\mathbf{r}')\ N(\mathbf{r}')\ P(\mathbf{r})\ . \tag{12.37}$$

To obtain a formal solution of Eq. (12.34) in steady state, it turns out to be helpful to use a coordinate independent representation. We write

$$\hbar i\ \frac{\partial}{\partial t}\ |P\rangle = \mathcal{H}_H\ |P\rangle - S\ V_s\ |P\rangle - d_{cv}\frac{\mathcal{E}_0(t)}{2}\ |f\rangle\ , \tag{12.38}$$

where the operator S is defined as

$$\langle\mathbf{r}|\ S|\xi\rangle = \xi(\mathbf{r}) - \int \frac{d\mathbf{r}'}{L^3}\ \xi(\mathbf{r}')\ N(\mathbf{r}-r')\ . \tag{12.39}$$

The Hilbert space vector $|f\rangle$ in Eq. (12.38) in position representation is given by

$$\langle\mathbf{r}|f\rangle = f(\mathbf{r}) = \delta(\mathbf{r})L^3 - N(\mathbf{r}).$$

The solution of Eq. (12.34) is complicated by the fact, that even though both operators S and V_s are Hermitian, their product is not, since

$$[\ S\ ,\ V\ ] \neq 0\ .$$

To continue we make the following assumptions:

i) the operator $\mathcal{H}_H - S\ V_s$ has complete sets of both right-hand and left-hand eigenvectors;
ii) all eigenvalues are real.

We now assume that we know the solutions to the eigenvalue problems

$$(\ \mathcal{H}_H - S\ V_s\ )\ |\xi_\lambda\rangle = \hbar e_\lambda\ |\xi_\lambda\rangle \tag{12.40}$$

and

$$\langle\eta_\lambda| \, (\mathcal{H}_H - S\, V_s) = \hbar e_\lambda \, \langle\eta_\lambda| \, . \tag{12.41}$$

The solutions are normalized such that

$$\langle\eta_\lambda | \xi_\mu\rangle = \delta_{\lambda\mu} \, . \tag{12.42}$$

Next we multiply Eq. (12.41) from the right by S and take the complex conjugate. Since the operator S commutes with $\mathcal{H}_H$, we can write the result in the form

$$(\mathcal{H}_H - S\, V_s) \, S \, |\eta_\lambda\rangle = \hbar e_\lambda \, S \, |\eta_\lambda\rangle \, . \tag{12.43}$$

A comparison of Eqs. (12.40) and (12.43) shows that it is possible to choose the left-hand and right-hand eigenvectors such that

$$\mathcal{N}_\lambda \, S \, |\eta_\lambda\rangle = |\xi_\lambda\rangle \, , \tag{12.44}$$

where $\mathcal{N}_\lambda$ is a normalization factor. Multiplying Eq. (12.44) by $\langle\eta_\lambda|$ and using Eq. (12.42), we obtain

$$\mathcal{N}_\lambda = \langle\eta_\lambda | S | \eta_\lambda\rangle^{-1} \tag{12.45}$$

Since the left-hand and right-hand eigenvectors can be normalized independently, we can choose the normalization for η such that

$$| \langle\eta_\lambda | S | \eta_\lambda\rangle | = 1 \tag{12.46}$$

and

$$\langle\eta_\lambda | S | \eta_\mu\rangle = \mathrm{sign} \, \langle\eta_\lambda | S | \eta_\lambda\rangle \, \delta_{\lambda\mu} \, . \tag{12.47}$$

The sign function is necessary because the operator S is not positive definite. Physically, the sign describes gain (minus) or absorption (plus) at the corresponding states. Using Eqs. (12.47) and (12.45) we obtain from Eq. (12.44)

$$|\xi_\lambda\rangle = \mathrm{sign} \, \langle\eta_\lambda | S | \eta_\lambda\rangle \, S \, |\eta_\lambda\rangle \, . \tag{12.48}$$

Since the right-hand eigenvectors form a complete set of functions, we can expand

$$|P\rangle = \sum_{\lambda} b_\lambda \, |\xi_\lambda\rangle \tag{12.49}$$

with

$$b_\lambda = \langle \eta_\lambda | P \rangle \, . \tag{12.50}$$

Inserting Eq. (12.49) into (12.38), multiplying from the left by $\langle \eta_\lambda |$ and using Eq. (12.50) yields

$$\hbar i \frac{\partial}{\partial t} b_\lambda = \hbar e_\lambda b_\lambda - d_{cv} \frac{\mathcal{E}_0(t)}{2} \langle \eta_\lambda | f \rangle \, . \tag{12.51}$$

The solution of this equation is

$$b_\lambda(t) = - d_{cv} \frac{\mathcal{E}_0(t)}{2} \frac{\langle \eta_\lambda | f \rangle}{\hbar(\omega + i\delta - e_\lambda)} \, , \tag{12.52}$$

where the field has been written as $\dfrac{\mathcal{E}_0(t)}{2} = \dfrac{\mathcal{E}_0}{2} \exp[i(\omega+i\delta)t]$. Inserting this solution into the expansion of Eq. (12.49) and multiplying by $\langle \mathbf{r} |$, we obtain

$$P(\mathbf{r}) = - d_{cv} \frac{\mathcal{E}_0(t)}{2} \sum_{\lambda} \frac{\langle \eta_\lambda | f \rangle \, \xi_\lambda(\mathbf{r})}{\hbar(\omega + i\delta - e_\lambda)} \, . \tag{12.53}$$

To evaluate $\langle \eta_\lambda | f \rangle \, \xi_\lambda(\mathbf{r})$ we use Eqs. (12.48) and (12.39) to write

$$\xi_\lambda(\mathbf{r}) = \text{sign} \, \langle \eta_\lambda | S | \eta_\lambda \rangle \, \langle \mathbf{r} | S | \eta_\lambda \rangle \tag{12.54}$$

and

$$\langle \eta_\lambda | f \rangle = \int \frac{d\mathbf{r}}{L^3} \langle \eta_\lambda | \mathbf{r} \rangle \langle \mathbf{r} | f \rangle = \int \frac{d\mathbf{r}}{L^3} \langle \eta_\lambda | \mathbf{r} \rangle \, [\delta(\mathbf{r}) L^3 - N(\mathbf{r})]$$

$$= \langle \eta_\lambda | S | \mathbf{r}{=}0 \rangle \, , \tag{12.55}$$

where Eq. (12.39) has been used. Repeating the steps given in Eqs. (10.87) - (10.91) and using Eq. (12.48) we finally obtain the total polarization as

$$P(t) = -|d_{cv}|^2 \frac{\mathcal{E}_0(t)}{2} L^3 \sum_\lambda sign\langle\eta_\lambda|S|\eta_\lambda\rangle \frac{|\langle \mathbf{r}=0|S|\eta_\lambda\rangle|^2}{\hbar(\omega + i\delta - e_\lambda)} + c.c. \quad (12.56)$$

and the corresponding absorption coefficient is

$$\boxed{\alpha(\omega) = \frac{4\pi^2\omega}{n_b c} |d_{cv}|^2 \sum_\lambda sign\langle\eta_\lambda|S|\eta_\lambda\rangle \; |\langle \mathbf{r}=0|S|\eta_\lambda\rangle|^2 \; \delta(\hbar e_\lambda - \hbar\omega)}$$

generalized Elliott formula (12.57)

where Eq. (1.53) has been used.

The result in Eq. (12.57) is very general and contains, both for *2d* and *3d*, the appropriate limiting cases, such as the free-carrier absorption discussed in Chap. 4 and the Elliott formula for the attracting electron-hole transitions. When we ignore the Coulomb potential V the problem is diagonalized in a plane wave basis, i.e.,

$$\langle \mathbf{r}|\eta_\mathbf{k}\rangle = \sqrt{\mathcal{N}_\mathbf{k}} \; e^{i\mathbf{k}\cdot\mathbf{r}} \quad (12.58)$$

and Eq. (12.45) yields

$$\mathcal{N}_k = |1 - f_{e,k} - f_{h,k}|^{-1} \equiv |1 - N_k|^{-1} . \quad (12.59)$$

Therefore,

$$\alpha(\omega) = \frac{4\pi^2\omega}{n_b c} |d_{cv}|^2 \sum_{\mathbf{k}} sign\langle\eta_k|S|\eta_k\rangle \; |\langle x{=}0|S|\eta_k\rangle|^2 \;\; \delta(\hbar e_k - \hbar\omega)$$

$$= \frac{4\pi^2\omega}{n_b c} |d_{cv}|^2 \sum_{\mathbf{k}} sign(1-N_k)\, \mathcal{N}_k \; |1-N_k|^2 \;\; \delta(\hbar e_k - \hbar\omega)$$

$$= \frac{4\pi^2\omega}{n_b c} |d_{cv}|^2 \sum_{\mathbf{k}} (1-N_k)\; \delta(\hbar e_k - \hbar\omega) \tag{12.60}$$

which is identical to the free-carrier absorption spectrum of Eq. (4.28).

For the case of an unexcited semiconductor when the electron and hole populations are zero, S becomes the unit operator and the whole problem is Hermitian. Then Eqs. (12.40) and (12.41) are just the Wannier equation (10.45) and the eigenfunctions η_λ are the corresponding eigenfunctions. Eq. (12.57) becomes

$$\alpha(\omega) = \frac{4\pi^2\omega}{n_b c} |d_{cv}|^2 \sum_{\lambda} |\eta_\lambda(\mathbf{r}{=}0)|^2 \;\; \delta(\hbar e_\lambda - \omega) \;, \tag{12.61}$$

which is the ordinary Elliott formula, Eq. (10.93) or (10.98), for the absorption spectrum of attracting electron-hole transitions in *3d* or *2d*, respectively. For the case of finite carrier densities and non-negligible Coulomb interaction, we have to introduce some additional approximations in the explixit evaluation of Eq. (12.57), which will be discussed in the following chapter.

REFERENCES

Discussions of the two-level Bloch equations can be found in:

L. Allen and J. H. Eberly, *Optical Resonances and Two-Level Atoms*, Wiley, New York (1975)

M. Sargent III, M.O. Scully, and W.E. Lamb, *Laser Physics*, Addison-Wesley, New York (1974).

For a detailed derivation of the semiconductor Bloch equations and for further references see:

M. Lindberg and S.W. Koch, Phys. Rev. **B38**, 3342 (1988).

For further reading on the material presented in this chapter see:

H. Haug and S. Schmitt-Rink, Progr. Quantum Electron. **9**, 3 (1984)

Optical Nonlinearities and Instabilities in Semiconductors, H. Haug ed., Academic, New York (1988)

A. Stahl and I. Balslev, *Electrodynamics of the Semiconductor Band Edge*, Springer Tracts in Modern Physics 110, Springer Verlag, Berlin (1987).

R. Zimmermann, *Many-Particle Theory of Highly Excited Semiconductors*, Teubner, Leipzig (1988)

PROBLEMS

Problem 12.1: Derive the electron-hole Hamiltonian, Eq. (12.4).

Problem 12.2: Work out the details in the derivation of the generalized Elliott formula (12.57) and show that it reproduces the limiting results for free carriers (for $V_s \to 0$) and the Elliott formula (for $f_e = f_h = 0$).

Chapter 13
OPTICAL QUASI-EQUILIBRIUM NONLINEARITIES

In Chap. 10 we solved the interband polarization equation for an unexcited crystal and found an absorption spectrum consisting of sharp absorption lines due to the bound states and a broad absorption band due to the ionization continuum. Now we analyze the changes of this band-edge spectrum when an electron-hole plasma exists in the crystal. In the free-carrier approximation this plasma spectrum has already be calculated and discussed in Chap. 4, but the derivation of the polarization equation in Sec. 12.3, which includes the relevant many-body corrections, allows us to calculate the modifications of the band edge spectrum due to the Coulomb interaction.

13-1. Numerical Matrix Inversion

The thermal distribution of electrons and holes is described by a Fermi function

$$f_{i,k} = \frac{1}{\exp(\beta(\hbar e_{i,k} - \mu_i)) + 1} , \tag{13.1}$$

where the value of the quasi-chemical potential μ_i is determined by the concentration of carriers, which can still be a slowly varying function in time and space. In homogeneous situations the concentration is determined by a rate equation for the total number of electrons or holes. Most often we can assume charge neutrality with $N_e = N_h$, because a locally different electron and hole concentration would create strong restoring forces. There are however frequently situations where the electron-hole plasma undergoes an ambipolar drift, i.e., the whole neutral plasma drifts with a wavevector $\mathbf{k}_D$. In such a situation the Fermi distribution is

$$f_{i,|\mathbf{k}-\mathbf{k}_D|} = \frac{1}{\exp(\beta(\hbar e_{i,|\mathbf{k}-\mathbf{k}_D|} - \mu_i)) + 1} \; . \tag{13.2}$$

With the thermal distribution (13.1) the polarization equation is

$$\hbar \left[\omega - e_{e,k} - e_{h,k} + i\delta\right] P_{\mathbf{k}} = -\,(1 - f_{e,k} - f_{h,k}) \left[d_{cv,k} \frac{\mathcal{E}_0}{2} + \sum_{\mathbf{k}'} V_{s|\mathbf{k}-\mathbf{k}'|} P_{\mathbf{k}'}\right] \tag{13.3}$$

where the single-particle energies $\hbar e_{i,k}$ are the renormalized energies, Eq. (9.20*a*). As discussed in Chap. 4, we have absorption and gain depending on the sign of the inversion factor

$$(1 - f_e - f_h) \; . \tag{13.4}$$

The cross-over between gain and absorption occurs at

$$\hbar\omega = E'_g + \mu_e + \mu_h, \tag{13.5}$$

if μ_i is counted from the actual band extrema, which are separated by the renormalized gap E'_g.

The second important influence of the plasma is the screening of the Coulomb potential. The attractive electron-hole potential is weakened by the plasma screening causing a reduction of the excitonic effects in the spectra. Actually, at a critical plasma density the combined effects of the screening and of the occupation of k-states by the plasma result in a vanishing exciton binding energy. Above this critical density, also called Mott density, only ionized states exist, however as we will see the attractive potential still modifies the plasma spectrum considerably.

The screened potential gives at the same time rise to a renormalization of the single-particle energies which results in the band gap shrinkage $\Delta E_g = E'_g - E_g$. The band gap shrinkage and the reduction of the exciton binding energy are of similar size, so that normally no shift of the exciton resonance occurs as the plasma density is increased. The influence of the plasma is only seen in a reduction of the exciton oscillator strength due to the increasing exciton Bohr radius with increasing plasma density, until the band edge reaches the exciton level at the Mott density. Above the Mott density the exciton resonance does no longer exist, however the attractive Coulomb potential still increases the probability to find the electron and hole at the same position, which causes an excitonic enhancement of the plasma absorption or gain spectrum.

As discussed in Sec. 12.3, the solution of the integral equation (13.3) for the interband polarization is complicated by the fact that the inversion factor (13.4) can change its sign, which means that the corresponding homogeneous eigenvalue equation is non-Hermitian. In the following we describe first a numerical method for obtaining rather accurate solutions by a matrix inversion, and than two approximate solutions which are considerably simpler and can be used above the Mott density, i.e. in the high-density regime. In the final section of this chapter we then discuss the so-called plasma theory, which provides a simple analytical approximation scheme for $3d$ semiconductors.

The first step to simplify Eq. (13.3) is to replace the vector sum by a scalar sum

$$\sum_{\mathbf{k}'} V_{s,|\mathbf{k}-\mathbf{k}'|} P_{\mathbf{k}'} \rightarrow \sum_{k'} \overline{V}_{s,k,k'} \; P_{k'} \; , \tag{13.6}$$

where the angle averaged potential is given by

$$\overline{V}_{s,k,k'} = \frac{1}{2} \int_{-1}^{+1} d(\cos\theta) \; V_{s,(k^2+k'^2-2kk'\cos\theta)^{1/2}} \qquad \text{in } 3d \tag{13.7a}$$

and

$$\overline{V}_{s,k,k'} = \frac{1}{2\pi} \int_{0}^{2\pi} d\phi \; V_{s,(k^2+k'^2-2kk'\cos\phi)^{1/2}} \; . \qquad \text{in } 2d \tag{13.7b}$$

With this approximation we take into account only s-wave scattering and the integral equation becomes one-dimensional. Because Eq. (13.3) is linear in the field (the polarization becomes nonlinear only via the plasma density, which is a function of the light intensity), we can introduce a susceptibility function χ_k by

$$P_k = \chi_k \, \frac{\mathcal{E}_0}{2} \, , \tag{13.8}$$

which obeys the integral equation

$$\chi_k = \chi_k^0 \left[1 + \frac{1}{d_{cv,k}} \sum_{k'} \bar{V}_{s,k,k'} \chi_{k'} \right], \tag{13.9}$$

where χ_k^0 is the free-carrier susceptibility function of Chap. 4

$$\chi_k^0 = - d_{cv,k} \frac{1-f_{e,k}-f_{h,k}}{\hbar(\omega+i\delta - e_{e,k} - e_{h,k})} . \tag{13.10}$$

In contrast to Chap. 4, however, the single-particle energies in Eq. (13.10) are the renormalized energies calculated in the quasi-static approximation of Chap. 9. Hence, Eq. (13.10) includes the band gap renormalization effects due to the electron-hole plasma. Next we introduce a vertex function Γ_k which describes just the deviations of χ from χ^0

$$\chi_k = \Gamma_k \chi_k^0 . \tag{13.11}$$

It obeys the equation

$$\boxed{\Gamma_k = 1 + \frac{1}{d_{cv,k}} \sum_{k'} \chi_{k'}^0 \bar{V}_{s,k,k'} \Gamma_{k'}} .$$

vertex integral equation (13.12)

The integral equation (13.12) will now be solved by various methods.

The integral over k' is approximated by a discrete sum with about 100 terms

$$\sum_{k'} \rightarrow \int \frac{dk'}{2\pi} \rightarrow \sum_i \frac{\Delta k'_i}{2\pi} . \tag{13.13}$$

The points k_i can be taken equidistantly. Better accuracy is obtained if the k_i are taken as the points of support of a Gaussian quadrature. At low densities the angle averaged potential becomes singular for $k'=k$. This singularity has to be removed before the numerical matrix inversion is performed. One adds and subtracts a term

$$\Big[\sum_{k'} F_{k,k'}\bar{V}_{s,k,k'}\Big]\,\Gamma_k\chi_k^0 + \sum_{k'}\bar{V}_{s,k,k'}\Big[\chi_{k'}^0\,\Gamma_{k'} - F_{k,k'}\chi_k^0\Gamma_k\Big]\,, \tag{13.14}$$

where $F_{k,k'}$ is chosen such that $F_{k,k} = 1$ and that the sum in the first bracket can be evaluated analytically, e.g., in *3d*

$$F_{k,k'} = \frac{2k^4}{k'^2(k'^2+k^2)}\ . \tag{13.15}$$

The bracket of the difference term in (13.14) vanishes at $k=k'$, where $\bar{V}_{s,k,k'}$ is singular, so that the total term is a smooth function at $k=k'$. With such a procedure, compensation terms can be found for the plasmon-pole approximation both in *3d* and *2d*.

In order to get more realistic optical spectra, one still has to account for the finite damping γ of the interband polarization which in the framework of Bloch equations is often called $1/T_2$. The population lifetime is T_1. Because of the equivalence of a two-level atom system and of a spin-one-half system, in which the polarization corresponds to the transverse magnetization and the population to the longitudinal magnetization, T_2 and T_1 are also called the transverse and longitudinal relaxation times, respectively. Because there are many processes which destroy the phase of the polarization, but not the population, T_2 is normally much shorter than T_1.

If one simply replaces the infinitesimal damping δ in χ_0 by γ, the cross-over between gain and absorption is no longer precisely determined by Eq. (13.5). Therefore, one often uses a spectral representation

$$\chi_k^0 = d_{cv,k}\int_{-\infty}^{+\infty}\frac{d\omega'}{2\pi}\,\frac{2\gamma/\hbar}{(\omega'-e_{e,k}-e_{h,k})^2+\gamma^2}\,\frac{1-f_e(\hbar\omega'-e_{h,k})-f_h(e_{h,k})}{\omega-\omega'+i\delta}\ . \tag{13.16}$$

A still better description of the band-tail absorption can be obtained with a frequency dependent $\gamma_k(\omega)$. Such a dynamical damping results naturally if it is calculated as the imaginary part of a frequency-dependent self-energy. Here we take it in the form

$$\gamma_k(\omega) = \gamma^0 \frac{1}{e^{(E_k - \hbar\omega)/E_\alpha} + 1}$$

in order to describe the exponential absorption tail. The energy E_k is given by $E_k = \hbar^2 k^2/2m$ and E_α is a numerical constant.

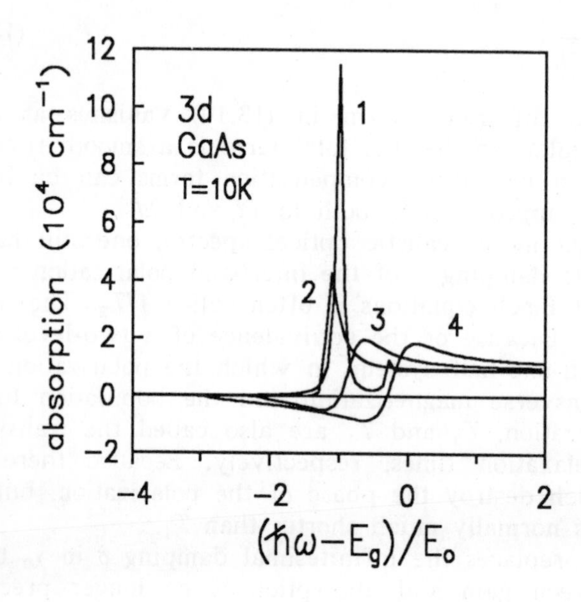

Fig. 13.1-a: Computed absorption spectra for bulk GaAs at T=10 K using the matrix-inversion procedure. The used parameters are: m_e=0.0665, m_h=0.457, ϵ_0=13.74, ϵ_∞=10.9, a_0=125 Å, E_0=4.2 meV, and the densities N are 0 (1), $5\cdot10^{15}$ cm^{-3} (2), $3\cdot10^{16}$ cm^{-3} (3), and $8\cdot10^{16}$ cm^{-3} (4), respectively.

Fig. 13.1 and Fig. 13.2 show the optical spectra which are obtained by numerical matrix inversion for the examples of bulk GaAs and bulk InSb for various plasma densities. The resulting complex susceptibility

$$\chi(\omega) = \frac{1}{L^3}\sum_k d^*_{cv,k}\chi_k \tag{13.17}$$

defines the complex optical dielectric function $\epsilon(\omega) = 1 + 4\pi\chi(\omega)$ which in turn determines the spectra of absorption $\alpha(\omega)$ (1.50) and refraction $n(\omega)$ (1.51). The GaAs spectra of Fig. 13.1 are calculated

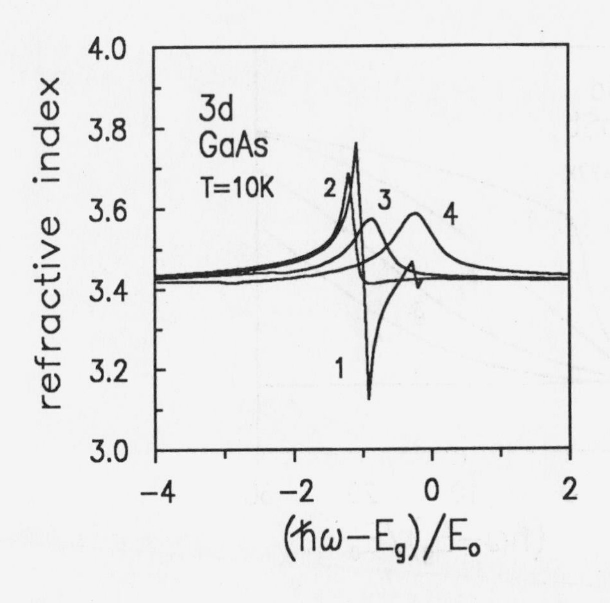

Fig. 13.1-b: Computed refractive index spectra for bulk GaAs using the same parameters as in Fig. 13.1-*a*.

with a damping $\hbar\omega = 0.05\ E_0$. The absorption spectrum in Fig. 13.1-*a* shows at zero plasma densities the well-resolved *1s* and *2s* exciton resonances followed by an ionization continuum enhanced by excitonic effects. The plasma density of $5 \cdot 10^{15}\ cm^{-3}$ is close to the Mott density where the exciton bound states cease to exists. The excitonic enhancement still causes the appearance of a maximum around the original exciton ground state position. At still higher densities a band gap reduction far below the position of the original exciton ground state is seen, and simultaneously a build-up of optical gain. The vanishing of the exciton resonance in the absorption spectrum causes also considerable changes in the refractive index, as shown in Fig. 13.1-*b*. Particularly if the laser beam is tuned below the exciton resonance, where the absorption is relatively weak, the index of refraction decreases with increasing plasma density. This effect is often exploited in optical switching devices which are based on dispersive optical bistability, see Chap. 14.

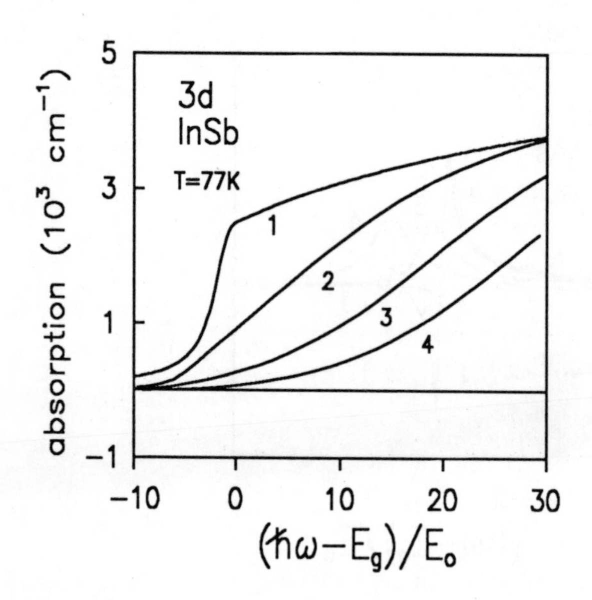

Fig. 13.2: Computed absorption spectra for bulk InSb at T=77 K using the matrix-inversion procedure. The used parameters are: m_e=0.0145, m_h=0.4, ϵ_0=17.05, ϵ_∞=15.7, a_0=644.8 Å, E_0=0.665 meV, and the densities N are 0 (1), $8\cdot10^{15}$ cm^{-3} (2), $2\cdot10^{16}$ cm^{-3} (3), and $3\cdot10^{16}$ cm^{-3} (4), respectively.

The bulk InSb spectra of Fig. 13.2 have been calculated with a damping $\hbar\gamma = 2E_0 = 1\text{meV}$, which is twice as large as the exciton Rydberg. This case is typical for narrow band-gap semiconductors. Here the exciton is not resolved in the unexcited medium and causes

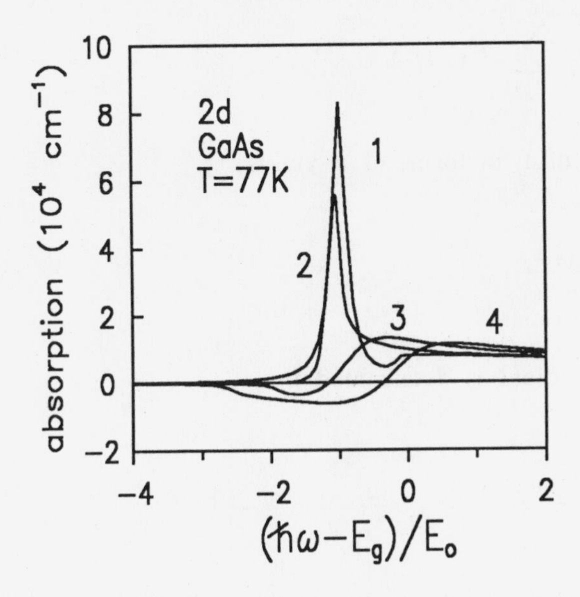

Fig. 13.3: Computed absorption spectra for *2d* GaAs at T=77 K using the matrix-inversion procedure. The used parameters are: a_0=62.5 Å and E_0=16.8 meV, all the other parameters are the same as in Fig. 13.1. The densities N are 0 (1), $1\cdot10^{11}$ cm^{-2} (2), $5\cdot10^{11}$ cm^{-2} (3), and $1\cdot10^{12}$ cm^{-2} (4), respectively.

only a step-like absorption edge.

The corresponding absorption spectra for ideal *2d* GaAs are shown in Fig. 13.3. Only one hole band was taken into account. Experimentally in quantum-well structures, one typically sees superimposed the contributions of the heavy- and the light-hole bands.

13-2. High-Density Approximations

The integral equation (13.12) can be solved approximately if the attractive electron-hole potential is not to strong, i.e., in the high-density limit, where plasma screening and phase-space occupation have reduced the strength of the Coulomb potential sufficiently. We rewrite (13.12) by introducing a formal interaction parameter σ, which will be assumed to be small

$$\Gamma_k = 1 + \frac{\sigma}{d_{cv,k}} \sum_{k'} \bar{V}_{s,k,k'} \, \chi^0{}_{k'} \, \Gamma_{k'} \quad , \tag{13.18}$$

A power expansion of Γ in terms of σ yields

$$\Gamma_k = \sum_n q_n \, \sigma^n \quad , \tag{13.19}$$

where the first coefficient is determined by

$$q_1 = \frac{1}{d_{cv,k}} \sum_{k'} \bar{V}_{s,k,k'} \, \chi^0{}_{k'} \ . \tag{13.20}$$

In general one can express Eq. (13.19) as the ratio of two polynomials

$$P_k^{(N,M)} = \frac{\sum_{n=0}^{N} r_n \sigma^n}{\sum_{n=0}^{M} s_n \sigma^m} \quad , \tag{13.21}$$

which represents the (N,M)-*Padé* approximation. The coefficients r_n and s_n can be obtained by a comparison with the original expansion (13.19). We use the simplest nontrivial $(0,1)$-*Padé* approximation

$$\Gamma_k = P_k^{(0,1)} = \frac{1}{1 - q_{1k}} \quad , \tag{13.22}$$

which results in an optical susceptibility of the form

$$\boxed{\chi(\omega) = \frac{1}{L^3} \sum_k \frac{d^*_{cv,k}\chi^0{}_k}{1-q_{1k}}}$$

Padé approximation (13.23)

where we used Eqs. (13.11) and (13.17). The denominator in Eq. (13.23) expresses the influence of the multiple electron-hole scattering due to their attractive interaction potential V_s, which causes an excitonic enhancement.

Another approximation for the integral equation (13.3) is obtained by inserting a dominant momentum, which we take as the Fermi momentum, into the angle-averaged screened Coulomb potential. The Fermi momentum is

$$k_F = (3\pi^2 n)^{1/3} \qquad \text{in } 3d \tag{13.24a}$$

$$k_F = (2\pi n)^{1/2} \qquad \text{in } 2d \ , \tag{13.24b}$$

see Eqs. (7.22) and (7.39), respectively. Then the integral equation (13.9) becomes

$$\chi_k = \chi^0{}_k \left[1 + \frac{1}{d_{cv,k_F}} \sum_{k'} \chi_{k'} \, \overline{V}_{s,k_F,k'} \right] = \chi^0{}_k \; (1 + S(\omega)), \tag{13.25}$$

and for

$$S(\omega) = \frac{1}{d_{cv,k_F}} \sum_{k'} \chi_{k'} \, \overline{V}_{s,k_F,k'} \tag{13.26}$$

we get the self-consistency equation

$$S(\omega) = \frac{1}{d_{cv,k_F}} \sum_k \chi^0{}_k \, \overline{V}_{s,k_F,k} \, (1 + S(\omega)) \tag{13.27}$$

with the solution

$$1 + S(\omega) = \frac{1}{1 - \dfrac{1}{d_{cv,k_F}} \sum \chi^0{}_k \, \overline{V}_{s,k_F,k}} \; . \tag{13.28}$$

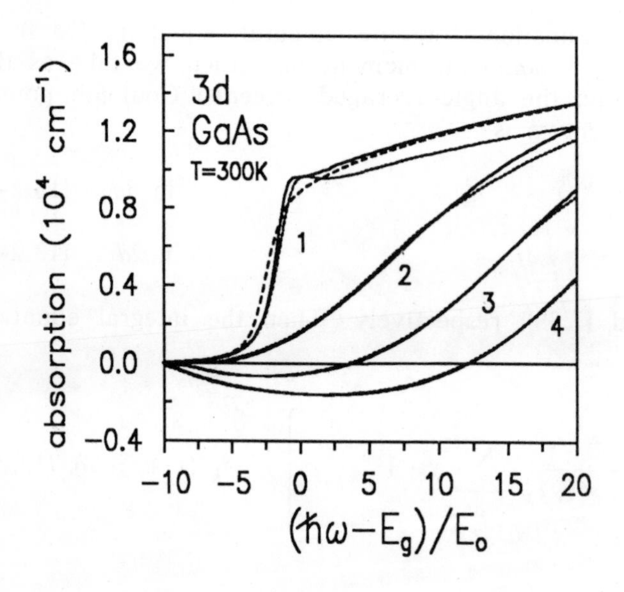

Fig. 13.4: *3d* absorption spectra at 300 *K* for the densities $1 \cdot 10^{16}\ cm^{-3}$ (1), $1 \cdot 10^{18}\ cm^{-3}$ (2), $2 \cdot 10^{18}\ cm^{-3}$ (3), and $3 \cdot 10^{18}\ cm^{-3}$ (4), respectively. Matrix inversion (full lines), *Padé* approximation (dashed lines), high-density approximation (dotted lines).

Eq. (13.28) yields the optical susceptibility

$$\chi(\omega) = \frac{\sum d_{cv,k}\ \chi^0{}_k}{L^3 - \frac{1}{d_{cv,k_F}} \sum_k \chi^0{}_k\ \bar{V}_{s,k_F,k}} \tag{13.29}$$

expressing again, in a different approximation, the effect of the excitonic enhancement, however with a simple momentum-independent enhancement factor $1 + S(\omega)$.

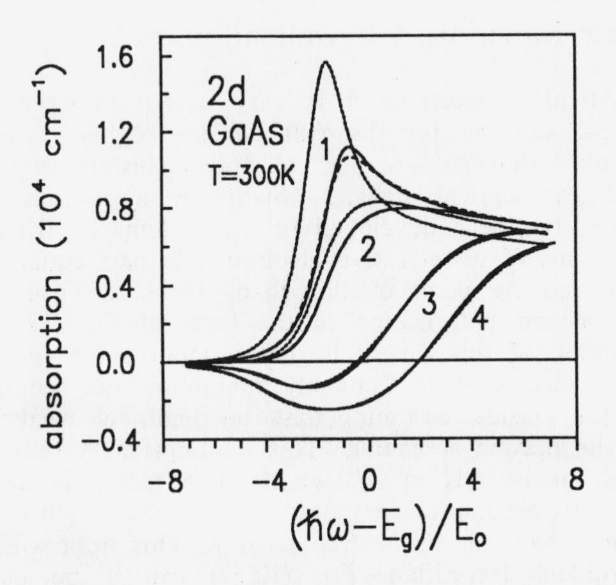

Fig. 13.5: *2d* absorption and gain spectra at 300 *K* for the densities $1\cdot10^{10}$ cm^{-2} (1), $1\cdot10^{11}$ cm^{-2} (2), $2\cdot10^{12}$ cm^{-2} (3), and $3\cdot10^{12}$ cm^{-2} (4), respectively. The full lines are the result of the matrix inversion, the dashed line is the *Padé* approximation and the dotted line represents the high-density approximation.

Taking the spectral representation only for the imaginary part, we compare the absorption spectra for *3d* and *2d* semiconductors in Figs. 13.4 and 13.5 *a*) by solving Eq. (13.12) by numerical matrix inversion (full curves), by using *b*) the (0,1)-*Padé* approximation (13.23) (dashed curves) and *c*) the high-density approximation (13.29) (dotted curves). The agreement of both approximations *b*) and *c*) with the numerical solution *a*) becomes very good, if a pronounced optical gain exists. At lower plasma-densities where the gain vanishes, but where the densities are still above the Mott density, the (0,1)-*Padé* approximation yields a slightly better description of the excitonic enhancement than the high-density approximation, particularly in the *2d* case. At low densities, where bound states of the exciton exist, both approximations are unable to give a reliable description of the bound-state resonances in the absorption spectra.

13-3. Effective Pair-Equation Approximation

For many practical applications it is very useful to have an approximate analytical solution for the full density regime, which can then be used in further studies, e.g., of optical bistability (see Chap. 14) or nonlinear optical devices. Such an approximation scheme for *3d* systems has been developed for *3d* systems by Banyai and Koch (1986) in terms of an effective electron-hole pair equation.

As a possible starting point of this theory one can use the equation for the interband polarization in the form of Eq. (12.38). The main approximation of this theory is that it ignores the reduction of the attractive electron-hole Coulomb interaction via occupation of k-states by the plasma, assuming that the dominant weakening comes through the plasma screening. This assumption is reasonable in *3d* systems, but it fails in *2d*, where state filling is more important. Formally, this assumption amounts to replacing S with the unit operator but not changing the source term f. This approximation makes the problem Hermitian. Eq. (12.57) cannot be used directly because it was derived using the explicit relation between f and S. Therefore we start from Eq. (12.38), replace S by the unit operator and repeat all the successive steps in Sec. 12.3. The population factors are reformulated according to Eq. (4.48) and the prefactor of the tanh-term is put equal to unity. The resulting absorption coefficient is then

$$\alpha(\omega) = \frac{4\pi^2\omega}{n_b c} |d_{cv}|^2 \tanh\left[\frac{\beta}{2}(\hbar\omega - E_g - \mu)\right] \sum_\nu |\psi_\nu(\mathbf{r}=0)|^2 \, \delta(E_\nu - \hbar\omega). \quad (13.30)$$

The remaining problem for the analytic evaluation is to obtain the eigenfunctions ψ_ν and the corresponding eigenvalues for the screened Coulomb potential. Unfortunately, we do not known an analytic solution for the Wannier equation with the Yukawa potential. However, there is a very good approximation to the screened Coulomb potential, which is the so-called *Hulthén* potential

$$V_H(r) = -\frac{2e^2\kappa/\epsilon_0}{e^{2\kappa r} - 1}, \quad (13.31)$$

for which one can solve the Wannier equation (10.45). As in Chap. 10, we make the ansatz

$$\psi_\nu(\mathbf{r}) = f_{n\ell}(r) Y_{\ell m}(\theta, \phi), \quad (13.32)$$

where $\nu = n, \ell, m$ represents the relevant set of quantum numbers. We obtain the radial equation

$$-\frac{\partial^2 f_{n\ell}}{\partial r^2} - \frac{2}{r}\frac{\partial f_{n\ell}}{\partial r} + \left[\frac{\ell(\ell+1)}{r^2} - \frac{g\lambda^2}{e^{\lambda r} - 1}\right] f_{n\ell} = \epsilon_{n\ell}\, f_{n\ell} \quad (13.33)$$

with

$$\lambda = 2\kappa, \; g = \frac{1}{a_0\kappa}, \text{ and } \epsilon_{n\ell} = \frac{E_{n\ell}}{E_0 a_0^2}. \quad (13.34)$$

Since we are only interested in those solutions which do not vanish in the origin,

$$f_{n\ell}(r=0) \neq 0,$$

we restrict the analysis of Eq. (13.33) to $\ell = 0$ and drop the index of f and ϵ. For this case we obtain from Eq. (13.33)

$$-\frac{\partial^2 u}{\partial r^2} - \frac{g\lambda^2}{e^{\lambda r} - 1} u = \epsilon u, \quad (13.35)$$

where we introduced

$$u(r) = rf(r) \quad . \tag{13.36}$$

Defining

$$z = 1 - e^{-\lambda r} \quad , \quad w(r) = u(r)/z(1-z)^{\beta} \tag{13.37}$$

with

$$\beta = \sqrt{-\epsilon/\lambda^2} \tag{13.38}$$

and inserting these definitions into Eq. (13.35) yields

$$z(1-z)\frac{\partial^2 w}{\partial z^2} + (2 - (2\beta+3)z)\frac{\partial w}{\partial z} - (2\beta+1 - g)w = 0 \quad . \tag{13.39}$$

This is the hypergeometric differential equation, which has the convergent solution

$$F(a,\, b,\, c;\, z) = 1 + \frac{ab}{c}\frac{z}{1!} + \frac{a(a+1)\ b(b+1)}{c(c+1)}\frac{z^2}{2!} + \ldots \quad c \neq 0, -1, -2, \ldots \tag{13.40}$$

where in our case

$$a = 1 + \beta + \sqrt{g + \beta^2}$$

$$b = 1 + \beta - \sqrt{g + \beta^2}$$

$$c = 2 \quad . \tag{13.41}$$

A) Bound states

For the bound-state solutions we have $\epsilon_\nu < 0$. Since f has to be a normalizable function, we request

$$f(r \to \infty) \to 0 \quad , \tag{13.42}$$

which yields the condition that

$$b = 1 - n \qquad n = 1,\, 2,\, 3,\, \ldots$$

and therefore

$$\beta = \frac{1}{2n}(g - n^2) \equiv \beta_n \text{ and } \epsilon_n = -\lambda^2\beta_n^2 \ ; \quad n = 1, 2, \ldots \tag{13.43}$$

The energetically lowest bound state (*1s*-exciton) is ionized for

$$g \to 1 \ , \text{ i.e., } \ a_0\kappa \to 1 \ ,$$

which is the Mott criterion for the *Hulthèn* potential, Eq. (13.31). Inverting the transformations (13.36), (13.37), and using Eq. (13.40), we obtain the bound-state wavefunctions as

$$\psi_\nu(r) = N_n \frac{z(1-z)^{\beta_n}}{r} F(1-n, 1+g/n, 2; z) Y_{\ell m} \ , \tag{13.44}$$

where the normalization constant is

$$N_n = \sqrt{\frac{1}{8\pi} g^3 \frac{1}{n}\left[\frac{1}{n^2} - \frac{n^2}{g^2}\right]} \ . \tag{13.45}$$

The relevant factor for the optical response is therefore

$$|\psi_n(r=0)|^2 = \frac{\lambda^3}{32\pi^2} g^3 \frac{1}{n}\left[\frac{1}{n^2} - \frac{n^2}{g^2}\right] , \tag{13.46}$$

where a factor $1/4\pi$ comes from the spherical harmonics.

B) Continuum states

For the unbound solutions of Eq. (13.33) we have $\epsilon_\nu > 0$ and

$$\beta_\nu = i\sqrt{\epsilon_\nu}/\lambda \ . \tag{13.47}$$

Inserting (13.47) into Eqs. (13.40) and (13.41) and normalizing the resulting wavefunction in a sphere of radius R, i.e.,

$$4\pi \int_0^R dr \ |u_\nu(r)|^2 = 1 \ , \tag{13.48}$$

we obtain $\epsilon_\nu = \nu^2\pi^2/R^2$ and

$$f_\nu = N_\nu \frac{z}{r} (1-z)^{\beta_\nu} F\left[1+\beta_\nu+\sqrt{g-|\beta_\nu|^2}\,,\ 1+\beta_\nu-\sqrt{g-|\beta_\nu|^2}\,,\ 2;\ z\right] \tag{13.49}$$

with

$$N_\nu = \frac{1}{\sqrt{R}} \left[\frac{|\beta_\nu|\ \sinh(2\pi|\beta_\nu|)\ g}{\cosh(2\pi|\beta_\nu|) - \cos\left[2\pi\sqrt{g-|\beta_\nu|^2}\right]}\right]^{1/2}, \tag{13.50}$$

where $\cos(ix) = \cosh(x)$. Finally, for the optical response, we need

$$\sum_\nu |\psi_\nu(r=0)|^2 f(\beta_\nu) =$$

$$\frac{1}{E_0 2\pi^2 a_0^3} \int_0^\infty dx \frac{\sinh(\pi g\sqrt{x})}{\cosh(\pi g\sqrt{x}) - \cos(\pi\sqrt{4g-xg^2})} f(x), \tag{13.51}$$

where the spin summation together with the spherical harmonics contribute a factor $1/2\pi$.

C) Optical spectra

Combining Eqs. (13.46) and (13.51) with Eq. (13.30) yields

$$\alpha(\omega) = \alpha_0 \tanh\left[\frac{\beta}{2}(\hbar\omega-E_g-\mu)\right] \frac{\hbar\omega}{E_0}\left\{\sum_n \frac{1}{n}\left[\frac{1}{n^2}-\frac{n^2}{g^2}\right]\delta_\Gamma\left(\frac{\hbar\omega-E_n}{E_0}\right)\right.$$

$$\left.+\int_0^\infty dx \frac{\sinh(\pi g\sqrt{x})}{\cosh(\pi g\sqrt{x})-\cos(\pi\sqrt{4g-xg^2})}\,\delta_\Gamma(\hbar\omega/E_0-x)\left[1-f_e\left(\frac{x}{\tilde{m}_e}\right)-f_h\left(\frac{x}{\tilde{m}_h}\right)\right]\right\} \tag{13.52}$$

where α_0 is defined in Eq. (4.47), the n-summation runs over all bound states,

$$\delta_\Gamma(x) = \frac{1}{\pi\Gamma\cosh(x/\Gamma)} , \quad \tilde{m}_\alpha = \frac{m_\alpha}{m_r} , \tag{13.53}$$

and a factor 2 in the exciton part results from the spin summation. Eq. (13.52) yields semiconductor absorption spectra that vary with carrier density $N = \Sigma f(k)$. The theoretical results agree very well with experimental observations for many different semiconductor materials. An example of such a comparison with experiments is shown in Fig. 13.6. From the absorptive changes

$$\Delta\alpha(\omega) = \alpha(\omega, N_2) - \alpha(\omega, N_1) , \tag{13.54}$$

one obtains the corresponding dispersive changes

$$\Delta n(\omega) = n(\omega, N_2) - n(\omega, N_1) \tag{13.55}$$

through the Kramers-Kronig transformation

$$\Delta n(\omega) = \frac{c}{\pi} P \int_0^\infty d\omega' \frac{\Delta\alpha(\omega')}{\omega'^2 - \omega^2} , \tag{13.56}$$

where P again indicates the principle value. Examples of the dispersive changes in bulk GaAs are shown in Fig. 13.6-*d*.

It is worthwhile to note at this point that the Kramers-Kronig transformation (13.56) is valid even though we are dealing with a nonlinear system, as long as $\Delta\alpha$ depends only parameters that are temporally constant. We have to be sure that the carrier density N does not vary in time, or that N varies sufficiently slowly so that it is justified to treat its time-dependence adiabatically.

REFERENCES

For details on the matrix-inversion technique see:

H. Haug and S. Schmitt-Rink, Progr. Quantum Electron. **9**, 3 (1984) and the given references

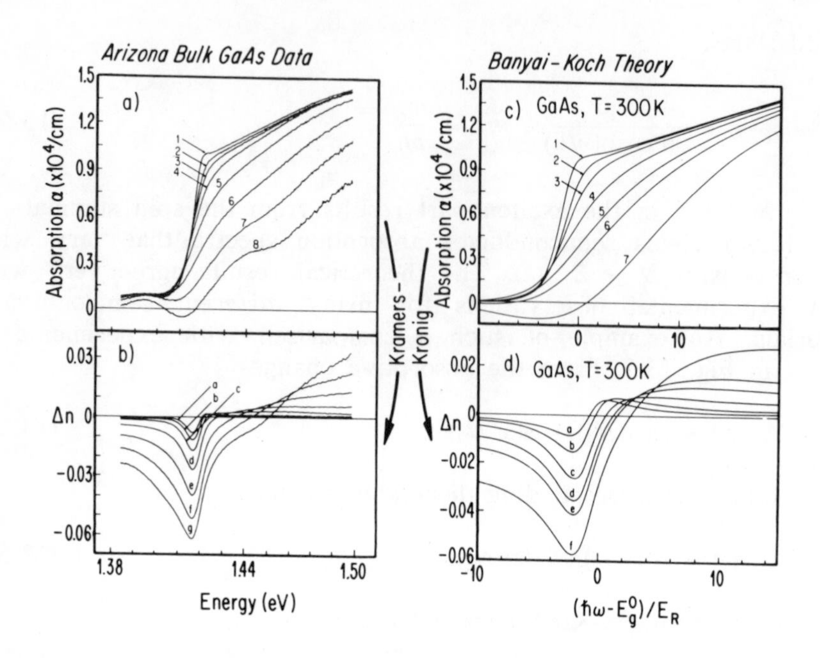

Fig. 13.6: Experimental and theoretical absorption and refractive index spectra are compared for room temperature GaAs. (*a*) The experimental absorption spectra are obtained for different excitation intensities I (mW): 1) 0; 2) 0.2; 3) 0.5; 4) 1.3; 5) 3.2; 6) 8; 7) 20; 8)50 using quasi-*cw* excitation directly into the band and a 15 μm excitation spot size. The oscillations in curve 8) are a consequence of imperfect antireflection coating. (*b*) The dispersive changes Δn are obtained through a Kramers-Kronig transformation of the absorptive changes $\alpha(I{=}0)$ - $\alpha(I)$ (*a*). The agreement with direct measurements of dispersive changes has been tested for the same conditions using a 299-Å multiple-quantum-well sample. (*c*) Calculated absorption spectra using the plasma theory for the densities N (cm^{-3}): 1) 10^{15} (linear spectrum); 2) $8{\cdot}10^{16}$; 3) $2{\cdot}10^{17}$; 4) $5{\cdot}10^{17}$; 5) $8{\cdot}10^{17}$; 6) $1{\cdot}10^{18}$; 7) $1.5{\cdot}10^{18}$. (*d*) Kramers-Kronig transformation of the calculated absorption spectra.

J.P. Löwenau, S. Schmitt-Rink, and H. Haug, Phys. Rev. Lett. **49**, 1511 (1982)

S. Schmitt-Rink, C. Ell and H. Haug, Phys. Rev. **B33**, 1183 (1986).

The *Padé* and high-density approximations are discussed in:

C. Ell, R. Blank, S. Benner, and H. Haug, Journ. Opt. Soc. Am. **B** (1989)

H. Haug and S.W. Koch, Phys. Rev. **A39**, 1887 (1989).

The effective pair-equation approximation (plasma theory) is derived and applied in:

L. Banyai and S.W. Koch, Z. Physik **B63**, 283 (1986).

See also:

S.W. Koch, N. Peyghambarian, and H. M. Gibbs, Appl. Phys. (Reviews) **63**, R1 (1988)

S.W. Koch, N. Peyghambarian, and M. Lindberg, J. Phys. *C* (Reviews) 21, 5229 (1988)

N. Peyghambarian and S.W. Koch, in *Nonlinear Photonics*, eds. H.M. Gibbs *et al.*, Springer, Berlin (1990).

PROBLEMS

Problem 13.2: Derive the expression for the optical susceptibility using the (1,2) *Padé* approximation.

Problem 13.2: Derive the self-consistency equation (13.27).

Problem 13.3: Discuss the *Hulthén* potential for small and large radii and compare it to the Yukawa potential.

Problem 13.4: Verify that Eq. (13.51) yields the correct free-particle result for $a_0 \to \infty$ and the Wannier result for $\kappa \to 0$.

Chapter 14
OPTICAL BISTABILITY

Optical instabilities in semiconductors can occur if one combines the strong material nonlinearities with additional feedback. The simplest example of such an instability is *optical bistability*, in which one has situations with two (meta-) stable values for the light intensity transmitted through a nonlinear material for one value of the input intensity I_0. The actually realized transmission intensity depends on the excitation history. A different state is reached, if one either decreases the incident intensity I_0 from a sufficiently high original level, or if one increases I_0 from zero.

The possibility to switch a bistable optical device between its two states allows the use of such a device as binary optical memory. The fact that bistable elements can be addressed simultaneously by many laser beams, has led many scientists to think about the possibility of parallel optical data processing.

A proper analysis of the optical instabilities in semiconductors requires a combination of the microscopic theory for the material nonlinearities with Maxwell's equations for the light field, including the appropriate boundary conditions. The polarization relaxes in very short times, determined by the carrier-carrier and carrier-phonon scattering, to its quasi-equilibrium value which is governed by the momentary values of the field $E(t)$ and the carrier density $n(t)$. Therefore we can use the quasi-equilibrium results of Chap. 13 for the optical susceptibility as the material equation, which implies that the polarization dynamics has been eliminated adiabaticaly.

The process of carrier generation through light absorption couples the electron-hole-pair density to the electromagnetic field. The electromagnetic field in turn is described by the macroscopic Maxwell equations, in which the polarization field depends on the value of the electron-hole-pair density through the equation for the susceptibility. This set of equations, i.e., the microscopic equation for the susceptibility together with the macroscopic equations for the carrier density and for the light field, constitutes the combined microscopic and macroscopic approach to consistently describe semiconductor nonlinearities.

14-1. The Coupled Propagation Equations

Before we discuss two examples of optical bistability in semiconductors, we present a systematic derivation of the relevant equations. Wave propagation in dielectric media is described by

$$\left[\nabla^2 - \text{grad div} - \frac{1}{c^2}\frac{\partial^2}{\partial t^2}\right] \mathcal{E} = \frac{4\pi}{c^2}\frac{\partial^2 \mathcal{P}}{\partial t^2}, \tag{14.1}$$

where Maxwell's equations request div $\mathbf{D} = 0$, since we assume no external charges. However, $div\mathcal{E} = -\mathcal{E}\cdot\nabla\ln\epsilon$, where ϵ is the medium dielectric constant, is generally not zero. Therefore, the electromagnetic field is no longer purely transverse but also has a longitudinal component.

To obtain the equations for the longitudinal and transverse variations of the field amplitudes, we assume a Gaussian incident light beam

$$\mathcal{E} = \mathcal{E}_0\, e^{-r^2/w_0^2}, \quad r^2 = x^2 + y^2,$$

with a characteristic (transverse) width w_0 which propagates in z direction. Using the ansatz

$$\nabla = \mathbf{e}_T\, \nabla_T + \mathbf{e}_z \frac{\partial}{\partial z}$$

$$\mathcal{E} = e^{-i(\omega t - kz)}(\mathbf{e}_T\, \mathcal{E}_T + \mathbf{e}_z\, \mathcal{E}_z)$$

$$\mathcal{P} = e^{-i(\omega t - kz)}(\mathbf{e}_T\, \mathcal{P}_T + \mathbf{e}_z\, \mathcal{P}_z) \tag{14.2}$$

we subdivide the wave equation (14.1) into a longitudinal and a transverse part. $\mathbf{e}_z$ and $\mathbf{e}_T$ in Eq. (14.2) are the unit vectors in z-direction and in the transverse directions, respectively. The transverse equation is

$$\nabla_T \left[\nabla_T \mathcal{E}_T + \left[ik + \frac{\partial}{\partial z} \right] \mathcal{E}_z \right] - \left[-k^2 + 2ik \frac{\partial}{\partial z} + \frac{\partial^2}{\partial z^2} + \nabla_T{}^2 \right] \mathcal{E}_T$$

$$= \frac{1}{c^2} \left[\omega^2 + 2i\omega \frac{\partial}{\partial t} - \frac{\partial^2}{\partial t^2} \right] [\, \mathcal{E}_T + 4\pi \, \mathcal{P}_T] \, , \tag{14.3}$$

and the longitudinal equation is

$$\left[ik + \frac{\partial}{\partial z} \right] \nabla_T \mathcal{E}_T - \nabla^2{}_T \mathcal{E}_z =$$

$$\frac{1}{c^2} \left[\omega^2 + i2\omega \frac{\partial}{\partial t} - \frac{\partial^2}{\partial t^2} \right] [\mathcal{E}_z + 4\pi \mathcal{P}_z] \, . \tag{14.4}$$

As usual, the polarization is related to the optical susceptibility or the dielectric function via

$$\mathcal{P}_{z/T} = \chi \, \mathcal{E}_{z/T} = \frac{\epsilon - 1}{4\pi} \; \mathcal{E}_{z/T} \, . \tag{14.5}$$

It is now convenient to split the susceptibility into the linear part χ_0 and the nonlinear, density-dependent part χ_{nl}

$$\chi = \chi_0 + \chi_{nl}(N) = \frac{\epsilon_0 - 1}{4\pi} + \frac{\epsilon_{nl}(N)}{4\pi} \, . \tag{14.6}$$

Even though we explicitly deal with a density-dependent nonlinearity, this treatment can easily be generalized to also include, e.g., thermal and other nonlinearities. To derive from Eqs. (14.3) and (14.4) the coupled equations for the longitudinal and transverse field components we use the so-called paraxial approximation, see Lax *et al.* (1975). Scaled variables are introduced as

$$x = \bar{x} \, w_0 \; ; \quad y = \bar{y} \, w_0 \; ; \quad z = \bar{z} \, l$$

$$t = \bar{t} \, \tau_R \; ; \quad \omega = \frac{\bar{\omega}}{\tau_R} \, , \tag{14.7}$$

where

$$l = k\, w_0^2 \tag{14.8}$$

is the diffraction length of the beam in z direction and $\tau_R = ln_b/c$, $n_b = \sqrt{\epsilon_0}$, is the characteristic propagation time over that distance. The dimensionless number

$$\mathcal{f} = \frac{w_0}{l} << 1 \tag{14.9}$$

serves as small parameter allowing to expand all quantities in powers of $\mathcal{f}$. Consistent equations are obtained using

$$\mathcal{E}_T = \mathcal{E}_T{}^{[0]} + \mathcal{f}^2\, \mathcal{E}_T{}^{[2]} + ...$$

$$\mathcal{E}_z = \mathcal{f}\, \mathcal{E}_z{}^{[1]} + \mathcal{f}^3\, \mathcal{E}_z{}^{[3]} + ...$$

$$\chi_{nl} = \mathcal{f}^2\, \chi_{nl}{}^{[2]} + \mathcal{f}^4\, \chi_{nl}{}^{[4]} + ... \tag{14.10}$$

Inserting Eqs. (14.9) and (14.10) into Eqs. (14.3) and (14.4) yields a set of coupled equations in the orders $O(\mathcal{f})$, $O(\mathcal{f}^3)$, ... Restricting ourselves to $O(\mathcal{f})$ we obtain

$$k\mathcal{E}_z{}^{[1]} = i\, \nabla_T\, \mathcal{E}_T{}^{[0]} \tag{14.11}$$

and

$$\boxed{\left[\frac{c}{n_b}\frac{\partial}{\partial z} - i\frac{c}{2kn_b}\nabla_T{}^2 + \frac{\partial}{\partial t} + \frac{c\alpha(\omega,N)}{2n_b} - i\frac{\omega\Delta n(\omega,N)}{n_b}\right]\mathcal{E}_T{}^{[0]} = 0}$$

transverse field equation (14.12)

where we used Eqs. (1.52), (1.53) to introduce absorption and refractive-index change,

$$\alpha(\omega,N) = \frac{4\pi\omega}{n_b c}\chi''(\omega,N) \quad \text{and}\quad \Delta n(\omega) \cong \frac{2\pi\chi'_{nl}(\omega,N)}{n_b}\ , \tag{14.13}$$

respectively. Eq. (14.12) shows that the electromagnetic field is purely transverse in lowest order, but it may nevertheless depend on the transverse coordinate. Only in next order there is a small longitudinal component whose size depends on $\mathcal{f}$.

For simplicity of notation we now replace $\mathbf{e}_T \ \mathcal{E}_T^{[0]}$ by $\mathcal{E}$. Through $\alpha(\omega, N)$ and $\Delta n(\omega, N)$, the field amplitude $\mathcal{E}$ is nonlinearly coupled to the electron-hole-pair density N. This density in turn couples to the light intensity

$$I = \frac{|\mathcal{E}|^2 c n_b}{8\pi} = \frac{\mathcal{E}_0^2 c n_b}{2\pi} , \tag{14.14}$$

where $\mathcal{E}_0$ is the real field amplitude in rotating-wave approximation, compare Chap. 4, especially Eq. (4.10), and Chap. 12. For simplicity we restrict ourselves in this chapter to a simple rate equation for the carrier density N

$$\boxed{\frac{\partial N}{\partial t} = -\frac{N}{\tau} + \frac{\alpha(\omega, N)}{\hbar\omega} I + \nabla D \nabla N}$$

electron-hole-pair rate equation (14.15)

Note that we use N for the electron-hole-pair density in this chapter to distinguish it clearly from the refractive index n. In Eq. (14.15) τ is the carrier relaxation time and D is the electron-hole-pair diffusion coefficient. The term $\alpha I / \hbar\omega$ describes carrier generation through light absorption, compare Eq. (12.16).

Eqs. (14.12) - (14.15), together with an expression for the nonlinear susceptibility determine the quasi-equilibrium nonlinear optical response of semiconductors. Note, that this approach involves the adiabatic elimination of the polarization dynamics, which is well justified for the present purposes because of the extremely fast phase destroying processes in semiconductors.

14-2. Bistability in Semiconductor Resonators

Dispersive optical bistability or bistability through decreasing absorption may be obtained if a semiconductor is brought into an optical resonator, which introduces additional feedback for the light field. In the following, we consider a Fabry-Perot resonator (Fig. 14.1) consisting of two lossless mirrors of reflectivity R and transmissivity $T = 1 - R$.

The nonlinear semiconductor material fills the space between the mirrors. In many practical applications, the mirrors are actually the end faces of the semiconductor crystal itself, and R is just the natural reflectivity, or R is increased through additional high reflectivity coatings evaporated onto the semiconductor surfaces.

As shown schematically in Fig. 14.1, it is useful to decompose the complex transverse field amplitude $\mathcal{E}$ into the forward and backward propagating parts to treat the feedback introduced by the mirrors. We write

$$\mathcal{E} = \mathcal{E}_F + \mathcal{E}_B = \xi_F e^{-i\phi_F} + \xi_B e^{i\phi_B} , \tag{14.16}$$

where the amplitudes ξ and the phases ϕ are real quantities. Depending on the phase relation between $\mathcal{E}_F$ and $\mathcal{E}_B$ there can be either constructive or destructive interference, leading to a maximum or minimum of the light intensity transmitted through the etalon.

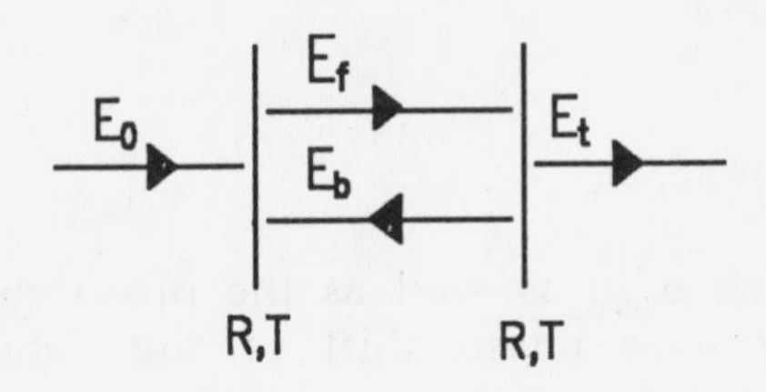

Fig. 14.1: Schematic drawing of a Fabry-Perot etalon. The mirrors have reflectivity R, transmissivity T=1-R, and E_0, E_f, E_b, E_t are incident, forward travelling, backward travelling, and transmitted field, respectively.

Dominantly dispersive nonlinearities can be observed in semiconductors, since the refractive index of the medium changes via the carrier density with the light intensity, causing the optical path between the mirrors of the etalon to change with intensity. These intensity-induced changes tune the etalon in or out of resonance with light of a fixed frequency. In reality, absorptive and dispersive changes occur simultaneously, but either one may be dominant in a particular frequency regime.

The boundary conditions of the Fabry Perot resonator can be written as

$$\mathcal{E}_F(z=0) = \sqrt{T}\mathcal{E}_0 + \sqrt{R}\mathcal{E}_B(z=0)$$

$$\mathcal{E}_B(z=0) = \sqrt{R}\ \mathcal{E}_F(z=L)\ e^{i\beta/2 - \alpha_{tot}L/2}$$

$$\mathcal{E}_F(z=L) = \frac{\mathcal{E}_t}{\sqrt{T}} = \mathcal{E}_F(z=0)\ e^{i\beta/2 - \alpha_{tot}L/2}\ , \tag{14.17}$$

yielding

$$\left|\frac{\mathcal{E}_t}{\mathcal{E}_0}\right|^2 = \frac{T^2}{(e^{\alpha_{tot}L/2} - \mathrm{R}e^{-\alpha_{tot}L/2})^2 + 4R\sin^2(\beta/2)}\ , \tag{14.18}$$

where we used

$$\cos(\beta) = 1 - 2\sin^2(\beta/2).$$

The effective absorption α_{tot}, as well as the phase shift β still have to be computed. The total phase shift of the light after passing through the resonator can be written as

$$\beta = \phi_F(z=L) - \phi_B(z=L) - 2\delta\ , \tag{14.19}$$

where 2δ contains all carrier-density-independent phase shifts of the linear medium and of the mirrors. If we ignore transverse variations, and insert Eq. (14.16) into Eq. (14.12), we obtain for the field amplitudes

$$\left[\frac{\partial}{\partial z} + \frac{n_b}{c}\frac{\partial}{\partial t} + \frac{\alpha(\omega,N)}{2}\right]\xi_{F/B} = 0\ , \tag{14.20a}$$

and for the phases

$$\left[\frac{\partial}{\partial z} + \frac{n_b}{c}\frac{\partial}{\partial t}\right]\phi_{F/B} = {}^{-}/_{+}\ \frac{\omega \Delta n(\omega, N)}{c}\ , \tag{14.20b}$$

where the minus sign is for ϕ_F .

Most semiconductor bistability experiments are done for resonator lengths $L \cong 0.5$ - $2\ \mu m$. Under these conditions, the round-trip time for the light in the resonator is substantially shorter than the carrier relaxation time τ and one is justified to adiabatically eliminate the dynamics of the light field (bad cavity limit),

$$\frac{\partial \xi}{\partial t} \cong 0 \quad \text{and} \quad \frac{\partial \phi}{\partial t} \cong 0\ . \tag{14.21}$$

Solving Eq. (14.20) for these conditions yields

$$\xi_{F/B}(z) = \xi_{F/B}(0)\ \exp\left[-\frac{1}{2}\int_0^z dz'\ \alpha[N(z')]\right] \tag{14.22a}$$

and

$$\phi_{F/B}(z) = {}^{-}/_{+}\ \frac{\omega}{c}\int_0^z dz'\ \Delta n[N(z')]\ , \tag{14.22b}$$

showing, that α_{tot} and β in Eq. (14.18) are given by

$$\alpha_{tot} = \int_0^L dz\ \alpha[N(z)] \tag{14.23}$$

and

$$\frac{\beta}{2} = -\left[\delta + \frac{\omega}{c}\int_0^L dz\ \Delta n[N(z)]\right]\ . \tag{14.24}$$

The spatial carrier distribution $N(z)$ has to be computed from Eq.

(14.15).

To describe typical semiconductor etalons it is often a good approximation to neglect all spatial density variations (diffusion dominated case). Then one can trivially evaluate the integrals in Eqs. (14.23) and (14.24), and Eq. (14.18) becomes

$$\boxed{I_t = \frac{I_0 T^2}{(e^{\alpha(\omega,N)L/2} - Re^{-\alpha(\omega,N)L/2})^2 + 4R\sin^2(\delta + \omega\Delta n(\omega,N)L/c)}}$$

transmission through a resonator (14.25)

Eq. (14.25) is the well-known equation for the transmission of a Fabry-Perot etalon. This transmission exhibits peaks whenever the argument of the $\sin^2$-term equals integer multiples of π.

Through α and Δn, Eq. (14.25) is coupled to the spatially averaged rate equation (14.15)

$$\frac{dN}{dt} = -\frac{N}{\tau} + \frac{\alpha(\omega,N)}{\hbar\omega} I , \tag{14.26}$$

where I has to be taken as the average intensity inside the resonator

$$I = I_t \frac{1 + R}{T} . \tag{14.27}$$

Note, that the incident intensity I_0 in Eq. (14.25) may still be time-dependent, but all variations have to be slow on the timescale of the resonator round-trip time (adiabatic approximation).

To explicitly solve Eqs. (14.25) - (14.27) for an example of practical interest, we use the band-edge nonlinearities of room-temperature GaAs (Fig. 13.6) as obtained from the plasma theory. Inserting the computed $\alpha(N)$ and $\Delta n(N)$ into Eqs. (14.25) - (14.27) allows us to directly study nonlinear optical device performance. We obtain bistable hysteresis curves by plotting transmitted intensity versus input intensity. In Fig. 14.2 we show some typical results for slightly different resonator lengths L, i.e., for different detunings of the excitation frequency ω with respect to the nearest resonator eigenfrequency $\omega_R < \omega$, where

$$\omega_R = m \frac{\pi c}{L n_b} , \; m = 0, 1, 2, ... \tag{14.28}$$

Curve 2 in Fig. 14.2 shows marginal and curve 3 shows well developed optical bistability for some range of input intensities, whereas curve 1 exhibits only nonlinear transmission.

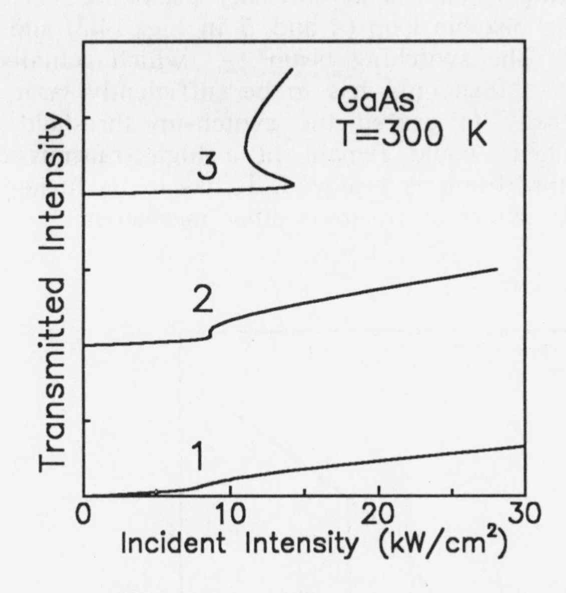

Fig. 14.2: Optical bistability for GaAs at T=300K, as computed from Eq. (14.25) using the absorption and refractive index data shown in Fig. 13.6. The mirror reflectivity R = 0.9, the resonator lengths are 2.0168 μm (1), 2.0188 μm (2), and 2.0233 μm (3), respectively, and the operating frequency ω is chosen such that $(\hbar\omega - E_g^0)/E_0 = -4$.

For a situation similar to the one in curve 3 of Fig. 14.2, a stability analysis shows that the intermediate branch with the negative slope is unstable and therefore not realized under usual experimental conditions. Hence, increasing the incident intensity I_0 from 0 leads to a transmitted intensity I_t that follows the lower bistable branch until I_0 reaches the switch-up intensity, denoted by B in Fig. 14.3, then I_t follows the upper branch. On the other hand, lowering I_0 from an original value $I_0 > B$ leads to a transmission following the upper branch, until I_0 reaches the switch-down intensity, denoted by A in Fig. 14.3.

Optically nonlinear or bistable semiconductor etalons may be used as all-optical logic or switching devices, see Gibbs (1985).

Since the bistable etalon maintains one of two discrete output states for some range of input intensity, it is possible to use it in the so-called *latched mode*. In that case one divides the total input beam into several beams, which may serve as *holding beam* or *switching* beam(s). The holding beam has an intensity positioned between the switch points of the bistable loop (*A* and *B* in Fig. 14.3) and is used to bias the device. The switching beam - which actually is the input logic beam - then only has to be sufficiently large for the total incident intensity to exceed the switch-up threshold B. The device, once switched, would remain in a high transmission state even if the switching beam is removed. It has to be turned off by interrupting its bias power or by some other mechanism.

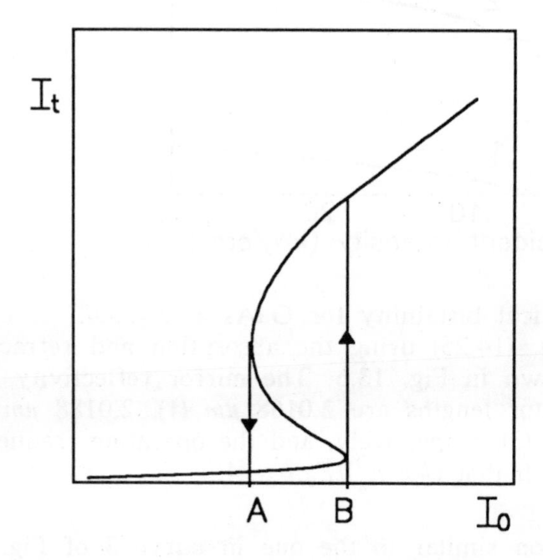

Fig. 14.3: Optical bistability for GaAs at *T*=300*K*, where *B*, *A* denote the switch-up and switch-down intensities.

In this latched mode of operation, one can obtain signal amplification. When a nonlinear etalon is operated as a passive device, it obviously has no overall gain relative to the total incident power. However, it is possible to achieve differential gain, in which case the device is able to transmit a larger signal than the signal used to switch it. Differential gain allows to use the output of one device as input (switching beam) for one or more other devices. This process is called *cascading* and the number of devices that can be switched with the output of a single device is usually referred to as *fan-out*. Using two beams as input in addition to the proper bias beam allows to realize all-optical gates which perform logic functions, such as AND, OR, or NOR.

14-3. Intrinsic Optical Bistability

From a conceptual point of view, the simplest example of

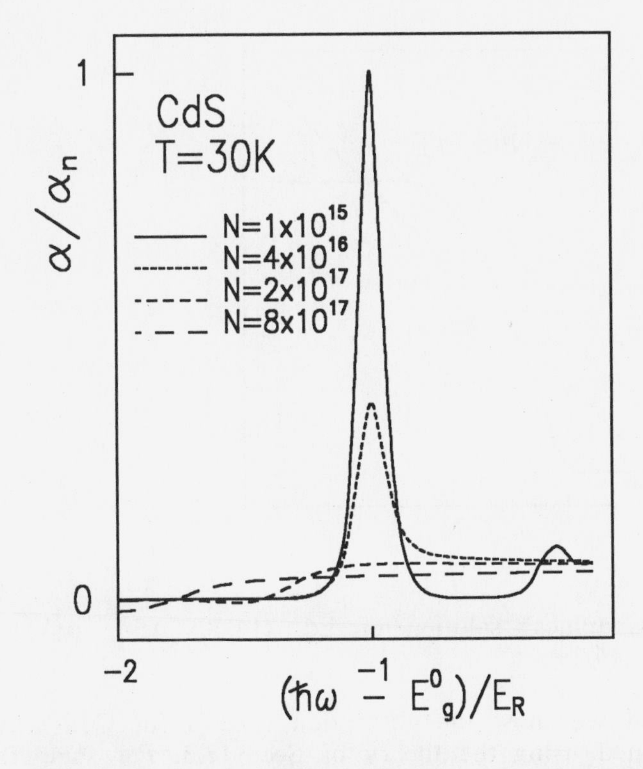

Fig. 14.4: Computed absorption spectra for CdS. These results have been obtained using the plasma theory, Sec. 13.3, with the parameters: m_e=0.235, m_h=1.35, ϵ_0=8.87, a_0=30.1Å, E_0=27meV, E_g^0=2.583eV, Γ=0.04E_0, and α_n=10^6/cm.

optical bistability is obtained, if one considers a medium whose absorption increases with increasing excitation density. Bistability in such a system may occur without any external feedback since the system provides its own internal feedback. Increasing the carrier density leads to an increasing absorption that causes the generation of even more carriers etc.. There are numerous mechanisms which may cause such an induced absorption in semiconductors and other systems. Here we concentrate on the induced absorption which is observed in semiconductors like CdS at low temperatures, as a consequence of the bandgap reduction (Koch *et al.*, 1985). However, most of the macroscopic features discussed below are quite general and may very well occur also in other systems.

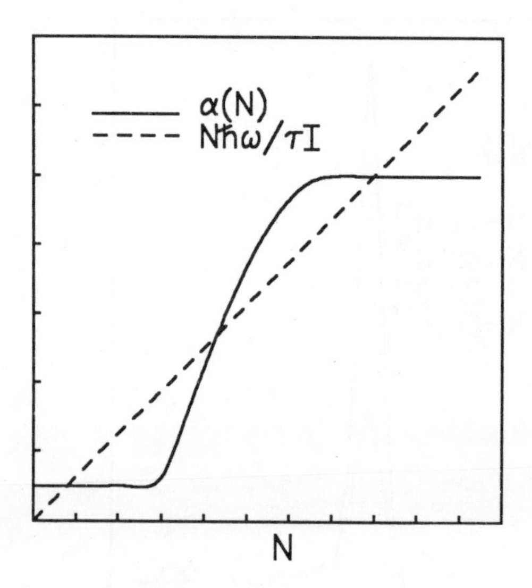

Fig. 14.5: Graphical solution of Eq. (14.30) using $\alpha(N)$ given by Eq. (14.29).

In Fig. 14.4 we show the absorption spectra for CdS which have been computed using the theory of Sec. 13.3. For some frequencies below the exciton resonance, we see that the absorption increases with increasing carrier density. Assuming now that the semiconductor is excited at such a frequency below the exciton, one has only weak absorption for low intensities. Nevertheless, if the exciting laser is sufficiently strong, even this weak absorption generates a density of electron-hole pairs which causes a reduction of the semiconductor bandgap. Eventually, the band edge shifts below the frequency of the exciting laser giving rise to a substantially increased one-photon absorption coefficient. Hence, one has an absorption, which increases with increasing carrier density.

Since we want to emphasize the general aspects of this type of optical bistability, we do not use the original CdS data, but a simple generic model for increasing absorption

$$\begin{aligned}\alpha(N) &= \alpha_L \;, \; N < N_1 \\ &= \alpha_L + (\alpha_H - \alpha_L) \sin\left[\frac{\pi(N-N_1)}{2(N_2-N_1)}\right] \;, \; N_1 < N < N_2 \\ &= \alpha_H, \; N_2 < N \;. \end{aligned} \tag{14.29}$$

Here α_L and α_H denote the low and high absorption values.

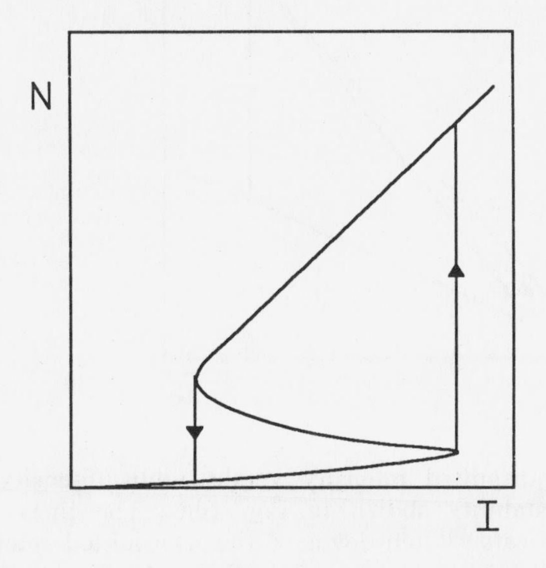

Fig. 14.6: Bistability of the carrier density N as function of intensity I, obtained as solution of Eq. (14.30). The lines with the arrows indicate switch-up of the carrier density N when I is increased from 0, and switch-down when I is decreased from a sufficiently high value.

The coupling between carrier density N and light intensity I is described by Eq. (14.15), which in the stationary, spatially homogeneous case leads to

$$\alpha(N) = \frac{N}{I}\frac{\hbar\omega}{\tau} . \tag{14.30}$$

This relation can be bistable, as indicated in Fig. 14.5, where we plot $\alpha(N)$, Eq. (14.29) and the straight line which is the RHS of Eq. (14.30). The slope of this straight line is inversely proportional to I. The intersection points with the curve $\alpha(N)$ are the graphical solutions of Eq. (14.30), clearly showing the occurrence of three simultaneous solutions indicating intrinsic optical bistability without resonator feedback (see Fig. 14.6).

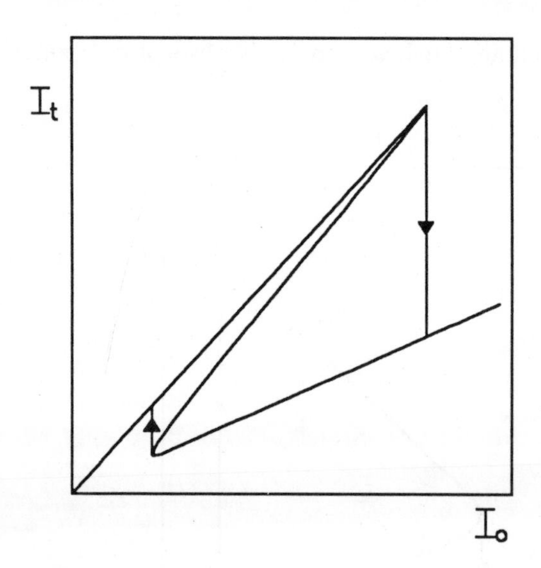

Fig. 14.7: Transmitted intensity versus input intensity for the carrier bistability shown in Fig. 14.6. The lines with the arrows indicate switch-down of the transmitted intensity I_t when the incident intensity I_0=$I(z$=0) is increased from 0, and switch-up when I_0 is decreased from a sufficiently high value.

It is worthwhile to stress at this point that for such an induced absorption bistability only a single pass of the light beam through the medium is required. Hence, one has no superposition of forward and backward travelling waves as, e.g., in the Fabry-Perot resonator. Therefore one may neglect nonlinear dispersive effects as long as one is only interested in the characteristic variation of the transmitted intensity and not in transverse beam-profile variations or in diffraction effects.

To obtain an equation for the light intensity, we multiply Eq. (14.12) by $\mathcal{E}^*$ and add the complex conjugate equation

$$\left[\frac{\partial}{\partial t} + \frac{c}{n_b}\frac{\partial}{\partial z} + \frac{c}{n_b}\,\alpha(N) \right] \mathcal{E}\,\mathcal{E}^* = -\,\frac{ic}{2kn_b}\,(\mathcal{E}\nabla_T{}^2\mathcal{E}^* - \text{c.c.}\,). \quad (14.31)$$

The RHS of Eq. (14.31) describes beam diffraction. This effect can be neglected if the length over which the light propagates is much smaller than the characteristic diffraction length. For a sufficiently thin sample, we can therefore approximate Eq. (14.31) as

$$\left[\frac{\partial}{\partial t} + \frac{c}{n_b}\,\frac{\partial}{\partial z} + \frac{c}{n_b}\,\alpha(N) \right] I = 0\ . \quad (14.32)$$

If N is constant, the steady state solution of Eq. (14.32) is Beer's law

$$I(z) = I(z{=}0)\ e^{-\alpha(N)z}\ , \quad (14.33)$$

where $z = 0,\ L$ are the sample front and end faces, respectively. Eq. (14.33) shows that the transmitted intensity $I_t = \mathrm{I}(z{=}L)$ is high (low) if $\alpha(N)$ is low (high).

If the carrier density exhibits optical bistability as shown in Fig. 14.6, the corresponding transmitted intensity shows the hysteresis loop plotted in Fig. 14.7. We see that the transmitted intensity follows the input intensity I_0 for low I_0. When I_0 exceeds the value at which the carrier density switches to its high value, the sample absorption also switches up and the transmitted intensity switches down.

The described situation is experimentally relevant, if $D\tau/L^2 \gg 1$, i.e., in the diffusion dominated case or for very thin semiconductor samples at the center of the input beam. Induced absorption bistability in CdS for these conditions has indeed been observed experimentally, and the experimental results are properly described by the outlined theory. In addition, there exist interesting longitudinal, transverse, and dynamic instabilities in such induced absorbers. However, the discussion of these effects goes beyond the scope of this book.

REFERENCES

For the paraxial approximation in two-level systems see:

M. Lax, W.H. Louisell, and W.N. McKnight, Phys. Rev. **A11**, 1365 (1975).

The theory of Lax *et al.* (1975) has been generalized for the case of semiconductor nonlinearities in:

S.W. Koch and E.M. Wright, Phys. Rev. **A35**, 2542 (1987).

For reviews of optical bistability and its applictions to optical logic see:

H.M. Gibbs *Optical Bistability: Controlling Light with Light*, Academic Press (1985)

M. Warren, S.W. Koch, and H.M. Gibbs, in the special issue on integrated optical computing of IEEE *Computer*, Vol. **20**, No. 12, *p.* 68 (1987)

From Optical Bistability Towards Optical Computing, eds. P. Mandel, S.D. Smith, and B.S. Wherrett, North Holland, Amsterdam (1987).

The theory of intrinsic optical bistability through bandgap reduction is discussed in:

S.W. Koch, H.E Schmidt and H. Haug, J. Luminescence **30**, 232 (1985).

PROBLEMS

Problem 14.1: Derive Eqs. (14.11), (14.12) and the corrections $O(\mathcal{f}^2)$ applying the procedure outlined in the text.

Problem 14.2: Show that the resonator formula (14.18) satisfies all the boundary conditions specified in Eqs. (14.17).

Problem 14.3: Discuss dispersive optical bistability for the case of a lossless Kerr medium with $\alpha = 0$ and $\Delta n = n_2 N$. Solve the coupled Eqs. (14.25) and (14.26) graphically and analyze the conditions for optical bistability.

Chapter 15
THE SEMICONDUCTOR LASER

From a technological point of view, semiconductor laser diodes are probably the most important electro-optical semiconductor devices. These lasers are used for many applications in data communication with optical fibers, for data recording and processing (CD players, laser printers), and for optical control and display devices. Today, low-power semiconductor laser diodes have become relatively cheap mass products. Modern crystal growth techniques are used to engineer lasers with well specified device properties. Most of the presently available diodes are made from binary, ternary and quaternary III–V compounds to get an active material with the desired band gap and laser frequency. Microstructures with low optical losses have been developed, and quantum confinement is also exploited to get low-threshold laser diodes. In addition to III–V compound lasers, narrow-gap semiconductor lasers, such as lead salt laser diodes, are used in the far infrared.

In this chapter we discuss the physical principles and the quantum mechanical equations which govern the action of semiconductor lasers. For the rich spectrum of different device designs we have to refer to special books on semiconductor lasers [e.g. Thompson (1980) or Agarwal and Dutta (1986)]. As in the description of passive bistable semiconductor devices in Chap. 14, we need the equations for the electrons in the semiconductor in combination with Maxwell's equations for the laser light. A classical treatment of the light field, if necessary supplemented with stochastic noise sources, is sufficient as long as we are not interested in the photon statistics and other quantum mechanical aspects of the laser light. The electric field $E(t)$ is driven by the dielectric polarization $P(t)$ which in term is determined by the inversion of the electrons, i.e., the electron-hole plasma density $N(t)$. The main difference to passive devices is that the field is generated by the laser itself, and the energy is supplied by a pump source, usually in form of an injection current. The resulting negative absorption provides the optical gain so that the spontaneously emitted light can grow into coherent laser light.

15-1. Material Equations

We showed in Chap. 13 that the electronic system of a semiconductor can be described with the reduced single-particle density matrix, whose off-diagonal elements yield the interband polarization. In laser diodes one has a relatively dense quasi-thermal electron-hole plasma, in which the relaxation of the polarization is determined by carrier-carrier collisions. The relaxation times are of the order of 100 *fs*, while the spontaneous lifetime in a direct-gap semiconductor is typically of the order of *ns*. On the other hand, the relaxation rate of the light field in the laser cavity is of the order of the cavity round-trip time, ≅ 10 *ps*, so that the polarization follows more or less instantaneously all changes of the electron-hole density and of the light field. Under these conditions the polarization dynamics can be eliminated adiabatically. i.e, we solve Eq. (12.15) for stationary situations and insert the result into the equations for the electron-hole density and for the laser field. If we assume spatial homogeneity in the semiconductor material, i.e., in the active laser region, we can express the polarization in the form

$$P(\omega,t) = \chi(\omega,t)\,\frac{\mathcal{E}_0(\omega,t)}{2}\,, \tag{15.1}$$

where $\mathcal{E}_0(\omega,t)$ is the field amplitude in rotating wave approximation, compare Chaps. 4 and 10. For the susceptibility we can use the results of Chap. 13, such as the Padé approximation (13.23),

$$\chi(\omega,t) = \frac{1}{L^3}\sum_k \frac{d^*_{cv,k}\chi^0{}_k}{1-q_{1k}}\,. \tag{15.2}$$

The time-dependence in Eqs. (15.1) and (15.2) stems from the relatively slow variations of the plasma density $N(t)$. In a simpler version of the theory one often neglects the many-body effects and uses only the free-carrier susceptibility χ^0 instead of the full χ.

In laser theory, $\chi(\omega,t)$ is often transformed into the complex gain function $g(\omega,t)$, where the gain per unit time is defined as

$$g(\omega,t) = i\,\frac{4\pi\omega}{\epsilon_0}\,\chi(\omega,t)\ . \tag{15.3}$$

The real part of the gain, g', is positive for negative absorption, $\chi'' < 0$, while the imaginary part, g'', is connected with the index of refraction.

In addition to the equation for the interband polarization, we also need dynamic equations for the diagonal elements of the reduced density matrix, i.e., for the populations (carrier densities). Again, because of the rapid carrier-carrier scattering, we can assume that the electrons and holes are described by quasi-equilibrium Fermi functions, Eq. (13.2). In principle, the parameters of the Fermi functions, i.e., the temperature T, the drift vector $\mathbf{k}_D$, and the chemical potential μ may still be slowly varying functions in time and space. However, to keep the present analysis as simple as possible, we assume that we have no drift, $\mathbf{k}_D = 0$, no temperature variations, and charge neutrality, $n_e = n_h = N$. We allow only slow temporal density variations and keep the chemical potential, which as usual is determined by

$$N = \frac{1}{L^d}\sum_{\mathbf{k}} f_{i,k}\ ,\quad i = e,\ h \tag{15.4}$$

as the only dynamic parameter in the Fermi distribution. Under these conditions, it is sufficient to consider the rate equation governing the time development of the total electron-hole density

$$\frac{dN}{dt} = r_p - r_{st} - r_{sp} - r_{nr}\ , \tag{15.5}$$

where the different rates on the RHS describe injection pumping of carriers (r_p), stimulated emission (r_{st}), spontaneous emission (r_{sp}), and the nonradiative transitions (r_{nr}), respectively. Note, that the adiabatic elimination of the polarization and the local equilibrium assumption are both justified under the condition of short inter-particle collision times.

The pump rate due to an injection current density j is given by

$$r_p = \frac{j\eta}{ed}\ , \tag{15.6}$$

where η is the quantum efficiency and d the transverse dimension of the active region in the laser. The loss rate due to stimulated emission is

$$r_{st} = \frac{2}{L^3}\,\mathrm{Im}\sum_{\mathbf{k}} d_{cv}\,\frac{\mathcal{E}_0(t)}{2}\,P^*_{\mathbf{k}} , \tag{15.7}$$

compare Eq. (14.15). Using Eqs. (13.8), (13.11), and (15.3) allows to write Eq. (15.7) as

$$r_{st} = \frac{1}{2L^3}\,\mathrm{Im}\sum_{\mathbf{k}} d_{cv}\,\chi^*_k\,\mathcal{E}_0^2(t) = \frac{1}{8\pi\omega}\,g'(\omega)\,\mathcal{E}_0^2(t) . \tag{15.8}$$

With the help of the *Padé* approximation and using the identity

$$[1-f_{e,k}-f_{h,k}]\ \delta(\omega-e_{e,k}-e_{h,k}) = [e^{\beta(\hbar\omega-\mu)}-1]\ f_{e,k}f_{h,k}\ \delta(\omega-e_{e,k}-e_{h,k}) ,$$

the imaginary part of the susceptibility becomes

$$\chi''(\omega) = [e^{\beta(\hbar\omega-\mu)}-1]\ \chi''_e(\omega) \equiv \chi''_a(\omega) - \chi''_e(\omega) . \tag{15.9}$$

Here the emission part of χ is defined as

$$\chi_e(\omega) = -\sum_{\mathbf{k}} \frac{|d_{cv,k}|^2\ f_{e,k}f_{h,k}}{1-q_1}\ \frac{1}{\hbar(\omega+i\delta-e_{e,k}-e_{h,k})} . \tag{15.10}$$

Using Eqs. (1.53) and (15.9), we can subdivide the absorption coefficient as

$$\alpha(\omega) = \frac{4\pi\omega}{cn_b}\,(\chi''_a - \chi''_e) \equiv \alpha_a - \alpha_e , \tag{15.11}$$

where α_a and α_e are the probabilities per unit length for absorbing and emitting a photon, respectively. Thus $\alpha_e c/n_b$ is the emission probability of a photon per unit time. The spontaneous emission rate into the continuum of all photon modes is with $\omega=\omega_{\mathbf{q},\lambda}$, where $\mathbf{q}$, λ label the photon wavevector and polarization,

$$r_{sp} = \frac{1}{L^3} \sum_{\mathbf{q},\lambda} \frac{c}{n_b} \alpha_e(\omega_{\mathbf{q},\lambda})$$

$$= \frac{2 \cdot 4\pi}{(2\pi)^3} \int dq\, q^2 \frac{4\pi\omega}{\epsilon_0} \chi_e''(\omega_{\mathbf{q}}) \tag{15.12}$$

or

$$r_{sp} = \int_0^\infty \frac{d\omega}{2\pi\epsilon_0} \left[\frac{2\omega n_b}{c}\right]^3 \chi_e''(\omega) , \tag{15.13}$$

where we have used $n_b \cong (\epsilon_0)^{1/2}$. The weight factor ω^3 shows that it is difficult for higher laser frequencies to overcome the losses due to spontaneous emission. This is one of the major problems in the development of an x-ray laser.

The non-radiative recombination rate has a linear term describing recombination under multi-phonon emission at deep traps levels, compare Eq. (14.15). Additionally, in narrow-gap semiconductors and at high plasma densities, one often has to consider also Auger recombination processes. These processes are proportional to the third power of the plasma density, so that the total non-radiative recombination rate can be written as

$$r_{nr} = \frac{N}{\tau} + CN^3 . \tag{15.14}$$

The influence of the Auger recombination rates increase with decreasing gap energy. In infrared semiconductor lasers this is often the dominant loss term.

15-2. Quantum Mechanical Langevin Equations

In a complete laser theory we have to include the fluctuations which are necessarily linked with dissipative processes. In the following we discuss quantum mechanical *Langevin equations*, which provide a method that allows to incorporate at least approximately dissipative processes and the connected fluctuations into the Heisenberg equations for a given quantum mechanical operator. With this method, we can include the noise terms in the equations of motion for the reduced density matrix of the electronic excitations and we get the fluctuation of the laser field.

As an example, we first show for the harmonic oscillator that it is quantum mechanically inconsistent to have only dissipative terms in addition to the usual Heisenberg equation. The equation of motion for the Boson annihilation operator of the harmonic oscillator is

$$\dot{b} = \frac{i}{\hbar}\,[\mathcal{H}, b] = -\,i\omega b\ . \tag{15.15}$$

To include damping in the traditional way, we write

$$\dot{b} = \left[-i\omega - \frac{\kappa}{2}\right] b\ . \tag{15.16}$$

The solution of this equation is $b(t) = b(0)\exp[-(i\omega+\kappa/2)t]$, and it is a simple exercise (see problem 15.1) to show that the commutator $[b(t), b^\dagger(t)] \neq 1$. The damping term causes a decay of the commutator. We can correct this inconsistency by adding a fluctuation operator $f(t)$ to the equation of motion

$$\boxed{\dot{b} = \left[-i\omega - \frac{\kappa}{2}\right] b(t) + f(t)}$$

Langevin equation (15.17)

For simplicity, we always assume in this book that the fluctuation operators can be described as Markoffian noise sources. This means that the stochastic fluctuations at different times are uncorrelated (no memory)

$$\boxed{\langle f(t)\, f^\dagger(s)\rangle = 2D\,\delta(t-s)}$$

Markoffian fluctuations (15.18)

The average in Eq. (15.18) is taken over a reservoir to which the system must be coupled in order to introduce irreversible behavior. D is called the diffusion constant, which is related via the dissipation-fluctuation theorem to the damping constant κ.

To obtain an explicit form for the fluctuation operator and the dissipation rate, let us discuss the coupling of the harmonic oscillator to a reservoir (also called bath), which consists of a large set of harmonic oscillators B_λ with a continuous energy spectrum $\hbar\Omega_\lambda$. We assume that the bath oscillators are all in thermal equilibrium and are not disturbed by the coupling to the harmonic oscillator which represents our system. In other words, we assume that the bath is infinitely large in comparison to the system of interest.

The total system-bath Hamiltonian is

$$\frac{\mathcal{H}}{\hbar} = \omega b^\dagger b + \sum_\lambda \Omega_\lambda \, B_\lambda^\dagger B_\lambda + \sum_\lambda g_\lambda \, (bB_\lambda^\dagger + B_\lambda b^\dagger) \qquad (15.19)$$

The equations of motion for the oscillators of the system and bath are

$$\dot{b} = - i\omega b - i\sum_\lambda g_\lambda B_\lambda \quad \text{and} \quad \dot{B}_\lambda = - i\Omega_\lambda B_\lambda - ig_\lambda b \ . \qquad (15.20)$$

With the ansatz $B(t) = \widetilde{B}(t)e^{-i\Omega_\lambda t}$ we find for the bath operator

$$B_\lambda(t) = B_\lambda(0)e^{-i\Omega_\lambda t} - ig_\lambda \int_0^t d\tau \ b(\tau) \ e^{-i\Omega_\lambda(t-\tau)} \ . \qquad (15.21)$$

Inserting this result into the equation for the system operator yields

$$\dot{b} = -i\omega b -i\sum_\lambda g_\lambda \left[B_\lambda(0)e^{-i\Omega_\lambda t} - ig_\lambda \int_0^t d\tau \ b(\tau) \ e^{-i\Omega_\lambda(t-\tau)} \right] , \qquad (15.22)$$

which we write in the form

$$\dot{b} = -i\omega b + f(t) - \frac{\kappa}{2} b(t) \ . \qquad (15.23)$$

Now, we identify

$$\frac{\kappa}{2} b(t) \cong \sum_{\lambda} g_{\lambda}^{2} b(0)\, e^{-i\Omega_{\lambda} t} \int_{0}^{t} d\tau\, e^{i(\Omega_{\lambda} - \omega)\tau}$$

$$\cong \sum_{\lambda} g_{\lambda}^{2}\, \pi\, \delta(\Omega_{\lambda} - \omega)\, b(t) \,, \qquad (15.24)$$

where we approximated $b(t) \cong b(0)\exp(-i\omega t)$. In Eq. (15.24) we recognize Fermi's golden rule for the transition rate per unit time of an energy quantum $\hbar\omega$ into the continuous spectrum of the bath. Using the density of states $\rho(\omega) \cong \rho$, assuming $g(\omega) \cong g$, and integrating over all bath energies, we get the damping constant

$$\kappa = g^{2} 2\pi\rho \,. \qquad (15.25)$$

The fluctuation operator $f(t)$ is given by the first-order term in Eq. (15.22) as

$$f(t) = -i \sum_{\lambda} g_{\lambda} B_{\lambda}(0) e^{-i\Omega_{\lambda} t} \,. \qquad (15.26)$$

Obviously, if we average $f(t)$ over the bath $\langle \ldots \rangle = tr\rho_{B} \ldots$, with

$$\rho_{B} = \frac{e^{-\beta H_{B}}}{tr e^{-\beta H_{B}}} \,, \qquad (15.27)$$

we find $\langle f(t) \rangle = 0$. However, for the second moments we get

$$\langle f^{\dagger}(t) f(s) \rangle = \sum_{\lambda} g_{\lambda}^{2} \langle B_{\lambda}^{\dagger}(0) B_{\lambda}(0) \rangle\, e^{i\Omega_{\lambda}(t-s)} \cong g^{2} g_{0}(\omega) \rho 2\pi \delta(t-s)$$

or

$$\langle f^\dagger(t)f(s)\rangle = \kappa\ g_0(\omega)\ \delta(t-s)\ , \tag{15.28}$$

$$\langle f(t)f^\dagger(s)\rangle = \kappa\ (g_0(\omega) + 1)\delta(t-s)\ , \tag{15.29}$$

$$\langle f(t)f(s)\rangle = 0 \text{ and } \langle f^\dagger(t)f^\dagger(s)\rangle = 0\ . \tag{15.30}$$

dissipation-fluctuation theorem for the harmonic oscillator

Here, $g_0(\omega) = (\exp(\beta\hbar\omega) - 1)^{-1}$ is the thermal Bose distribution for the bath quanta. Eqs. (15.28) - (15.30) are called dissipation-fluctuation relations because they link the correlations of the fluctuations and the damping constant κ, which describes the dissipation rate. We see that our simple model yields, at least approximately, Markoffian correlations for the fluctuations.

It turns out in the discussion of noise sources in the electron-hole system of the semiconductor laser that one needs a more general formulation of the quantum mechanical dissipation-fluctuation theorem. Following Lax (1966), we therefore write the general Langevin equation for a set of quantum mechanical variables O_μ as

$$\frac{d}{dt}\,O_\mu = A_\mu(\{O_\mu\},t) + F_\mu(\{O_\mu\},t)\ , \tag{15.31}$$

where the dissipation rates A_μ are calculated in second-order perturbation theory in the interaction of the system with its reservoirs, as shown above for the example of the harmonic oscillator. The Markoff assumptions for the fluctuations are then

$F_\mu(t)$ is independent of all $O_\mu(s)$ for $t>s$;

$\langle F_\mu(t)\rangle = 0$;

$$\langle F_\mu(t)F_\nu(s)\rangle = 2D_{\mu\nu}\ \delta(t-s). \tag{15.32}$$

In order to determine the generalized diffusion coefficients $D_{\mu\nu}$, we use the Langevin equation (15.31). We obtain $O_\mu(t+\Delta t)$, where Δt is a small time interval, from $O_\mu(t)$ as

$$O_\mu(t+\Delta t) - O_\mu(t) = \int_t^{t+\Delta t} ds\ A_\mu(\{O_\mu\},s) + \int_t^{t+\Delta t} ds\ F_\mu(\{O_\mu\},s)\ ,$$

or

$$\Delta O_\mu \cong A_\mu\ \Delta t + \int_t^{t+\Delta t} ds\ F_\mu(s). \tag{15.33}$$

For the time derivative of the bilinear expression $\langle O_\mu(t)O_\nu(t)\rangle$ we find

$$\begin{aligned}\frac{d}{dt}\langle O_\mu(t)O_\nu(t)\rangle &\cong \frac{1}{\Delta t}[\langle O_\mu(t+\Delta t)O_\nu(t+\Delta t)\rangle - \langle O_\mu(t)O_\nu(t)\rangle] \\ &\cong \frac{1}{\Delta t}[\langle (O_\mu(t) + \Delta O_\mu)(O_\nu(t) + \Delta O_\nu)\rangle - \langle O_\mu(t)O_\nu(t)\rangle] \\ &\cong \frac{1}{\Delta t}[\langle O_\mu(t)\Delta O_\nu\rangle + \langle \Delta O_\mu O_\nu(t)\rangle + \langle \Delta O_\mu \Delta O_\nu\rangle].\end{aligned}$$

Using Eq. (15.33) we get

$$\begin{aligned}\frac{d}{dt}\langle O_\mu(t)O_\nu(t)\rangle \cong\ & \langle O_\mu A_\nu\rangle + \langle A_\mu O_\nu\rangle \\ & + \langle O_\mu(t)\int_t^{t+\Delta t} ds\ F_\nu(s)\rangle + \langle \int_t^{t+\Delta t} ds\ F_\mu(s)O_\nu(t)\rangle \\ & + \langle A_\mu(t)\int_t^{t+\Delta t} ds\ F_\nu(s)\ \rangle + \langle \int_t^{t+\Delta t} ds\ F_\mu(s)A_\nu(t)\rangle \\ & + \frac{1}{\Delta t}\langle \int_t^{t+\Delta t} ds\ F_\mu(s) \int_t^{t+\Delta t} ds'\ F_\nu(s')\rangle\ .\end{aligned} \tag{15.34}$$

The terms in the second line of Eq. (15.34) do not contribute because $O_\mu(t)$ is not correlated with the fluctuations $F_\nu(s)$ for $t<s<t+\Delta t$ (see Eq. (15.32)). The terms in the third lines are of third order in the system-bath interaction and are neglected. Inserting the Markoff assumption, Eq. (15.36), into the last row of Eq. (15.34), we get the result

$$\boxed{2D_{\mu\nu} = \frac{d}{dt}\langle O_\mu O_\nu\rangle - \langle A_\mu O_\nu\rangle - \langle O_\mu A_\nu\rangle}$$

general dissipation-fluctuation theorem (15.35)

According to Eq. (15.35), one has to calculate the time-derivative of the bath-averaged term $\langle O_\mu O_\nu\rangle$ and subtract the expressions which contain the linear dissipation rate A_μ in order to get the diffusion coefficients $D_{\mu\nu}$.

To illustrate the use of Eq. (15.35) we now use the example of spontaneous emission to calculate the diffusion coefficient for $n_{i\mathbf{k}}$ and $P_\mathbf{k}$. The interaction Hamiltonian with the bath of photons is

$$\mathcal{H}_I = \sum \hbar d_{cv}\, (a^\dagger_{c,\mathbf{k}} a_{v\mathbf{k}} B_\lambda + B^\dagger_\lambda a^\dagger_{v\mathbf{k}} a_{c,\mathbf{k}}) , \tag{15.36}$$

where λ describes the continuum of all photon modes. We assume that the bath is in the vacuum state $|0_\lambda\rangle$ without photons, so that we get only spontaneous emission. The change of the population density in the state $c,\mathbf{k}$ calculated in second order perturbation theory is

$$\begin{aligned}\frac{d}{dt}\langle n_{c,\mathbf{k}}\rangle &= -\, d^2_{cv} \sum_\lambda \langle n_{c,\mathbf{k}}(1-n_{v\mathbf{k}})\rangle\, \pi\delta(\epsilon_{ck} - \epsilon_{vk} - \Omega_\lambda) \\ &= \langle \mathbf{A}_{n_{c,\mathbf{k}}}\rangle,\end{aligned} \tag{15.37}$$

in agreement with Fermi's golden rule. In order to calculate the diffusion coefficient $D_{n_{c,\mathbf{k}},n_{c,\mathbf{k}}}$ we have to evaluate

$$\frac{d}{dt}\langle n_{c,\mathbf{k}} n_{c,\mathbf{k}}\rangle = \frac{d}{dt}\langle n_{c,\mathbf{k}}\rangle = \langle A_{n_{c\mathbf{k}}}\rangle,$$

so that

$$2D_{n_{c\mathbf{k}},n_{c\mathbf{k}}} = d^2_{cv} \sum_\lambda \langle n_{c,\mathbf{k}}(1-n_{v\mathbf{k}})\rangle\, \pi\delta(\epsilon_{ck} - \epsilon_{vk} - \Omega_\lambda) \tag{15.38}$$

The result is the typical shot noise for population variables. In general, shot noise consists of the sum of all transition rates in and out of the considered state. Because we assumed that no photons are present in the bath, only emission processes, i.e., transitions out of the state $c,\mathbf{k}$ are possible. In local equilibrium, where $\langle n_{i\mathbf{k}} \rangle = f_{ik}$, we have to consider only the total number of electrons in one band. Because optical transitions at different $\mathbf{k}$-values are not correlated, one gets for the fluctuations of the total electron density N_c the result

$$2D_{N_c,N_c} = d_{cv}^2 \sum_{\lambda,\mathbf{k}} f_{ck}(1-f_{vk})\ \pi\delta(\epsilon_{ck} - \epsilon_{v\kappa} - \Omega_\lambda)\ , \tag{15.39}$$

which is just the total spontaneous transition rate. Similarly, one calculates the auto-correlation of the fluctuations of the polarization P_k

$$2D_{P_\mathbf{k},P_\mathbf{k}^\dagger} + 2D_{P_\mathbf{k}^\dagger,P_\mathbf{k}} = 2\gamma_k\{\ f_{ck}(1-f_{vk}) + f_{vk}(1-f_{ck})\ \}\ . \tag{15.40}$$

The polarization fluctuations (15.40) determine, as will be seen, mainly the coherence properties of the semiconductor laser light.

The various dissipative in the carrier rate equation (15.5) can all be modeled by a coupling to a suitable bath. The resulting noise sources all have shot noise character [see e.g. Haug, 1967, 1969]

$$\frac{dN}{dt} = r_p - r_{st} - r_{sp} - r_{nr} + F_p + F_{st} + F_{sp} + F_{nr}\ . \tag{15.41}$$

The correlations of the fluctuations are discussed below.

Diode lasers are pumped by an injection current, which is driven by a voltage source via a serial resistor R_s. If the serial resistor is larger than the differential resistance of the diode, R_s suppresses the current fluctuations in the diode. The resulting pump noise is, as Yamamoto and Machida (1987) have shown, simply the Nyquist noise in the serial resistor

$$\langle F_p(t)\ F_p(s)\rangle = \frac{4\ kT}{e^2L^3R_s}\ \ \delta(t - s)\ . \tag{15.42}$$

Obviously by increasing R_s one can essentially suppress the pump noise.

For the stimulated emission we obtain from the generalized dissipation fluctuation theorem, Eq. (15.42),

$$\langle F_{st}(t)\, F_{st}(s)\rangle = \frac{|\mathcal{E}_0(t)|^2}{2L^3}\, [\chi''_a(\omega) + \chi''_e(\omega)]\, \delta(t-s) \,, \tag{15.43}$$

where we used Eqs. (15.3) and (15.9). In the evaluation of the various terms in Eq. (15.43) we had to take the expression for χ before the expectation values were taken, i.e.,

$$(1 - f_{e,k} - f_{h,k}) \rightarrow a_{h,k} a^{\dagger}_{h,k} - a^{\dagger}_{e,k} a_{e,k} \,. \tag{15.44}$$

A close inspection of Eq. (15.43) shows that $\langle F_{st}(t)\, F_{st}(s)\rangle$ is proportional to

$$f_e f_h + (1-f_e)\,(1-f_h) \,,$$

which describes the sum of the interband transition rates due to the laser action. These contributions arise from the polarization fluctuations through the adiabatically eliminated interband-polarization.

The noise terms due to spontaneous emission have the correlation

$$\langle F_{sp}(t)\, F_{sp}(s)\rangle = \frac{r_{sp}}{L^3}\, \delta(t - s) \,, \tag{15.45}$$

compare Eq. (15.39), and the nonradiative transition noise yields

$$\langle F_{nr}(t)\, F_{nr}(s)\rangle = \frac{r_{nr}}{L^3}\, \delta(t - s) \,. \tag{15.46}$$

15-3. Stochastic Field Equations

In the semiclassical description of the Langevin equations for the laser light we follow essentially the discussion of Vahala and Yariv (1983). We start from the wave equation (14.1) and include an additional term

$$-\frac{4\pi\sigma\epsilon_0}{c^2}\,\frac{\partial \mathcal{E}}{\partial t} \,,$$

where σ is the conductivity, on the LHS of Eq. (14.1) to model the losses of the laser cavity. In the simplest way, the cavity of a semiconductor laser is the resonator formed by the two parallel cleaved end faces of the crystal.

Next, we expand the field in terms of orthonormal spatial modes $\mathbf{u}_n(\mathbf{r})$, i.e.,

$$E(r,t) = \sum_n \mathcal{E}_n(t)\, u_n(r) \ , \tag{15.47}$$

and equivalently for the polarization. The spatial eigenmodes fulfill the condition

$$\nabla^2\, \mathbf{u}_n(\mathbf{r}) = -\, \frac{\omega_n^2 \epsilon_0}{c^2}\, \mathbf{u}_n(\mathbf{r}) \ , \tag{15.48}$$

where ω_n is the eigenfrequency of the nth resonator mode.

Expressing the polarization through the carrier-density-dependent susceptibility and the field,

$$P_n = \chi(N)\, \mathcal{E}_n \ , \tag{15.49}$$

we obtain

$$\frac{d^2}{dt^2} \left\{ \left[1 + \left(\frac{4\pi}{\epsilon_0} \right) \chi(N) \right] \mathcal{E}_n \right\} + \kappa\, \frac{d\mathcal{E}_n}{dt} + \omega_n^2\, \mathcal{E}_n = 0 \ , \tag{15.50}$$

where we introduced $\kappa = 4\pi\sigma/\epsilon_0$ as the cavity loss rate and $\chi(N{=}0) = 0$, since the background susceptibility has been included in ϵ_0.

Now we express all quantities in Eq. (15.50) in terms of their mean values and slowly varying amplitude and phase perturbations, i.e.,

$$\mathcal{E}_n = \frac{1}{2}\,[A_0 + A(t)]\; e^{-i(\omega_m t\; -\; \phi(t))}$$

$$N = N_0 + n(t)$$

$$\chi(N) = \chi(N_0) + \xi n(t), \tag{15.51}$$

where

$$\xi = \left.\frac{\partial\chi}{\partial N}\right|_{N=N_0} .$$

We insert (15.51) into Eq. (15.50) and linearize the resulting equation in terms of the perturbations $A(t)$, $\phi(t)$ and $n(t)$. As described in problem (15.3), we obtain

$$i\omega_m(\dot{A}+iA_0\dot{\phi}) + i\omega_m A_0\zeta\,\dot{n} + \omega_m^2\frac{\zeta A_0}{2}\,n$$

$$+ \left[\omega_m^2 - \omega_n^2 + i\omega_m\kappa + \frac{4\pi\omega_m^2}{\epsilon_0}\,\chi(N_0)\right]\frac{A_0}{2} = F(t)\;, \tag{15.52}$$

where $\zeta = 4\pi\xi/\epsilon_0$. On the RHS of Eq. (15.52) we added the classical Markoffian noise term $F(t) = F'(t) + i\,F''(t)$ with

$$\langle F(t)\rangle = 0$$

$$\langle F'(t)F'(t')\rangle = \langle F''(t)F''(t')\rangle = W\;\delta(t-t')$$

$$\langle F'(t)F''(t)\rangle = 0\;, \tag{15.53}$$

where W is given by the rate of spontaneous emission into the laser mode. The expression for W can be determined from the quantum mechanical theory of Sec. 15-2.

Eq. (15.52) describes the field and phase perturbations which are coupled to the carrier density perturbations $n(t)$. Introducing the expansion (15.51) into the Langevin equation (15.41) for the carrier density yields

$$\dot{n} + \frac{\xi''}{2} A_0^2 \, n - \chi''(N_0) A_0 A + \frac{n}{\tau}$$

$$- \frac{\chi''(N_0)}{2} A_0^2 + \frac{N_0}{\tau} - r_p = F_n \, , \tag{15.54}$$

where, for simplicity we omitted the spontaneous emission and the nonlinear nonradiative recombination term CN^3. F_n is the sum of the relevant noise contributions.

The perturbations of the carrier density and the field amplitude and phase have zero mean values. Therefore, taking the bath average of the coupled Eqs. (15.52) and (15.54) yields

$$\kappa = - \frac{4\pi\omega_m}{\epsilon_0} \chi''(N_0) \, , \tag{15.55}$$

$$\omega_m^2 = \frac{\omega_n^2}{1 - 4\pi\chi'(N_0)/\epsilon_0} \tag{15.56}$$

$$N_0 = \tau \left[r_p + \frac{\chi''(N_0)}{2} A_0^2 \right] . \tag{15.57}$$

These equations determine the operating point of the laser. Especially Eq. (15.55) describes the laser threshold, i.e., the point at which the gain equals the losses,

$$\kappa = g'(\omega_m) \, , \tag{15.58}$$

where Eq. (15.3) has been used. Assuming an originally unexcited system without electrons or holes, we increase the plasma density by increasing the pump current density j. The increasing plasma density then leads to growing gain g', until the gain equals the losses and the laser threshold is reached. Clearly, the condition (15.58) is fulfilled first at that frequency which corresponds to the gain maximum. However, since the cavity modes ω_n are discretely spaced, one may not have an eigenmode directly at the gain maximum. In this case the mode closest to the gain maximum starts to lase. The actual laser frequency ω_m is determined from Eq. (15.56).

In our simple spatially homogeneous model, in which spectral hole burning is impossible due to the rapid carrier-carrier scattering, the plasma an support only one stable laser mode. Once the threshold of the first laser mode is reached, increased pumping does not increase the plasma density, but rather leads to an increase of the intensity of the lasing mode. At this point it should be noted that the laser equations derived in this chapter are valid only if the field E is not too strong, since we did not include any high-field effects. Intensive laser beams may very well contribute to the energy renormalization for the electrons and holes. These radiative self-energy corrections describe the spectral hole burning, which for lasers gives rise to an extra gain saturation at high fields.

From Eqs. (15.52) and (15.54) we can now obtain three coupled equations which link the amplitude, phase and carrier-density perturbations to the Langevin noise terms

$$\dot{\rho} + \zeta' \, \dot{n} + \frac{\omega_m \zeta''}{2} \, n = \frac{F''}{\omega_m A_0} \tag{15.59}$$

$$\dot{\phi} + \xi'' \, \dot{n} - \frac{\omega_m \zeta'}{2} \, n = - \frac{F'}{\omega_m A_0} \tag{15.60}$$

$$\dot{n} + \frac{n}{\tau_R} - \frac{2\omega_R^2}{\omega_m \zeta''} \, \rho = F_n \; . \tag{15.61}$$

Here we have defined

$$\rho = \frac{A}{A_0} \; , \; \frac{1}{\tau_R} = \frac{1}{\tau} - 2A_0^2 \, \xi'' \; , \; \omega_R^2 = \frac{1}{2} \, A_0^2 \, \omega_m \, \zeta'' \chi''(N_0) \; ,$$

where ω_R and τ_R are the frequency and damping time of the carrier-density relaxation oscillations.

Eqs. (15.59) - (15.61) are coupled first-order differential equations which can be solved by Laplace transformation, see problem (15.4). From the resulting expressions for $\rho(t)$, $\phi(t)$, and $n(t)$ we can compute the autocorrelation functions, such as $\langle \rho(t+\tau)\rho(t) \rangle$. We obtain for the phase autocorrelation function, see problem (15.5),

$$\langle \phi(t+\tau)\phi(t) \rangle = \frac{W}{\omega_m^2 \, A_0^2} (\, 1 + \alpha^2) \, t$$
$$+ \left[\, B_1 cos(\beta\tau) + B_2 sin(\beta|\tau|) \, \right] e^{-|\tau|/2\tau_R} \; , \tag{15.62}$$

where B_1 and B_2 are combinations of the constants entering Eqs.

(15.59) - (15.61),

$$\beta^2 = \left[\omega_R^2 - \frac{1}{4\tau_R^2}\right]$$

and

$$\alpha = \frac{\zeta'(N_0)}{\zeta''(N_0)} = \frac{\partial\chi'/\partial N_0}{\partial\chi''/\partial N_0} \,. \tag{15.63}$$

is the so-called linewidth enhancement factor. In semiconductor lasers α is particularly large, because at the gain maximum the density dependent dispersive changes are larger than the gain changes. The laser linewidth is obtained from

$$\begin{aligned}
\langle \mathcal{E}^*(t+\tau)\, \mathcal{E}(t)\rangle &\cong \frac{A_0^2}{4}\, e^{i\omega_m\tau}\; i\langle \exp\{i[\phi(t+\tau)-\phi(t)]\}\rangle \\
&\cong \frac{A_0^2}{4}\, e^{i\omega_m\tau}\; \exp\left\{-\frac{1}{2}\langle[\phi(t+\tau)-\phi(t)]^2\rangle\right\} \\
&\cong \frac{A_0^2}{4}\, e^{i\omega_m\tau}\; \exp\left[\,-\frac{W}{2\omega_m^2 A_0^2}\,(1+\alpha^2)\,|\tau| \,+\right. \\
&\qquad \left. +\, e^{-|\tau|/\tau_R}\,[B_1 cos(\beta\tau) + B_2 sin(\beta|\tau|)] - B_1\right] .
\end{aligned} \tag{15.64}$$

Assuming for the moment $|\tau| \gg \tau_R$, we obtain from Eq. (15.64)

$$\langle \mathcal{E}^*(t+\tau)\, \mathcal{E}(t)\rangle \cong \frac{A_0^2}{4}\, e^{i\omega_m\tau} \exp\left[\,-\frac{W}{2\omega_m^2 A_0^2}\,(1+\alpha^2)\,|\tau|\,\right] . \tag{15.65}$$

Fourier transformation of Eq. (15.65) yields the field spectrum

$$S_E(\omega) = \int_{-\infty}^{\infty} d\tau \; e^{-i\omega\tau} \; \langle \mathcal{E}^*(\tau) \; \mathcal{E}(0) \rangle$$

$$= \frac{A_0^2}{4} \frac{\Delta\omega}{(\omega-\omega_m)^2 + (\Delta\omega/2)^2} , \tag{15.66}$$

where

$$\Delta\omega = \frac{W}{\omega_m^2 A_0^2} (1 + \alpha^2) . \tag{15.67}$$

Eq. (15.67) is the Schawlow-Townes linewidth formula, except for the additional term α^2. The semiconductor laser linewidth (15.67) has been derived first by Haug and Haken (1967), but only later has it been recognized that the linewidth factor α^2 in semiconductors is much larger than unity, because the density-dependent refractive changes of the complex gain function are quite large in semiconductors as compared to atomic systems.

Fourier transformation of the full correlation function (15.64) is not possible analytically. In order to show the basic features, we follow Vahala and Yariv (1983). We ignore the term proportional B_2 and approximate $\exp(-|\tau|/\tau_R) \cong 1$, $\beta \cong \omega_R$, assuming weakly damped relaxation oscillations. We use the associated series (Abramowitz and Stegun, 1972) for the modified Bessel functions I_n to write

$$\exp\{B_1[\cos(\omega_R\tau)-1]\} = I_0(B_1) + 2 \sum_{n=1}^{\infty} I_n(B_1) \cos(n\omega_R\tau) . \tag{15.68}$$

Inserting our approximations and Eq. (15.68) into Eq. (15.64), and evaluating Eq. (15.66) yields

$$S_E(\omega) = \frac{A_0^2}{4} \Delta\omega \sum_{n=-\infty}^{\infty} \frac{e^{-B_1} I_n(B_1)}{(\omega-\omega_m-n\omega_R)^2 + (\Delta\omega/2)^2} , \tag{15.69}$$

showing that the laser spectrum consists of a series of lines at $\omega=\omega_m+n\omega_R$. An example of the spectrum is shown in Fig. 15.1. The

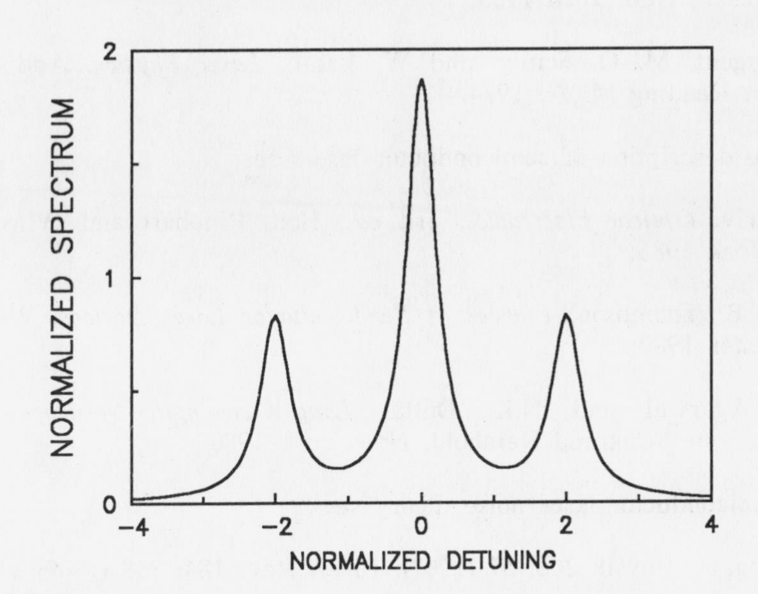

Fig. 15.1: Plot of the main mode and the first two sidemodes of the normalized laser spectrum, Eq. (15.66). For illustration we choose the frequency of the relaxation oscillations $\omega_R = 2$, $B_1 = 1$, $\Delta\omega/\omega_R = 0.25$.

main peak is at ω_m and the sidebands occur at multiples of the carrier relaxation oscillation frequency. Such spectra, usually with stronger damped sidemodes, are indeed observed in semiconductor lasers above threshold, see, e.g., Thompson (1980) or Agarwal and Dutta (1986).

REFERENCES

M. Abramowitz and I.A. Stegun, *Handbook of Mathematical Functions* (Dover Publ., 1972)

For the general laser theory see:

H. Haken, *Laser Theory*, Handbuch der Physik Vol. *XXV/2c*, Springer, Berlin 1970;

M. Lax, in 1966 Brandeis University Summer Institute of Theoretical

Physics Vol. II, eds. M. Chretien, E.P. Gross, and S. Deser, Gordon and Breach, New York 1968;

M. Sargent, M. O. Scully, and W. Lamb, *Laser Physics*, Addison-Wesley, Reading M. A. 1974.

For the description of semiconductor lasers see:

A. Yariv, *Optical Electronics*, 3rd *ed.*, Holt, Rinehart and Winston, New York 1985;

G. H. B. Thompson, *Physics of Semiconductor Laser Devices*, Wiley, Chichester 1980;

G.P. Agarwal and N.K. Dutta, *Long-Wavelength Semiconductor Lasers*, Van Norstrand Reinhold, New York 1986.

For semiconductor laser noise theory see

H. Haug, Z. Physik 200, 57 (1967); Phys. Rev. 184, 338 (1969)

H. Haug and H. Haken, Z. Physik 204, 262 (1967)

K. Vahala and A. Yariv, IEEE-QE 19, 1096 and 1102 (1983)

Y. Yamamoto and S. Machida, Phys. Rev. A 35, 5114 (1987)

H. Haug and S.W. Koch, Phys. Rev. A 39, 1887 (1989).

PROBLEMS

Problem 15.1: Show that the commutator relation of the harmonic oscillator is violated if one includes only dissipation in the Heisenberg equation for b and $b^\dagger$.

Problem 15.2: Use the general fluctuation-dissipation theorem, Eq. (15.35), to derive the diffusion coefficients, Eqs. (15.38) - (15.40), of the harmonic oscillator.

Problem 15.3: Use the expansion (15.51) to derive Eq. (15.52) from

Maxwell's wave equation including the loss term $- 4\pi\sigma\epsilon_0/c^2 \; \partial\mathcal{E}/\partial t$. Hint: Neglect all second-order derivatives of the perturbations, all products of perturbations, terms proportional to $(\omega_m^2-\omega_n^2)A$, σA, $\alpha\phi$, and use

$$[1 + 4\pi\chi(N_0)/\epsilon_0 \;] \begin{bmatrix} A \\ \phi \end{bmatrix} \cong \begin{bmatrix} A \\ \phi \end{bmatrix} .$$

Problem 15.4: Solve Eqs. (15.59) - (15.61) using Laplace transformations.

Problem 15.5: Use the solutions of problem (15.4) to compute $\langle\phi(t+\tau)\phi(t)\rangle$. Keep decaying terms $\propto \exp(-\omega|\tau|)$, but neglect all terms proportional $\exp(-\omega t)$, for any $\omega > 0$.

Chapter 16
OPTICAL STARK EFFECT IN SEMICONDUCTORS

In Sec. 2-3. we present an elementary treatment of the optical Stark shift in a two-level system. This effect is caused by a coherent light field which mixes the wavefunctions of the two states introducing the so-called dressed states. While these effects are well-known in atomic systems, they have been observed only recently in semiconductors since the dephasing times in semiconductors are much shorter than in atomic systems. Therefore, one has to use very short laser pulses in order to study these coherent optical phenomena in semiconductors.

The dephasing times are usually considerably larger for a non-resonantly driven polarization. In our discussion of the optical Stark effect in semiconductors we therefore concentrate on the case of nonresonant excitation of the exciton, where we can ignore absorption and generation of real carriers. In the following sections, we first discuss the stationary properties of the optical Stark effect and then we analyze the dynamic modifications.

16-1. Stationary Results

In this section we present the analysis of the stationary optical Stark effect. The stationary condition applies only when the amplitude variations of the light field are so slow, that we can make an adiabatic approximation. For femtosecond experiments this procedure is not valid. Nevertheless, we start our discussion with the stationary case in order to understand the similarities and differences of this coherent phenomenon in atomic and semiconductor systems. Particularly, we analyze analytically the role of the many-body effects in semiconductors. The dynamical solutions for pulsed excitation are discussed in the following section.

As an introduction, we first reformulate the treatment of Chap. 2 of the optical Stark shift of a two-level atom in terms of the Bloch equations for the polarization and the population. In second quantization the Hamiltonian for the two-level system is

$$H = \sum_{j=1,2} \hbar\epsilon_j a_j^\dagger a_j - [d_{21}\mathcal{E}(t) a_2^\dagger a_1 + \text{h.c.}], \tag{16.1}$$

with the coherent pump field $\mathcal{E}(t) = \mathcal{E}_p e^{-i\omega_p t}$. Via the Heisenberg equation for the operators we get the following equation for the polarization $P = \langle a_1^\dagger a_2 \rangle$ and the density in the upper state $n = n_2 = \langle a_2^\dagger a_2 \rangle = 1 - n_1$:

$$i\dot{P} = \epsilon P - (1-2n) d_{21}\mathcal{E}_p \tag{16.2}$$

$$\dot{n} = \frac{i}{\hbar}(d_{21}\mathcal{E}_p P^* - \text{h.c.}) \tag{16.3}$$

with $\epsilon = \epsilon_2 - \epsilon_1$. These two completely coherent equations (no damping terms) have a conserved quantity (see problem 16.1)

$$K = (1-2n)^2 + 4|P|^2$$

with the initial condition n=0 and P=0 we get K=1, or

$$n = \frac{1}{2}(1 \pm \sqrt{1-4|P|^2}) \; . \tag{16.4}$$

Eq. (16.4) shows that the density is completely determined by the polarization. This is naturally only true for a fully coherent process, or in the language of quantum mechanics for *virtual excitations*. These excitations of the atom vanish if the field is switched off, whereas real excitations would stay in the system and would decay on a much longer timescale determined by the carrier lifetime. From Eqs. (16.2) - (16.4) one can again derive the results of Chap. 2 (see problem 16.2).

Next, we turn to the coherent Bloch equations of the semiconductor which have been derived in Chap. 12, see Eqs. (12.15) - (12.17). Here we include no damping or collision terms, which is clearly not valid for resonant excitation, where real absorption occurs. With $n_k = n_{c,k} = 1 - n_{v,k}$ we can write Eqs. (12.15) - (12.17) as

$$i\dot{P}_k = e_k P_k - (1-2n_k)W_k \tag{16.5}$$

$$\dot{n}_k = i\,(W_k P_k^* - W_k^* P_k)\ , \tag{16.6}$$

where e_k is the pair energy renormalized by the exchange energy

$$\hbar e_k = \hbar(e_{c,k} - e_{v,k}) = E_g + \frac{\hbar k^2}{2m} - 2\sum_{k'} V_{k-k'} n_{k'} \tag{16.7}$$

and W_k is the renormalized Rabi frequency

$$W_k = \frac{1}{\hbar}\left[d_{cv}\mathcal{E}(t) + \sum_{k'} V_{k-k'} P_{k'} \right]. \tag{16.8}$$

Note that there is a complete formal analogy between the two-level atom equations (16.2) - (16.3) and the semiconductor equations (16.5) - (16.6) for each k-state, except for the renormalizations of the pair energy, Eq. (16.7), and of the Rabi frequency, Eq. (16.8), which mix the k-states in a complicated way. From this analogy we get immediately the conservation law

$$n_k = \frac{1}{2}\,(1 \pm \sqrt{1-4|P_k|^2}) \tag{16.9}$$

If the fields are switched on adiabatically only the minus sign of (16.9) can be realized and the relations $0 \le n_k \le 1/2$ and $0 \le |P_k|^2 \le 1/2$ hold.

Solving Eqs. (16.5) and (16.9) adiabatically, i.e., neglecting all possible slow amplitude variations, we find for $\mathcal{E}(t) = \mathcal{E}_p \exp(-i\omega_p t)$

$$P_k = \frac{(1-2n_k)\, W_k}{\hbar(e_k - \omega_p)}\ . \tag{16.10}$$

Eq. (16.10) together with Eqs. (16.6) - (16.9) form a complicated system of nonlinear integral equations, which can be solved numerically or which has to be simplified by further approximations.

In experiments, one usually applies a weak test beam to measure the effects which the strong pump beam introduces in the semiconductor (see Sec. 16-2. and Fig. 16.3 for more details). To study this situation, we add to the pump beam a weak test beam $\mathcal{E}_t \exp(-i\omega_t t)$ which induces an additional small polarization δP_k. Linearizing Eqs. (16.5) - (16.9) yields

$$i\delta\dot{P}_k = \delta e_k P_k + e_k \delta P_k + 2\delta n_k W_k - (1-2n_k)\delta W_k \;, \tag{16.11}$$

where

$$\delta e_k = -\frac{2}{\hbar}\sum_{k'} V_{k-k'}\delta n_{k'}, \tag{16.12}$$

and

$$\delta n_k = \frac{P_k \delta P_k^* + P_k^* \delta P_k}{1-2n_k}, \tag{16.13}$$

and

$$\delta W_k = \frac{1}{\hbar}\left[d_{cv}\mathcal{E}_t e^{-i\omega_t t} + \sum_{k'} V_{k-k'}\delta P_{k'} \right]. \tag{16.14}$$

We now eliminate the time dependence of the pump field by splitting off a factor $\exp(-i\omega_p t)$, e.g., $P_k = p_k \exp(-i\omega_p t)$, similarly with δP_k and W_k, etc., where we always use lower case letters to indicate that the time variation with the pump field frequency has been split off. This way we obtain

$$i\delta\dot{p}_k = \delta e_k p_k + (e_k-\omega_p)\delta p_k + 2\delta n_k \omega_k - (1-2n_k)\delta\omega_k \;, \tag{16.15}$$

with

$$\delta\omega_k = \frac{1}{\hbar}\left[d_{cv}\mathcal{E}_t e^{-i\Delta t} + \sum_{k'} V_{k-k'}\delta p_{k'} \right], \tag{16.16}$$

where $\Delta = \omega_p - \omega_t$ is the frequency difference of pump and test beam. This system of equations can be solved if the solutions of P_k and n_k under the influence of the strong pump beam alone are known. To get a stationary problem we have to choose the form

$$\delta p_k = \delta p_k^+ e^{+i\Delta t} + \delta p_k^- e^{-i\Delta t} \tag{16.17}$$

Once δp_k^+ is known we get in the usual way the susceptibility and the absorption spectrum of the test beam. Particularly, one is interested to see how the exciton absorption spectrum is influenced by a pump beam which is detuned far below the lowest exciton resonance. In order to get a realistic absorption spectrum, one takes a finite damping for the test-beam induced polarization δP_k into account. Note again, that a purely coherent equation for the pump-beam induced P_k, and a dissipative equation for δP_k are a physically justified model, because of the rapidly decreasing damping with increasing detuning, i.e., the frequency-dependent dephasing. In Fig. 16.1 we show the results of a numerical evaluation of the stationary equations for a quasi-two-dimensional GaAs quantum-well structure. The detuning of the pump beam with respect to the *1s*-exciton was chosen as 10 exciton Rydberg energies. One sees clearly the blue shift of the exciton resonance with increasing pump intensity, but surprisingly the band edge shifts similarly to higher energies and the oscillator strength of the exciton does not decrease.

In order to analyze these numerical results we investigate the theory up to linear order in the pump intensity analytically. Thus $\mathcal{E}_p$, P_k and n_k can be considered as small expansion parameters. From Eq. (16.9) we find

$$n_k = |P_k|^2 \tag{16.18}$$

The quasi-stationary equation for the polarization induced by the test beam reduces with $\delta p_k^+ >> \delta p_k^-$ to (see problem 16.3)

$$\sum_{k'} \left[H^0_{kk'} + \Delta H_{kk'} - \hbar(\omega_t + i\delta)\delta_{kk'} \right] \delta p^+_{k'} = (1-2|p_k|^2)\, d_{cv}\mathcal{E}_t, \tag{16.19}$$

where H^0 is the unperturbed pair Hamiltonian

$$H^0_{kk'} = \left[E_g + \frac{\hbar^2 k^2}{2m} \right] \delta_{kk'} - V_{k-k'}\,, \tag{16.20}$$

and

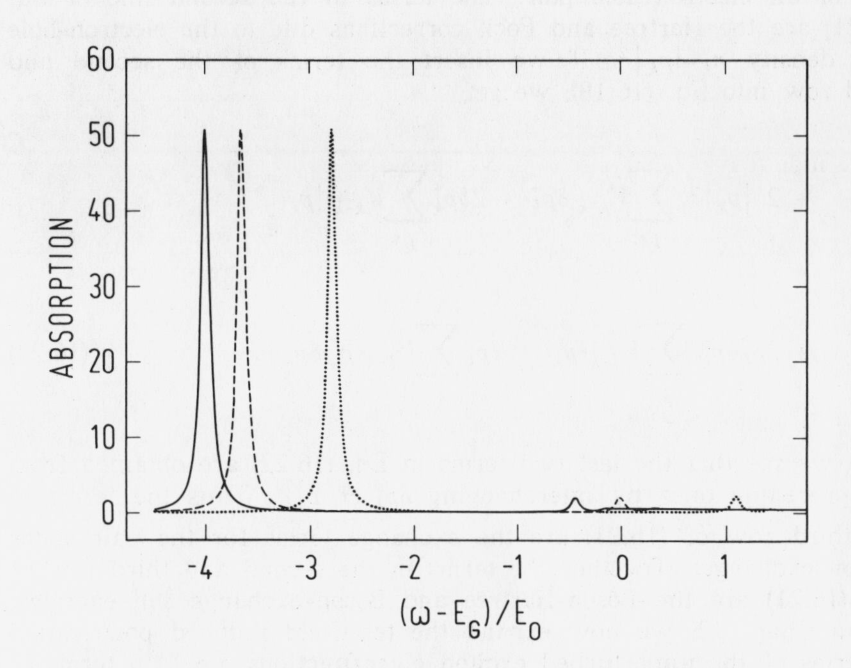

Fig. 16.1: Calculated *2d* absorption spectrum according to Ell *et al.* for $(E_g - \omega_p)/E_0 = 10$ and the pump intensities $I_p = 0$ (full line), 7.5 (---), and 30 MW/cm^2 (····).

$$\Delta H_{kk'} = 2\delta_{kk'} d_{cv} \mathcal{E}_p p_k^*$$

$$+ 2|p_k|^2 V_{k-k'} - 2\delta_{kk'} \sum_{k''} V_{k-k''} |p_{k''}|^2$$

$$+ 2\delta_{kk'} p_k^* \sum_{k''} V_{k-k''} p_{k''} - 2p_k V_{k-k'} p_{k'}^* . \tag{16.21}$$

The first term of the perturbation Hamiltonian ΔH describes the optical Stark shift in a non-interacting two-level system. $\mathcal{E}_p p^*_k$ corresponds to the annihilation of a pump photon and the virtual creation of an electron-hole pair. The terms in the second line of Eq. (16.21) are the Hartree and Fock corrections due to the electron-hole pair density $n_k = |p_k|^2$. If we insert the terms of the second and third row into Eq. (16.19), we get

$$+ 2\,|p_k|^2 \sum_{k'} V_{k-k'} \delta p^+_{k'} - 2\delta p^+_k \sum_{k'} V_{k-k'} |p_{k'}|^2$$

$$+ 2p^*_k \delta p^+_k \sum_{k'} V_{k-k'} p_{k'} - 2p_k \sum_{k'} V_{k-k'} p^*_{k'} \delta p_{k'} \; . \tag{16.22}$$

Now we see that the last two terms in Eq. (16.22) are obtained from the preceding ones by interchanging $\delta p^+_k \leftrightarrows p_k$. Thus the terms in the third row of (16.21) are the exchange terms for the pair states (Boson exchange). Together the terms in the second and third row of Eq. (16.21) are the Boson-Hartree and Boson-exchange self-energies. As in Chap. 12, we now expand the test-field induced polarization in terms of the unperturbed exciton eigenfunctions, i.e. , in terms of the eigenfunctions of $\mathcal{H}^0$, Eq. (16.20),

$$\delta p^+_k = \sum_\lambda \delta p^+_\lambda \, \psi_{\lambda k} \tag{16.23}$$

and get in the exciton representation by multiplying (16.19) with $\psi^*_{\lambda k}$ from the LHS and summing over all k vectors:

$$\sum_\lambda \hbar \left[(\omega_\lambda - \omega_t - i\delta)\delta_{\lambda\lambda'} + \Delta H_{\lambda\lambda'}\right] \; \delta p^+_{\lambda'} = \sum_k \psi_{\lambda k}(1 - 2|\,p_k|^2) d_{cv} \mathcal{E}_p . \tag{16.24}$$

The perturbation Hamiltonian is given by

$$\Delta H_{\lambda\lambda'} = \Pi_{\lambda\lambda'} + \Delta_{\lambda\lambda'} \; , \tag{16.25}$$

where $\Pi_{\lambda\lambda'}$ is the anharmonic exciton - pump-field interaction

$$\Pi_{\lambda\lambda'} = 2\, \mathcal{E}_p \sum_k \psi^*_{\lambda k} d_{cv}\; p^*_k\; \psi_{\lambda' k} \;, \tag{16.26}$$

and $\Delta_{\lambda\lambda'}$ is the exciton-exciton interaction

$$\Delta_{\lambda\lambda'} = 2 \sum_{kk'} V_{k-k'} \psi^*_{\lambda k} (p^*_k \; - \; p^*_{k'})(p_k \; \psi_{\lambda' k'} \; + \; p_{k'} \; \psi_{\lambda' k}) \; . \tag{16.27}$$

We can rewrite Eq. (16.24) as

$$\hbar \delta p^+_\lambda = \frac{\sum_k \psi^+_{\lambda k}(1-2|\,p_k\,|^2) d_{cv} \mathcal{E}_p \; - \; \sum_{\lambda' \neq \lambda} \Delta H_{\lambda\lambda'} \delta p^+_{\lambda'}}{\overline{\omega}_\lambda - \omega_t - i\delta} \; , \tag{16.28}$$

where $\overline{\omega}_\lambda$ is the renormalized exciton frequency

$$\overline{\omega}_\lambda = \omega_\lambda + \Delta H_{\lambda\lambda} \; . \tag{16.29}$$

Eq. (16.28) can be solved iteratively with

$$\hbar \delta p^{+(1)}_\lambda = \frac{\sum_k \psi^+_{\lambda k}(1-2|\,p_k\,|^2) d_{cv} \mathcal{E}_p}{\overline{\omega}_\lambda - \omega_t - i\delta} \; . \tag{16.30}$$

Note that in this procedure even the first order contains already the shifted exciton energies $\hbar\overline{\omega}_\lambda$, i.e. , the Stark shift as well as the phase space filling. In the next order one finds

$$\hbar \delta p^{+(2)}_\lambda = \hbar \delta p^{+(1)}_\lambda + \frac{\sum_{\lambda'} \Delta H_{\lambda\lambda'} \delta p^{+(1)}_{\lambda'}}{\overline{\omega}_{\lambda'} - \omega_t - i\delta} \; . \tag{16.31}$$

The linear optical susceptibility of the test beam is finally obtained as

$$\chi_t(\omega_t) = 2 \sum_\lambda \frac{d_\lambda \, \delta p_\lambda^+}{\mathcal{E}_t} , \tag{16.32}$$

where

$$d_\lambda = \sum_k d_{cv} \psi_{\lambda k}^* . \tag{16.33}$$

The resulting susceptibility has the form (see problem 16.4)

$$\chi_t(\omega_t) = \frac{2}{\hbar} \sum_\lambda \frac{\bar{f}_\lambda}{\omega_\lambda - \omega_t - i\delta} , \tag{16.34}$$

where $\bar{f}_\lambda$ is the renormalized exciton oscillator strength:

$$\bar{f}_\lambda = |d_\lambda|^2 - 2d_\lambda^* \sum_k \psi_{\lambda k}^* \, |p_k|^2 \, d_{cv} - \sum_{\lambda' \neq \lambda} \frac{d_\lambda^* \Delta H_{\lambda\lambda'} d_{\lambda'} + (\lambda \rightleftarrows \lambda')}{\hbar(\omega_\lambda - \omega_{\lambda'})} . \tag{16.35}$$

Note that the result (16.34) is put into the usual form of the exciton susceptibility of chap. 10, but it contains renormalized exciton energies ω_λ and renormalized oscillator strengths $\bar{f}_\lambda$. In this low-intensity regime the optical Stark effect can be described completely in terms of shifts of the exciton levels and in terms of changes of the exciton oscillator strengths. The first correction term of the oscillator strength in Eq. (16.35) describes the reduction due to phase space filling, while the terms due to the perturbation ΔH describe the corrections due to the anharmonic exciton-photon and due to the exciton-exciton interaction.

In the linear approximation in the pump field the polarization p_k can be written in terms of an exciton Green's function

$$p_k(\omega_p) = - d_{cv} \, \mathcal{E}_p \, G^r(k, \omega_p) , \tag{16.36}$$

with

$$G^r(k,\omega_p) = \sum_\lambda \frac{\psi^*_{\lambda,k}\psi_\lambda(r{=}0)}{\hbar(\omega_\lambda-\omega_p-i\delta)} , \tag{16.37}$$

For small detuning $\omega_p - \omega_{1s} \ll E_0$ the Green's function simplifies to

$$G^r(k,\omega_p) = \frac{\psi^*_{1s,k}\psi_{1s}(r{=}o)}{\hbar(\omega_{1s}-\omega_p-i\delta)} , \tag{16.38}$$

so that

$$\Pi_{\lambda\lambda} = \psi^*_{1s}(r{=}0) \sum_k \psi_{1s,k} |\psi_{\lambda,k}|^2 \frac{2\,|d\mathcal{E}_p|^2}{\hbar(\omega_{1s}-\omega_p)} , \tag{16.39}$$

The Stark shift due to the anharmonic exciton-photon interaction is equal to the usual two-level Stark shift $(2\,|d\mathcal{E}_p|^2)/\hbar(\omega_{1s}-\omega_p)$ times an enhancement factor due to the electron-hole correlation

$$\rho_\lambda = \psi^*_{1s}(r{=}0) \sum_k \psi_{1sk} |\psi_{\lambda k}|^2 . \tag{16.40}$$

For the band edge we find with $|\psi_{\lambda=\infty,k}|^2 = \delta_{k,0}$

$$\rho_\infty = \psi^*_{1s}(r{=}0)\ \psi_{1s,k=0} . \tag{16.41}$$

Using the two and three dimensional exciton wave functions (see Chap. 10)

$$\psi_{1sk} = \frac{\sqrt{2\pi}\ a_0}{(1 + (ka_0/2)^2)^{3/2}}$$

$$\psi_{1sk} = \frac{8\sqrt{\pi a_0^3}}{(1 + (ka_0)^2)^2}$$

we find (see problem 16.4)

$$\rho_{1s} = \begin{bmatrix} 16/7 \\ 7/2 \end{bmatrix} ; \quad \rho_\infty = \begin{bmatrix} 4 \\ 8 \end{bmatrix} \quad \text{for} \quad \begin{bmatrix} 2d \\ 3d \end{bmatrix}. \tag{16.42}$$

For small detuning we obtain the surprising result that the contribution of the anharmonic exciton-photon interaction to the Stark shifts is larger for the band gap than for the exciton.

Similarly, the exciton-exciton interaction $\Delta_{\lambda\lambda}$ can be written for small detuning as

$$\Delta_{\lambda\lambda} = 2|\psi_{1s}^*(r=0)d\mathcal{E}_p|^2 \sum_{kk'} \frac{V_{k-k'}\psi_{\lambda k}^*(\psi_{1sk}^*-\psi_{1sk'}^*)(\psi_{1sk}\psi_{\lambda k'}-\psi_{1sk'}\psi_{\lambda k})}{\hbar^2(\omega_{1s}-\omega_p)^2} \tag{16.43}$$

$$= \nu_\lambda \frac{2|d\mathcal{E}_p|^2}{\hbar(\omega_{1s}-\omega_p)} . \tag{16.44}$$

The enhancement factor ν_λ diverges as $1/(\omega_{1s} - \omega_p)$ for $\omega_p \to \omega_{1s}$. Again the integrals can be evaluated analytically and yield

$$\nu_{1s} = \frac{E_0}{\hbar(\omega_{1s}-\omega_p)} \begin{bmatrix} 64(1-315\pi^2/2^{12}) \\ 26/3 \end{bmatrix} \cong \begin{bmatrix} 15.4 \\ 8.66 \end{bmatrix} \quad \text{for} \quad \begin{bmatrix} 2d \\ 3d \end{bmatrix} \tag{16.45}$$

$$\nu_\infty = \frac{E_0}{\hbar(\omega_{1s}-\omega_p)} \begin{bmatrix} 64(1-3\pi/16) \\ 24 \end{bmatrix} \cong \begin{bmatrix} 26.3 \\ 24 \end{bmatrix} \quad \text{for} \quad \begin{bmatrix} 2d \\ 3d \end{bmatrix} \tag{16.46}$$

Again we see that these contributions to the Stark shifts are larger for the band gap than for the exciton ground state! For general values of the detuning the shifts have to be evaluated numerically. Fig. 16.2 shows the resulting shifts of the exciton and the band gap for varying detuning. For all values of the detuning the blue shift of the band gap is larger than that of the exciton ground state, i.e. the exciton binding energy increases.

Now we are in a position to understand the surprising numerical results for the quasi-stationary Stark effect shown in Fig. 16.1. The phase-space filling due to the action of the pump beam which would result in a reduction of the oscillator strength of the exciton is overcompensated by the anharmonic exciton-photon and

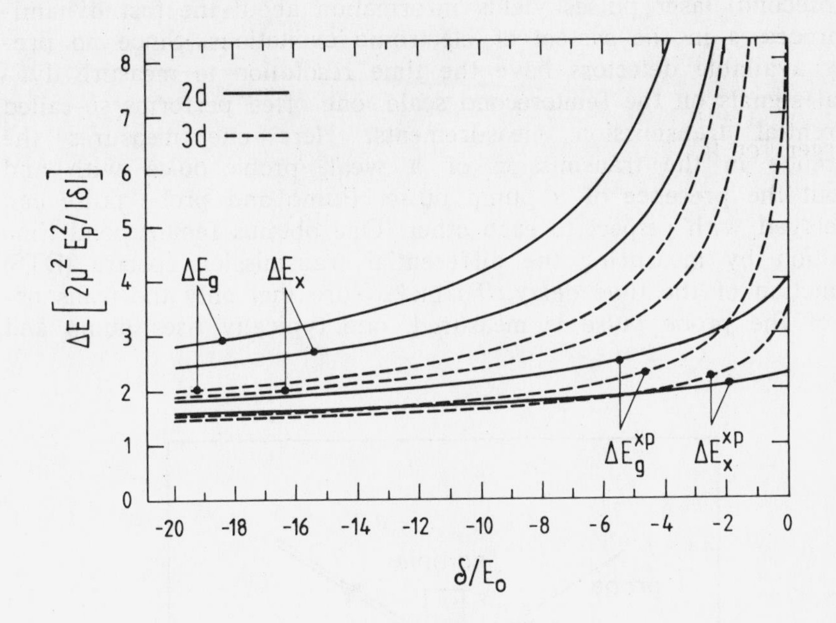

Fig. 16.2: Calculated shifts of the exciton ΔE_x and of the band gap ΔE_g versus the detuning $\delta = \omega_p - E_{1s}$; as well as shifts due to the anharmonic exciton-photon interaction alone, ΔE_x^{xp} and ΔE_g^{xp} respectively. From Haug *et al.* (1988).

exciton-exciton interaction which increase the binding energy and thus the oscillator strength. These conclusions can indeed be verified by evaluating the oscillator strength (16.35) explicitly.

16-2. Dynamic Results

Probing the response of a semiconductor with ultra-short (femtosecond) laser pulses yields information about the fast dynamical processes in the system of electronic excitations. Since no presently available detectors have the time resolution to measure dynamical signals on the femtosecond scale, one often performs so-called differential transmission measurements. Here one measures the difference in the transmission of a weak probe pulse with and without the presence of a pump pulse. Pump and probe pulse can be delayed with respect to each other. One obtains femtosecond time resolution by measuring the differential transmission spectra (DTS) as function of the time delay. To make sure that only the transmission of the probe pulse is measured, one typically uses pump and

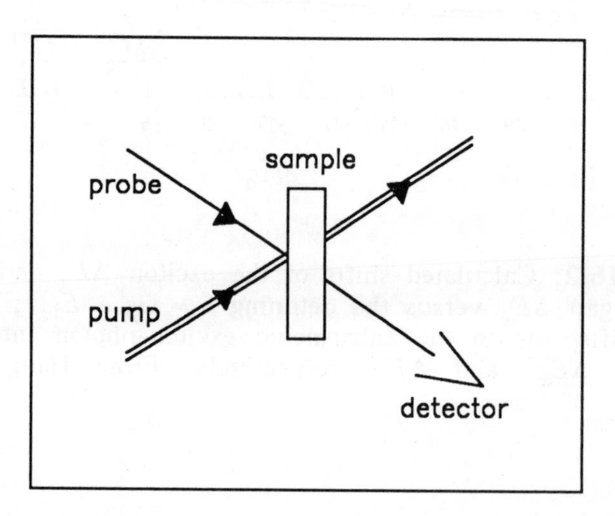

Fig. 16.3: Schematic plot of a femtosecond pump-probe experiment.

probe pulses which do not *co*-propagate (see Fig. (16.3).

The differential transmission spectrum can be computed as

$$DTS(\omega) = \frac{T(\omega) - T_0(\omega)}{T_0(\omega)} \cong - \delta\alpha(\omega)L$$

differential transmission spectrum (16.47)

where L is the sample length, $T(\omega)$ and $T_0(\omega)$ are the probe transmission in the presence and absence of the pump, and $\delta\alpha$ is the pump-induced absorption change, respectively. To relate absorption and transmission in Eq. (16.47), we have used Beer's law

$$T(\omega) = e^{-\alpha(\omega)L} \ . \tag{16.48}$$

Due to the presence of the nonlinear medium, the probe transmission is modified by the pump.

In this section we analyze some of the general features observed in femtosecond pump-probe spectroscopy which are relevant for the optical Stark effect in semiconductors. In the dynamic regime the full semiconductor Bloch equations can only be solved numerically. Therefore, in order to present some analytical insights and to keep the theory as simple as possible, we start from the transformed semiconductor Bloch equations (12.30) in the low excitation coherent regime. We add the appropriate phenomenological damping rates to write the equations for the polarization and density as

$$\frac{\partial}{\partial t} \tilde{P}_\lambda = - (i\epsilon_\lambda + \gamma) \tilde{P}_\lambda - \frac{id_{cv}}{\hbar} \mathcal{E}(t) (2\tilde{n}_\lambda - 1) \tag{16.49}$$

$$\frac{\partial}{\partial t} \tilde{n}_\lambda = - \Gamma\tilde{n}_\lambda + i \frac{d_{cv}}{\hbar} \mathcal{E}(t) \tilde{P}^*_\lambda - i \frac{d^*_{cv}}{\hbar} \mathcal{E}^*(t) \tilde{P}_\lambda \ . \tag{16.50}$$

As discussed above, the total field $\mathcal{E}$ consists of pump field $\mathcal{E}_p$ and test field $\mathcal{E}_t$

$$\mathcal{E}(\mathbf{r},t) = \mathcal{E}_p(\mathbf{r},t) + \mathcal{E}_t(\mathbf{r},t)$$

$$= \mathcal{E}_p(t)\, e^{-i(\mathbf{k}_p \cdot \mathbf{r} + \Omega t)} + \mathcal{E}_t(t)\, e^{-i(\mathbf{k}_t \cdot \mathbf{r} + \Omega t)} \tag{16.51}$$

with $\mathbf{k}_p \neq \mathbf{k}_t$. In order to eliminate the optical frequencies we introduce the notation

$$\tilde{P}_\lambda \equiv e^{-i\Omega t}\, p_\lambda \quad \text{and} \quad w_\lambda \equiv (1 - 2\tilde{n}_\lambda) \,. \tag{16.52}$$

Inserting these definitions into Eqs. (16.49) and (16.50) yields

$$\frac{\partial}{\partial t} p_\lambda = - i[(\epsilon_\lambda - \Omega) + \gamma]\, p_\lambda + i \frac{d_{cv}}{\hbar} [\mathcal{E}_p(t)\, e^{-i\mathbf{k}_p \cdot \mathbf{r}} + \mathcal{E}_t(t)\, e^{-i\mathbf{k}_t \cdot \mathbf{r}}]\, w_\lambda \tag{16.53}$$

$$\frac{\partial}{\partial t} w_\lambda = - \Gamma(w_\lambda - 1) - i \frac{2d_{cv}}{\hbar} [\mathcal{E}_p(t)\, e^{-i\mathbf{k}_p \cdot \mathbf{r}} + \mathcal{E}_t(t)\, e^{-i\mathbf{k}_t \cdot \mathbf{r}}]\, x_\lambda^* + i \frac{2d_{cv}^*}{\hbar} \left[\mathcal{E}_p^*(t)\, e^{i\mathbf{k}_p \cdot \mathbf{r}} + \mathcal{E}_t^*(t)\, e^{i\mathbf{k}_t \cdot \mathbf{r}}\right] p_\lambda \,. \tag{16.54}$$

In order to keep the theory as simple as possible, we assume that $\mathcal{E}_t(t)$ is short on all the relevant time scales, so we can approximate

$$\mathcal{E}_t(t) = \mathcal{E}_t\, \delta(t-t_t). \tag{16.55}$$

This corresponds to a broad frequency spectrum. The experiment measures only that part of the signal which propagates in probe direction. Therefore, we are interested only in that component of the solution for p_λ which has the spatial factor

$$\sim e^{i\mathbf{k}_t \cdot \mathbf{r}}. \tag{16.56}$$

Since we consider the case of an arbitrarily weak probe, we include only terms which are linear in $\mathcal{E}_t$. To obtain analytic results, we ignore all terms which are higher than second order in $\mathcal{E}_p(t)$.

Clearly, for $t < t_t$, we have no signal in the direction of the probe. The only contribution is

$$p_\lambda\,(t < t_t) = i \frac{d_{cv}}{\hbar}\, e^{-i\mathbf{k}_p \cdot \mathbf{r}} \int_{-\infty}^{t} dt'\, e^{-[i(\epsilon_\lambda - \Omega)+\gamma](t-t')}\, \mathcal{E}_p(t')\, w_\lambda(t') \,. \tag{16.57}$$

For that period of time during which the probe is incident on the sample, we can solve Eq. (16.54) as

$$\int_{t_{t_-}}^{t_{t_+}} dt \; \frac{\partial w_\lambda}{\partial t} = w_\lambda(t_{t_+}) - w_\lambda(t_{t_-})$$

$$\cong -i \; \frac{2d_{cv}}{\hbar} \; e^{-i\mathbf{k}_t \cdot \mathbf{r}} \mathcal{E}_t \; x^*_\lambda(t_{t_-}) + i \; \frac{2d^*_{cv}}{\hbar} \; e^{i\mathbf{k}_t \cdot \mathbf{r}} \mathcal{E}^*_t \; p_\lambda(t_{t_-}) + .. \, , \qquad (16.58)$$

where we denote by t_{t_-} and t_{t_+} the times just before and after the probe pulse, respectively. In Eq. (16.58) we used

$$p_\lambda(t_t) = p_\lambda(t_{t_-}) + O(\mathcal{E}_t), \qquad (16.59)$$

where the correction term of order $\mathcal{E}_t$ has been neglected, since we are interested only in terms linear in $\mathcal{E}_t$. The expression for $p_\lambda(t_{t_-})$ is given by Eq. (16.57). Inserting Eq. (16.58) into Eq. (16.57) shows that a grating

$$\sim e^{i(\mathbf{k}_t - \mathbf{k}_p) \cdot \mathbf{r}} \qquad (16.60)$$

is formed in the sample. Light from the pump pulse can scatter from the grating into the direction of the probe and can therefore be seen by the detector. In addition, the probe transmission is also modified through the saturation of the transitions.

To include all effects systematically, we now solve Eqs. (16.53) and (16.54) for the polarization and density variables by expanding them in powers of the fields

$$p_\lambda(t) = i \; \frac{d_{cv}}{\hbar} \int_{-\infty}^{t} dt' e^{-[i(\epsilon_\lambda - \Omega) + \gamma](t-t')} [\mathcal{E}_p(t') e^{-i\mathbf{k}_p \cdot \mathbf{r}} + \mathcal{E}_t(t') e^{-i\mathbf{k}_t \cdot \mathbf{r}}] w_\lambda(t')$$

$$\cong i \; \frac{d_{cv}}{\hbar} \; \mathcal{E}_t \; e^{-i\mathbf{k}_t \cdot \mathbf{r}} \; e^{-[i(\epsilon_\lambda - \Omega) + \gamma](t - t_t)} \; w_\lambda(t_{t_-}) \; \Theta(t - t_t)$$

$$+ \; i \; \frac{d_{cv}}{\hbar} \int_{-\infty}^{t} dt' \; e^{-[i(\epsilon_\lambda - \Omega) + \gamma](t-t')} \; \mathcal{E}_p(t') \; w_\lambda(t') \; e^{-i\mathbf{k}_p \cdot \mathbf{r}} \qquad (16.61)$$

and

$$w_\lambda(t) = 1 - \frac{i2d_{cv}}{\hbar}\int_{-\infty}^{t} dt'\; e^{-\Gamma(t-t')}\,[\mathcal{E}_p(t')\,e^{-i\mathbf{k}_p\cdot\mathbf{r}} + \mathcal{E}_t(t')\,e^{-i\mathbf{k}_t\cdot\mathbf{r}}]\,x_\lambda^*(t')$$

$$+\, i\,\frac{2d_{cv}^*}{\hbar}\int_{-\infty}^{t} dt'\; e^{-\Gamma(t-t')}\left[\mathcal{E}_p^*(t')\,e^{i\mathbf{k}_p\cdot\mathbf{r}} + \mathcal{E}_t^*(t')\,e^{i\mathbf{k}_t\cdot\mathbf{r}}\right] p_\lambda(t')$$

$$\cong 1 - i\,\frac{2d_{cv}}{\hbar}\,e^{-\Gamma(t-t_t)}\;\mathcal{E}_t\,e^{-i\mathbf{k}_t\cdot\mathbf{r}}\;x_\lambda^*(t_{t\_})\;\Theta(t-t_t)$$

$$+\, i\,\frac{2d_{cv}^*}{\hbar}\,e^{-\Gamma(t-t_t)}\;\mathcal{E}_t^*\,e^{i\mathbf{k}_t\cdot\mathbf{r}}\;p_\lambda(t_{t\_})\;\Theta(t-t_t)$$

$$-\, i\,\frac{2d_{cv}}{\hbar}\int_{-\infty}^{t} dt'\; e^{-\Gamma(t-t')}\;\mathcal{E}_p(t')\;x_\lambda^*(t')\;e^{-i\mathbf{k}_p\cdot\mathbf{r}}$$

$$+\, i\,\frac{2d_{cv}^*}{\hbar}\int_{-\infty}^{t} dt'\; e^{-\Gamma(t-t')}\;\mathcal{E}_p^*(t')\;p_\lambda(t')\;e^{i\mathbf{k}_p\cdot\mathbf{r}}\;, \tag{16.62}$$

where we used Eq. (16.55) for the probe pulse. Now we insert Eq. (16.62) into Eq. (16.61) and solve the resulting integral equation iteratively. This way, we obtain many terms, most of which do not contribute to our final result. In order to keep our equations as short as possible, we write only those terms which lead to a contribution in the final result that influences the probe transmission. We obtain

$$p_\lambda(t) = \frac{id_{cv}}{\hbar}\,\mathcal{E}_t\;e^{-i\mathbf{k}_t\cdot\mathbf{r}}\;e^{-[i(\epsilon_\lambda-\Omega)+\gamma](t-t_t)}\;w_\lambda(t_{t\_})\;\Theta(t-t_t)$$

$$+\, 2\left[\frac{d_{cv}}{\hbar}\right]^2\mathcal{E}_t e^{-i(\mathbf{k}_t+\mathbf{k}_p)\cdot\mathbf{r}}\,x_\lambda^*(t_{t\_})\int_{t_t}^{t} dt'\; e^{-[i(\epsilon_\lambda-\Omega)+\gamma](t-t')}e^{-\Gamma(t'-t_t)}\mathcal{E}_p(t')\Theta(t-t_t)$$

$$-\, 2\,\frac{|d_{cv}|^2}{\hbar^2}\int_{-\infty}^{t} \mathrm{dt'e}^{-[i(\epsilon_\lambda-\Omega)+\gamma](t-t')}\,\mathcal{E}_p(t')\int_{-\infty}^{t'} dt''\; e^{-\Gamma(t'-t'')}\mathcal{E}_p^*(t'')p_\lambda(t'')$$

$$+\;\dots\;. \tag{16.63}$$

The third line of this expression is of the order $(\mathcal{E}_p)^2$ and contains an integral over p_λ. Since we keep only terms up to the order $\mathcal{E}_t(\mathcal{E}_p)^2$ in our analysis, it is sufficient to solve this integral by inserting the first line of Eq. (16.63) to get

$$
\begin{aligned}
p_\lambda(t) &= i\,\frac{d_{cv}}{\hbar}\,\mathcal{E}_t\,e^{-i\mathbf{k}_t\cdot\mathbf{r}}\,e^{-[i(\epsilon_\lambda-\Omega)+\gamma](t-t_t)}\,w_\lambda(t_{t\_})\,\Theta(t-t_t) \\
&+ 2\left(\frac{d_{cv}}{\hbar}\right)^2 \mathcal{E}_t e^{-i(\mathbf{k}_t+\mathbf{k}_p)\cdot\mathbf{r}}\,x^*_\lambda(t_{t\_})\int_{t_t}^{t} dt'\;e^{-[i(\epsilon_\lambda-\Omega)+\gamma](t-t')}e^{-\Gamma(t'-t_t)}\mathcal{E}_p(t')\Theta(t-t_t) \\
&- i\,\frac{2d_{cv}}{\hbar}\frac{|d_{cv}|^2}{\hbar^2}\mathcal{E}_t e^{-i\mathbf{k}_t\cdot\mathbf{r}}\int_{t_t}^{t} dt'\;e^{-[i(\epsilon_\lambda-\Omega)+\gamma](t-t')}\mathcal{E}_p(t') \\
&\times \int_{t_t}^{t'} dt''\;e^{-\Gamma(t'-t'')}\,\mathcal{E}^*_p(t'')\,e^{-[i(\epsilon_\lambda-\Omega)+\gamma](t''-t_t)}\,\Theta(t-t_t) \\
&+ \ldots \,.
\end{aligned}
\tag{16.64}
$$

At the end of our calculations, we are interested in the optical susceptibility $\chi_\lambda(\omega)$ for the probe pulse. Therefore we study the Fourier transform of the polarization

$$
\begin{aligned}
\tilde{P}_\lambda(\omega) &= \int_{-\infty}^{\infty} dt\;e^{-i\omega t}\,\tilde{P}_\lambda(t) = \int_{-\infty}^{\infty} dt\;e^{-i(\omega-\Omega)t}\,p_\lambda(t) \\
&= \int_{t_t}^{\infty} dt\;e^{-i(\omega-\Omega)t}\,p_\lambda(t) = \int_{0}^{\infty} dt\;e^{-i(\omega-\Omega)t}\,p_\lambda(t+t_t)\,e^{-i(\omega-\Omega)t_t}\,.
\end{aligned}
\tag{16.65}
$$

Here we used the result of Eq. (16.64), that $p_\lambda(t)$ has a component proportional to (16.49) only for $t>t_t$, i.e., after the probe hit the sample. For the spectrum of the probe (test) pulse we can write

$$\mathcal{E}_t(\omega) = \int_{-\infty}^{\infty} dt\ \mathcal{E}_t(t)\ e^{i\omega t} = \int_{-\infty}^{\infty} dt\ \mathcal{E}_t(t)\ e^{i(\omega-\Omega)t}\ e^{-i\mathbf{k}_t\cdot\mathbf{r}}$$

$$\cong \mathcal{E}_t\ e^{i(\omega-\Omega)t_r}\ e^{-i\mathbf{k}_t\cdot\mathbf{r}} \tag{16.66}$$

Using Eqs. (16.52) and (16.64) - (16.66), we obtain

$$\begin{aligned}\widetilde{P}_\lambda(\omega) = i\ \frac{d_{cv}}{\hbar}\ \frac{\mathcal{E}_t(\omega)}{\gamma-i(\omega-\epsilon_\lambda)} \Bigg[\ & w_\lambda(t_{t_-}) \\ & - i\ \frac{2d_{cv}}{\hbar}\ e^{-i\mathbf{k}_p\cdot\mathbf{r}}\ x^*_\lambda(t_{t_-}) \int_0^\infty dt'\ e^{[i(\omega-\Omega)-\Gamma]t'}\ \mathcal{E}_p(t'+t_t) \\ & - 2\ \frac{|d_{cv}|^2}{\hbar^2} \int_0^\infty dt\ e^{i(\omega-\Omega)t}\ \mathcal{E}_p(t+t_t) \\ & \times \int_0^t dt'\ e^{-\Gamma(t-t')}\ e^{-[i(\epsilon_\lambda-\Omega)+\gamma]t'}\ \mathcal{E}^*_p(t'+t_t)\Bigg] \\ \equiv \chi_\lambda(\omega)\ \mathcal{E}_t(\omega). & \end{aligned} \tag{16.67}$$

Extracting the probe susceptibility, we find

$$\begin{aligned}\chi_\lambda(\omega) = i\ \frac{d_{cv}}{\hbar}\ \frac{1}{\gamma-i(\omega-\epsilon_\lambda)} \Bigg[\ & w_\lambda(t_{t_-}) \\ - 2\frac{|d_{cv}|^2}{\hbar^2} \int_{-\infty}^{t_t} dt'\ & e^{[i(\epsilon_\lambda-\Omega)-\gamma](t-t')}\mathcal{E}^*_p(t') \int_0^\infty dt\ e^{[i(\omega-\Omega)-\Gamma]t}\mathcal{E}_p(t+t_t) \\ - 2\frac{|d_{cv}|^2}{\hbar^2} \int_0^\infty dt e^{i(\omega-\Omega)t}\mathcal{E}_p(t+t_t) & \int_0^t dt' e^{-\Gamma(t-t')}e^{-[i(\epsilon_\lambda-\Omega)+\gamma]t'}\mathcal{E}^*_p(t'+t_t)\Bigg]\end{aligned} \tag{16.68}$$

where Eq. (16.57) has been used to express $x^*_\lambda(t_{t_-})$. The last term in Eq. (16.68) can also be written in the form

$$- 2 \frac{|d_{cv}|^2}{\hbar^2} \int_0^\infty dt' \int_{t'}^\infty dt \; e^{i(\omega-\Omega)t} \mathcal{E}_p(t+t_t) \; e^{-\Gamma(t-t')} e^{-[i(\epsilon_\lambda-\Omega)+\gamma]t'} \mathcal{E}_p^*(t'+t_t)$$

$$= - 2 \frac{|d_{cv}|^2}{\hbar^2} \int_0^\infty dt' \int_0^\infty dt \; e^{i(\omega-\Omega)t} e^{-\Gamma t} e^{-[i(\epsilon_\lambda-\omega)+\gamma]t'} \mathcal{E}_p(t+t'+t_t) \mathcal{E}_p^*(t'+t_t)$$

$$\cong - 2 \; \frac{|d_{cv}|^2}{\hbar^2} \; \frac{1}{\Gamma+i(\omega-\Omega)} \int_0^\infty dt \; e^{-[i(\epsilon_\lambda-\omega)+\gamma]t} \; |\mathcal{E}_p(t+t_t)|^2$$

$$\cong i \; 2 \; \frac{|d_{cv}|^2}{\hbar^2} \; \frac{1}{\omega-\Omega} \int_0^\infty dt \; e^{-[i(\epsilon_\lambda-\omega)+\gamma]t} \; |\mathcal{E}_p(t+t_t)|^2 \; . \tag{16.69}$$

Taking $\epsilon_\lambda = \omega_x$, with ω_x denoting the exciton resonance frequency, we can use Eq. (16.69) to discuss the excitonic optical Stark effect. In order to see the light induced shift more clearly, we consider the large detuning case,

$$|\omega-\Omega| >> \gamma \; , \; |\omega-\epsilon_\lambda| \; .$$

Inserting Eq. (16.69) into Eq. (16.68), we obtain the asymptotic behavior as

$$\delta\chi_\lambda(\omega) \cong 2 \; \frac{d_{cv}}{\hbar} \; \frac{|d_{cv}|^2}{\hbar^2} \; \frac{1}{\omega-\Omega} \int_0^\infty dt \; \frac{e^{-[i(\epsilon_\lambda-\omega)+\gamma]t}}{\gamma-i(\omega-\epsilon_\lambda)} \; |\mathcal{E}_p(t+t_t)|^2 \tag{16.70}$$

and for the pump-induced absorption change, or differential transmission spectrum, Eq. (16.47), we get

$$\delta\alpha(\omega) = - \; \mathrm{Im} \left[\frac{d_{cv}}{\hbar} \; \chi_\lambda(\omega) \right] =$$

$$\cong \frac{|d_{cv}|^4}{\hbar^4} \; \frac{2}{\Omega-\epsilon_\lambda} \; \mathrm{Im} \left[\int_0^\infty dt \; \frac{e^{-[i(\epsilon_\lambda-\omega)+\gamma]t}}{\gamma-i(\omega-\epsilon_\lambda)} \; |\mathcal{E}_p(t+t_t)|^2 \right] . \tag{16.71}$$

To analyze Eq. (16.71) let us assume for a moment that we excite the sample by a *cw*-beam, i.e., $\mathcal{E}_p(t) = \mathcal{E}_p =$ const. Then we obtain from Eq. (16.71)

$$\boxed{\delta\alpha(\omega) \propto -\frac{|\mathcal{E}_p|^2}{e-\Omega_p} \frac{2\gamma(\epsilon_\lambda - \omega)}{[(\epsilon_\lambda - \omega)^2 + \gamma^2]^2}}$$

optical Stark effect for *cw*-excitation (16.72)

which describes the absorption change caused by the shift of a Lorentzian resonance. This can be seen by looking at

$$\frac{\gamma}{\gamma^2+(\epsilon_\lambda-\omega-\delta)^2} - \frac{\gamma}{\gamma^2+(\epsilon_\lambda-\omega)^2} \cong -\frac{2\gamma(\epsilon_\lambda-\omega)}{(\gamma^2+(\epsilon_\lambda-\omega)^2)^2} , \tag{16.73}$$

where we assumed $\delta << |\epsilon_\lambda - \omega|$. Hence, Eq. (16.72) yields a dispersive shape around the resonance, $\omega = \epsilon_\lambda$, which describes decreasing and increasing absorption below and above the resonance, respectively.

For the case of pulsed excitation, the sample response is much more complex. Inserting the full Eq. (16.68) into the first line of Eq. (16.71), we obtain the results shown in Fig. 16.4 for different pump-probe delays. Fig. 16.2 shows, that for negative time delays, $t_t < 0$, i.e., when the probe pulse comes before the pump pulse maximum, the probe-transmission change shows oscillatory structures which evolve into the dispersive shape of the optical Stark effect.

Similar oscillations are obtained for the case of resonant interband excitation (Koch *et al.*, 1988). In this situation, the pump laser is tuned into the spectral regime of interband absorption, coupling an entire region of electron-hole transitions. The spectral extend of this region is given by the spectral width of the pump pulse. For negative pump-probe delays, the femtosecond experiments also show transient transmission oscillations. In contrast to the optical Stark effect, however, these oscillations then develop into a symmetric feature, called the *spectral hole*, which describes the saturation of the pump-laser coupled electron-hole transitions.

The general origin of the transient transmission oscillations is found in the grating, Eq. (16.60), which scatters parts of the pump pulse into the direction of the probe pulse. For $t<t_t$, the scattered pump interferes with the probe and causes the oscillations. Alternatively, one can also view the transient oscillations as perturbed free induction decay. The probe pulse excites the polarization, which decays on the time scale of the coherence decay time. The pump pulse then modifies the medium and perturbs (shifts) the resonances,

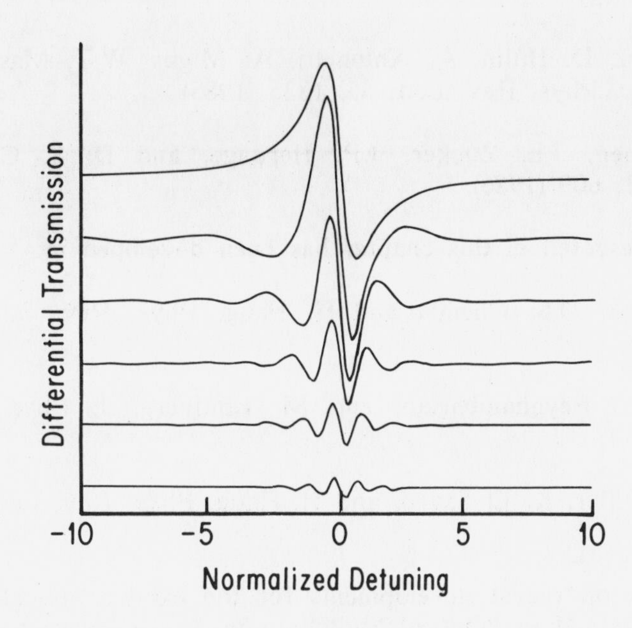

Fig. 16.4: Differential transmission spectra calculated in the spectral vicinity of the exciton resonance $\epsilon_\lambda = \omega_x$. The detuning is defined as $(\omega-\omega_x)/\sigma$, where σ^{-1} is the temporal width of the pump pulse. The FWHM of the pump pulse was assumed to be 120 *fs*, and the central pump frequency was detuned -10 below the resonance. The different curves are for different pump-probe delays t_t with 100 *fs* intervals, starting from the bottom at - 500 *fs* (probe before pump) to the top curve which is for 0 *fs* (pump-probe overlap). From Koch *et al.* (1988).

thus leading to the interference oscillations. For more details, see the review article by Koch *et al.* (1988) and the given references.

REFERENCES

The first reports on the observation of the nonresonant optical Stark effect are:

A. Mysyrowicz, D. Hulin, A. Antonetti, A. Migus, W.T. Masselink, and H. Morkoc, Phys. Rev. Lett. 55, 1335 (1985)

A. von Lehmen, J.E. Zucker, J.P. Heritage, and D. S. Chemla, Optics Lett. 11, 609 (1986).

The theory presented in this chapter has been developed in:

S. Schmitt-Rink, D.S. Chemla and H. Haug, Phys. Rev. *B37*, 941 (1988)

S.W. Koch, N. Peyghambarian, and M. Lindberg, J. Phys. C21, 5229 (1988).

C. Ell, J.F. Müller, K. El Sayed, and H. Haug, Phys. Rev. Lett. 62, 306 (1989).

Several papers on recent developments for the exciton optical Stark effect are contained in "Optical Switching in Low-Dimensional Systems", eds. H. Haug and L. Banyai, Plenum ASI Series Vol. 149, New York 1989.

PROBLEMS

Problem 16.1: Show for a two-level system that $(1-2n)^2 + |P|^2$ is a conserved quantity.

Problem 16.2: Use. Eqs. (16.2) - (16.4) to derive the optical Stark shift results of Sec. 2-3. for the two-level atom.

Problem 16.3 : Prove that $\delta p_k^+ >> \delta p_k^-$ and derive the linearized Eq. (16.19).

Problem 16.4 : Calculate the enhancement factors (16.41) using the exciton wavefunctions in *2d* and *3d*.

Chapter 17
FREE-CARRIER ELECTROABSORPTION

An important tool in solid-state spectroscopy is the application of static electric or magnetic fields. These fields give rise to changes in the optical spectra, which allow to draw conclusions on the nature of the optical transitions. In this chapter we discuss the effects of dc-electric fields on the free-carrier absorption of bulk (*3d*) and quantum-well (quasi-*2d*) semiconductors and we show that the field effects in the respective absorption spectra are remarkably different.

Applying an electric field to atomic systems causes a reduction of the overall symmetry which leads to the splitting of degenerate levels and to field-dependent level shifts. This effect is called the (dc) Stark effect. If one applies an external electric field to a semiconductor, this field has a pronounced influence on the optically active electron-hole pairs. In comparison to these effects, it is often justified to disregard field-induced changes in the atomic orbits. Following this philosophy, we therefore describe the atomic orbits by the unperturbed Bloch functions and study the influence of the field on the relative motion of the electron-hole pair using effective mass approximation.

17-1. Bulk Semiconductors

In Chap. 10 we compute the semiconductor band-edge absorption spectrum for direct allowed optical transitions. Using Eqs. (10.91) and (1.53) we can write the result as

$$\alpha(\omega) = \alpha_b \sum_n |\psi_n(\mathbf{r}=0)|^2 \, \delta(E_n - \hbar\omega) \,, \tag{17.1}$$

where

$$\alpha_b = \frac{8\pi^2 |d_{cv}|^2 \omega}{n_b c} , \tag{17.2}$$

including a factor 2 from the spin summation. In Eq. (17.1) ψ_n and E_n are the eigenfunctions and energy eigenvalues of the electron-hole pair, respectively. Note, that E_n in Eq. (17.1) includes the band-gap energy E_g.

If we disregard the Coulomb effects altogether, the stationary Schrödinger equation for the relative motion of the electron-hole pair in the presence of an electric field F (parallel to the z-axis) can be written as

$$\left[-\frac{\hbar^2 \Delta}{2m_r} - \mathrm{ezF} - E_n \right] \psi_n(\mathbf{r}) = 0 . \tag{17.3}$$

To solve this equation, we make the ansatz

$$\psi_n(\mathbf{r}) = \frac{1}{L} e^{i(k_x x + k_y y)} \psi_n(z) \tag{17.4}$$

where $L = \mathcal{V}^{1/3}$ is the linear extension of the system. We write the energy eigenvalue as

$$E_n = \frac{\hbar^2}{2m_r} (k_{||}^2 + \kappa_n^2) \equiv E_{n,k_{||}} , \tag{17.5}$$

with $k_{||}^2 = k_x^2 + k_y^2$. Inserting Eqs. (17.4) and (17.5) into Eq. (17.3) we find

$$\left[\frac{d^2}{dz^2} + fz + \kappa_n^2 \right] \psi_n(z) = 0, \tag{17.6}$$

where

$$f = eF \frac{2m_r}{\hbar^2} = \frac{eF}{E_0 a_0^2} , \tag{17.7}$$

and E_0 and a_0 are the usual excitonic units, defined in Eqs. (10.70) and (10.71), respectively. Eq. (17.7) shows that fa_0^3 is the ratio between the dipole energy in the field and the exciton Rydberg energy, ea_0F/E_0. Introducing the dimensionless variable Z by

$$Z = f^{1/3} z \tag{17.8}$$

and substituting

$$\zeta_n = Z + \kappa_n^2 f^{-2/3} = Z + (a_0\kappa_n)^2 \left[\frac{E_0}{a_0 eF}\right]^{2/3} , \tag{17.9}$$

we find

$$\psi_n''(\zeta_n) = - \zeta_n \psi_n(\zeta_n) , \tag{17.10}$$

where $\psi_n''(\zeta) = d^2\psi_n/d\zeta^2$. The solution of Eq. (17.10) is

$$\psi_n(z) = a_n \; Ai(-\zeta_n) . \tag{17.11}$$

Here, $Ai(x)$ is the Airy function (Abramowitz and Stegun, 1972)

$$Ai(x) = \frac{1}{\pi} \int_0^\infty \mathrm{du} \cos\left[\frac{u^3}{3} + ux\right] \tag{17.12}$$

and a_n is the normalization constant. The Airy function decays exponentially for positive arguments

$$\lim_{x\to\infty} Ai(x) = \frac{1}{2\sqrt{\pi}x^{1/4}} e^{-\frac{2}{3}x^{3/2}} \left[1 - \frac{3c_1}{2x^{2/3}}\right] , \tag{17.13}$$

with $c_1 = 15/216$. For negative arguments the Airy function oscillates,

$$\lim_{x\to\infty} Ai(-x) = \frac{1}{\sqrt{\pi}x^{1/4}} \sin\left[\frac{2}{3}x^{3/2} + \frac{\pi}{4}\right] , \tag{17.14}$$

expressing the accelerating action of the field. The normalization constant a_n is determined by

$$a_n^{-2} = \int_{-\infty}^{\infty} dz \, |Ai(-\zeta_n)|^2 = \lim_{L\to\infty} \int_{-L}^{L} dz \, |Ai(-\zeta_n)|^2 , \tag{17.15}$$

or, using Eqs. (17.8) and (17.9),

$$\frac{f^{1/3}}{a_n^2} = \lim_{L\to\infty} \int_{-Lf^{1/3}}^{Lf^{1/3}} dx \, |Ai(x)|^2 . \tag{17.16}$$

With partial integration we obtain

$$\int_{-Lf^{1/3}}^{Lf^{1/3}} dx \, |Ai(x)|^2 = xAi^2(x)\Big|_{-Lf^{1/3}}^{Lf^{1/3}} - \int_{-Lf^{1/3}}^{Lf^{1/3}} dx \; 2x \; Ai(x) \; Ai'(x). \tag{17.17}$$

Because $Ai(x)$ satisfies the differential equation

$$Ai''(x) = xAi(x) , \tag{17.18}$$

we get

$$\int_{-Lf^{1/3}}^{Lf^{1/3}} dx \, |Ai(x)|^2 = xAi^2(x)\Big|_{-Lf^{1/3}}^{Lf^{1/3}} - \int_{-Lf^{1/3}}^{Lf^{1/3}} dx \; 2 \; Ai''(x) \; Ai'(x)$$

$$= xAi^2(x)\Big|_{-Lf^{1/3}}^{Lf^{1/3}} - \int_{-Lf^{1/3}}^{Lf^{1/3}} dx \; \frac{d(Ai'(x))^2}{dx}$$

$$= \left[xAi^2(x) + (Ai'(x))^2\right]\Big|_{-Lf^{1/3}}^{Lf^{1/3}} . \tag{17.19}$$

Inserting Eq. (17.19) into Eq. (17.16) yields

$$a_n^{-2} = \lim_{L\to\infty} \sqrt{\frac{L}{f^{1/3}}} \, \frac{1}{\pi} \,, \tag{17.20}$$

where we used the fact that $Ai(x)$ and $Ai'(x)$ vanish for $x \to \infty$. For $Ai(-x)$ and $Ai'(-x)$ we inserted the asymptotic expressions given by Eq. (17.14) and by

$$\lim_{x\to\infty} Ai'(-x) = \frac{x^{1/4}}{\sqrt{\pi}} \cos\left[\frac{2}{3}x^{3/2} + \frac{\pi}{4}\right] , \tag{17.21}$$

respectively.

The energy eigenvalues are computed from the boundary condition

$$\psi_n(z=L) = 0 \tag{17.22}$$

as

$$\frac{2}{3} \sqrt{f} \left[L + \frac{\kappa_n^2}{f}\right]^{3/2} = \left[n - \frac{1}{4}\right] \pi \,. \tag{17.23}$$

Solving Eq. (17.23) for κ_n and inserting the result into Eq. (17.5) yields

$$E_{n,k_{||}} = \frac{\hbar^2}{2m_r} \left\{ k_{||}^2 - Lf + \left[\frac{3\pi f}{2}(n-1/4)\right]^{2/3} \right\} . \tag{17.24}$$

Now we have all the ingredients needed to evaluate Eq. (17.1), which we write as

$$\alpha(\omega) = \alpha_b \int_{-\infty}^{+\infty} \frac{dk_x}{2\pi} \int_{-\infty}^{+\infty} \frac{dk_y}{2\pi} \sum_n \frac{\pi f^{1/3}}{\sqrt{L}} \delta(E_{n,k_{||}} + E_g - \hbar\omega)$$

$$\times \left| Ai\left[f^{-2/3}\left(k_{||}^2 - \frac{2m_r E_{n,k_{||}}}{\hbar^2} \right) \right] \right|^2 . \tag{17.25}$$

We change the sum over n to an integral over E

$$\sum_n \to \int_0^\infty dE \ \frac{dn}{dE} \tag{17.26}$$

and evaluate the density of states using Eq. (17.23)

$$\frac{dn}{dE} = \frac{2m_r}{\hbar^2 \pi \sqrt{f}} \left(L + \frac{\kappa_n}{f} \right)^{1/2} \cong \frac{2m_r}{\hbar^2 \pi \sqrt{f}} \sqrt{L} . \tag{17.27}$$

We dropped the additive term in $L + \kappa_n / f$, since we are finally interested in the limit $L \to \infty$, see above. Inserting Eqs. (17.26) and (17.27) into Eq. (17.25) yields

$$\alpha(\omega) = \alpha_b \int_0^\infty \frac{dk_{||}^2}{(2\pi)^2} \frac{2m_r}{\hbar^2 f^{1/3}} \left| Ai\left[f^{-2/3}\left(k_{||}^2 + \frac{2m_r(E_g - \hbar\omega)}{\hbar^2} \right) \right] \right|^2$$

$$= \frac{\alpha_b}{2\pi} \frac{m_r f^{1/3}}{\hbar^2} \int_\epsilon^\infty dx \ |Ai(x)|^2 , \tag{17.28}$$

where

$$\epsilon = \frac{2m_r(E_g - \hbar\omega)}{\hbar^2 f^{2/3}} = \frac{E_g - \hbar\omega}{E_0}\left[\frac{E_0}{ea_0F}\right]^{2/3} . \tag{17.29}$$

Again, we evaluate the integral in Eq. (17.28) by partial integration, following the steps in Eqs. (17.16) - (17.19). The result is

$$\int_\epsilon^\infty dx\, |Ai(x)|^2 = -\epsilon Ai^2(\epsilon) + (Ai'(\epsilon))^2 . \tag{17.30}$$

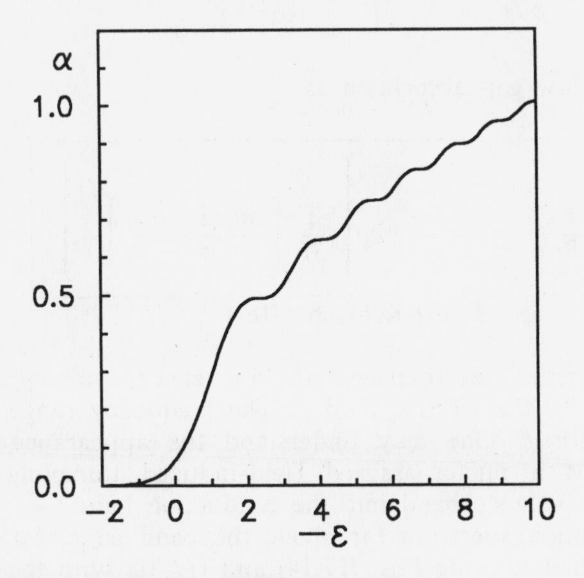

Fig. 17.1: Absorption spectrum for free carriers in an electric field according to Eq. (17.31). The absorption α is given in units of $\alpha' = \alpha_b m_r f^{1/3}/(2\pi\hbar^2)$ and ϵ is given by Eq. (17.29).

The total absorption spectrum is thus

$$\alpha(\omega) = \frac{\alpha_b}{2\pi} \frac{m_r f^{1/3}}{\hbar^2}[- \epsilon Ai^2(\epsilon) + (Ai'(\epsilon))^2] \quad .$$

electroabsorption for free carriers (17.31)

Fig. 17.1 shows the resulting absorption as function of ϵ. Due to the oscillatory character of the Airy functions for negative arguments, one gets oscillations in the absorption spectrum above the band gap. The amplitude of these oscillations decreases with increasing energy. We can also see from Fig. 17.1 that the absorption has a tail below the gap, i.e., for $\hbar\omega < E_g$ or $\epsilon > 0$.

Using the asymptotic form (17.13) and

$$\lim_{x\to\infty} Ai'(x) = \frac{x^{1/4}}{2\sqrt{\pi}} e^{-\frac{2}{3}x^{3/2}} \left[1 + \frac{21c_1}{10x^{2/3}}\right] , \tag{17.32}$$

we obtain the below-gap absorption as

$$\alpha(\omega) \cong \frac{\alpha_b}{8\pi^2} \frac{f}{E_g - \hbar\omega} \exp\left[-\frac{4}{3f}\left[\frac{2m_r(E_g-\hbar\omega)}{\hbar^2}\right]^{3/2}\right]$$

Franz-Keldysh effect (17.33)

Eq. (17.33) describes the exponential low energy absorption tail which is caused by the electric field f. The frequency range of this tail increases with f. One may understand the appearance of the absorption tail as a photon-assisted field-induced tunneling of an electron from the valence band into the conduction band.

The absorption spectrum far above the band edge, $\hbar\omega >> E_g$ or $\epsilon << 0$, can be estimated using Eqs. (17.14) and (17.21) with the result

$$\alpha(\omega) = \frac{\alpha_b}{(2\pi)^2} \left[\frac{2m_r}{\hbar^2}\right]^{3/2} \sqrt{\hbar\omega - E_g} . \tag{17.34}$$

Eq. (17.34) agrees with Eq. (4.46) describing free-carrier absorption in a *3d*-system.

17-2. Quantum Wells

If one applies the electric field perpendicular to the layer of a quantum well, the situation is quantitatively different from that in bulk material. Because of the opposite charges, the field pushes electron and hole toward the opposite walls of the well. Hence, the overlap between the corresponding particle-in-*a*-box wave functions is drastically modified. To discuss this effect, we again disregard for the time being the modifications caused by the electron-hole Coulomb interaction.

In a spatially inhomogeneous situation, such as in a quantum well, one has to use a two-point susceptibility function in real space representation $\chi(\mathbf{R},\mathbf{R}',\omega)$, which connects non-locally the polarization and the field according to

$$P(\mathbf{R},\omega) = \int d^3R' \;\chi(\mathbf{R},\mathbf{R}',\omega)E(\mathbf{R}',\omega) \; . \tag{17.35}$$

The optical susceptibility is given by a generalization of Eq. (17.1) as

$$\chi(\mathbf{R},\mathbf{R}',\omega) = \chi_0 \sum_{\mu} \frac{\psi_\mu^*(\mathbf{R},\mathbf{r}{=}0)\psi_\mu(\mathbf{R}',\mathbf{r}'{=}0)}{\hbar(\omega+i\delta) - E_\mu} \; . \tag{17.36}$$

Here, $\psi_\mu(\mathbf{R},\mathbf{r})$ is the wavefunction of an electron-hole pair, and $\mathbf{R}$, $\mathbf{r}$ are the center-of-mass and relative coordinates, respectively. In spatially homogeneous situations χ depends only on $\mathbf{R}$-$\mathbf{R}'$. The Fourier transform with respect to the difference of the center-of-mass coordinates yields the spatial dispersion, i.e., the wavevector dependence of the susceptibility discussed in Chap. 11. However, due to the spatially inhomogeneous situation in a quantum well one has

$$\chi(\mathbf{R},\mathbf{R}',\omega) \neq \chi(\mathbf{R}-\mathbf{R}',\omega) \; .$$

The light wavelength in the visible is of the order of 10^{-4} cm. This is much larger than the typical quantum-well width, which for GaAs is around 10^{-6}cm. Therefore, it is useful to introduce a susceptibility which is averaged over the quantum-well volume

$$\overline{\chi} = \frac{1}{\mathcal{V}} \int d^3R \int d^3R' \; \chi(\mathbf{R},\mathbf{R}',\omega) \; . \tag{17.37}$$

This averaged susceptibility locally connects $P(\mathbf{R})$ and $E(\mathbf{R})$.

Let us consider, for simplicity, a potential well of infinite depth extending over $-L/2 \leq z \leq L/2$. The pair wavefunction of a narrow quantum well can be taken as the product of particle-in-a-box wavefunctions for the electron and hole times the function describing the relative motion in the plane of the layer

$$\psi_\mu(\mathbf{R},\mathbf{r}) = \psi_{n_e}(z_e)\, \psi_{n_h}(z_h)\, \phi_{k_{||}}(\mathbf{r}_{||}) \; . \tag{17.38}$$

Electron and hole wavefunctions in the z direction obey the equation

$$\left[-\frac{\hbar^2}{2m_i} \frac{d^2}{dz_i^2} \pm eEz_i \right] \psi_{n_i}(z_i) = E_{n_i} \; \psi_{n_i}(z_i) \; , \tag{17.39}$$

where the + (-) sign is linked to i=e (h). The boundary conditions are

$$\psi_{n_i}(z=\pm L/2) = 0. \tag{17.40}$$

Without the field the wavefunctions are just the simple trigonometric functions with even and odd parity

$$\psi_n^0(z) = \begin{Bmatrix} \cos(k_n z) \\ \sin(k_n z) \end{Bmatrix} \text{with } k_n = \frac{\pi}{L} \begin{Bmatrix} 2n+1 \\ 2n \end{Bmatrix} \text{for } n=0,1,2,.. \tag{17.41}$$

The absorption spectrum resulting from Eqs. (17.36) - (17.41) is

$$\alpha(\omega) = \frac{\alpha_b}{L} \sum_{\mathbf{k}_{||}, n_e, n_h} \frac{\delta(\hbar\omega - E_g - E_{k_{||}} - E_{n_e} - E_{n_h})}{A_{n_e} A_{n_h}} \times \left| \int_{-L/2}^{+L/2} dz\, \psi_{n_e}(z)\, \psi_{n_h}(z) \right|^2$$

quantum-confined Franz-Keldysh spectrum (17.42)

with $E_{k_{||}} = \frac{\hbar^2 k_{||}^2}{2m}$ and the normalization

$$A_{n_i} = \int_{-L/2}^{+L/2} dz\, |\psi_{n_i}(z)|^2 \; .$$

The overlap integral in Eq. (17.42) results from the spatial average, Eq. (17.37), over $R_z = z_e = z_h = z$, because $r_z = z_e - z_h = 0$ according to (17.36), and in the same way over R_z'.

Eq. (17.39) is again solved in terms of Airy functions. However, in order to fulfill the boundary conditions, we have to use a linear combination of the two independent types of Airy functions, $Ai(\zeta)$ and $Bi(\zeta)$, where

$$Bi(x) = \frac{1}{\pi} \int_0^\infty du \left\{ e^{-u^3/3 + ux} + \sin\left[\frac{u^3}{3} + ux\right] \right\}, \tag{17.43}$$

see Abramowitz and Stegun (1972). The solution of Eq. (17.39) is thus

$$\psi_{n_i}(z_i) = a_i Ai(\zeta_i) + b_i Bi(\zeta_i) \; , \tag{17.44}$$

with

$$\zeta_{n_i} = \pm f^{1/3} z_i - \kappa_{n_i}^2 f^{-2/3}. \tag{17.45}$$

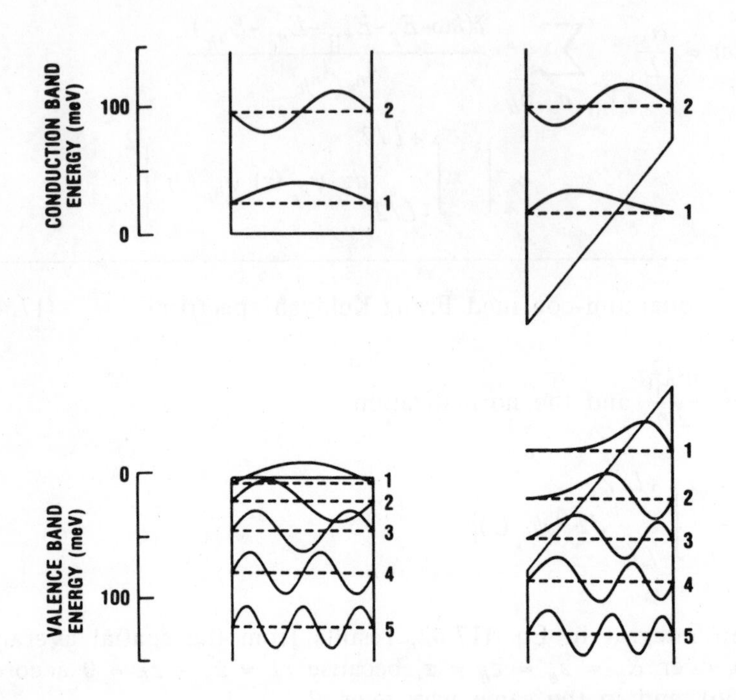

Fig. 17.2: Calculated wavefunctions and energy levels for a 150 Å thick GaAs-like quantum well at 0 and 10^5 Vcm^{-1}. (From Schmitt-Rink *et al.* , 1989)

The boundary condition yields the requirement for the existence of solutions

$$Ai(\zeta_{i+})Bi(\zeta_{i-}) = Ai(\zeta_{i-})Bi(\zeta_{i+}) \tag{17.46}$$

and

$$\frac{b_i}{a_i} = - \frac{Ai(\zeta_{i+})}{Bi(\zeta_{i+})} , \tag{17.47}$$

where $\zeta_{i\pm}$ corresponds to $z_i = \pm L/2$, respectively. Eq. (17.46) determines the energies $\kappa^2_{n_i}$ and Eq. (17.47) yields the relative weight of the Airy functions Bi and Ai. The summations over n_e and n_h in Eq. (17.42) are now replaced by integrations over the energies E_e

and E_h

$$\alpha(\omega) = \frac{\alpha_b}{L} \sum_{\mathbf{k}_{||}} \int_{-eEL/2}^{\infty} dE_e \int_{-eEL/2}^{\infty} dE_h \; \frac{dn_e}{dE_e} \frac{dn_e}{dE_e}$$

$$\times\delta(\hbar\omega - E_g - E_{k_{||}} - E_e - E_h) \; I_{eh}, \tag{17.48}$$

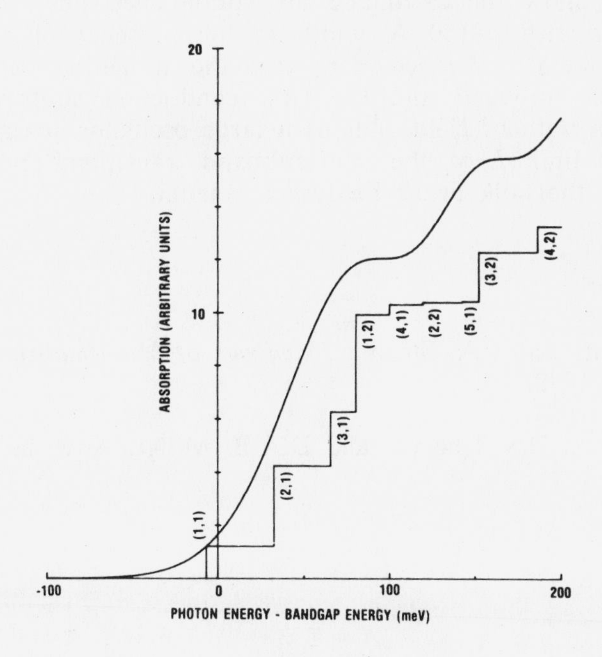

Fig. 17.3: Calculated absorption of a 150 Å thick GaAs-like quantum well at 10^5 Vcm^{-1}. The individual transitions are labelled $(n_v, \; n_c)$ where n_v (n_c) is the valence (conduction) subband number. The smooth line is the calculated Franz-Keldysh effect for bulk material, see Fig. 17.1. (From Schmitt-Rink *et al.* , 1989)

where I_{eh} is again the square of the normalized overlap integral between the electron and hole wavefunctions

$$I_{eh} = \frac{\left| \int_{-L/2}^{+L/2} dz \; \psi_{n_e}(z) \psi_{n_h}(z) \right|^2}{A_{n_e} \; A_{n_h}} \; . \tag{17.49}$$

Fig. 17.2 shows the calculated wavefunctions in the potential well with and without an electric field. The picture of the wavefunctions gives immediately the information how the overlap integral I_{eh} changes due to the field for the various inter-subband transitions. In Fig. 17.3 we show the calculated absorption spectrum for a GaAs quantum well with L=150 Å width in the presence of an electric field of 10^5 Vcm^{-1}. We see, e.g., that the transition between the second valence subband and the first conduction subband, which was forbidden without field, obtains a large oscillator strength in the field. For the limit $L \to \infty$, the inter-subband transitions approach the modulation of the bulk Franz-Keldysh spectrum.

REFERENCES

M. Abramowitz and I.A. Stegun, *Handbook of Mathematical Functions* (Dover Publ., 1972)

S. Schmitt-Rink, D.S. Chemla, and D.A.B. Miller, Adv. in Phys. **38**, 89 (1989)

PROBLEMS

Problem 17.1: Use first-order perturbation theory in the applied field to evaluate the absorption spectrum, Eq. (17.42), for the quantum-confined Franz-Keldysh effect. Use the basis functions (17.41) to show the reduction in oscillator strength for the transition between the lowest electron-hole subband, $0,h \to 0,e$. This transition is fully allowed without field, and it is reduced in the presence of the field. The same way show that the field makes the transition $0,h \to 1,e$ dipole-allowed.

Chapter 18
EXCITON ELECTROABSORPTION

In this chapter we treat the electroabsorption of the exciton, which means that in extension of the treatment in Chap. 17 we additionally include the attractive electron-hole Coulomb potential. Instead of Eq. (17.3) we then have to solve the basic pair equation

$$\left[- \frac{\hbar^2 \Delta}{2m_r} - ezF - \frac{e^2}{\epsilon_0 r} - E_\mu \right] \psi_\mu = 0 \; . \tag{18.1}$$

Again, we discuss the solution of this equation and the resulting optical spectra both for bulk and quantum-well semiconductors.

18-1. Bulk Semiconductors

The exciton in a bulk semiconductor looses its stability in the presence of an electric field, as can be seen easily by inspecting the total electron-hole potential in Eq. (18.1). Plotting this potential in z direction, Fig. 18.1, we see immediately that the exciton can be ionized if one of the carriers tunnels from z_1 to z_2 through the potential barrier. The tunneling causes a lifetime broadening of the exciton resonance. For example, for GaAs the exciton resonance vanishes completely for fields larger than 10^3V/cm. In addition to the broadening there is also a shift of the exciton resonance, the so-called (dc) Stark shift. Second-order perturbation shows immediately that the shift of the ground state is quadratic in the field and negative

$$\Delta E_0 \cong - \frac{(ea_0 F)^2}{E_0} \equiv - \mathcal{F}^2 E_0 \; , \tag{18.2}$$

which holds as long as the perturbation is sufficiently small, i.e.,

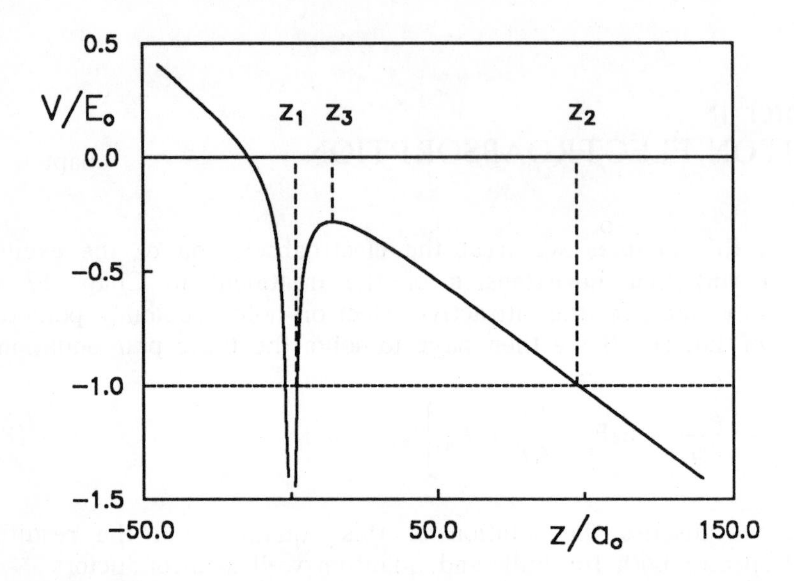

Fig. 18.1: Exciton potential in z-direction with an applied electric field ($\mathcal{F}$ = 0.01).

$\mathcal{F} \ll 1$. But still more interesting is the question how the Franz-Keldysh absorption tail will be modified by excitonic effects.

To study the region of the exciton absorption tail, we use the quasi-classical approximation introduced into quantum mechanics by Wentzel, Kramers and Brillouin, and often called the WKB method. This approach has been applied by Dow and Redfield (1970) to the present problem, we follow here the analytical approximations by Merkulov and Perel (1973) and Merkulov (1974).

In excitonic units, Eq. (18.1) becomes

$$\left[\Delta + \mathcal{F} z + \frac{2}{r} - \epsilon \right] \psi = 0 \ , \tag{18.3}$$

where all coordinates are scaled with the Bohr radius and all energies with the exciton Rydberg energy, respectively. The scaled pair energy is always negative in the tail region, therefore we have introduced the positive energy scale

$$\epsilon \equiv -\frac{E_\mu}{E_0} \equiv \frac{E_g - \hbar\omega}{E_0} > 1 \ . \tag{18.4}$$

Here, $\hbar\omega$ is the energy of the exciting photon, which is assumed below the band gap energy, so that ϵ is always positive and larger than unity.

The maximum of the potential (point z_3 in Fig. 18.1) has in z direction an energy of $-2\sqrt{2\mathcal{F}}$. For the analysis in this chapter we assume that the applied field is not too strong, such that a potential barrier still exists for the lowest exciton state, i.e.,

$$2\sqrt{2\mathcal{F}} < \epsilon_{min} = 1. \tag{18.5}$$

Under this condition, the exciton still exists as a quasi-bound state and we can essentially divide the solution of the problem into three steps: i) For the regime far away from the center of the exciton, we make a quasiclassical approximation and use $\hbar$ as formal expansion parameter. ii) For $z < z_1$, i.e., inside the Coulomb well, we neglect the electric field in comparison to the Coulomb potential and use the quantum mechanical solution (Chap. 10) for the exciton problem. iii) We match the solutions in the regime $z_1 < z < z_3$, where the quasi-classical approximation is still reasonably good and where the electric field is still small in comparison to the Coulomb potential.

First, we derive the quasiclassical solution for $z > z_2$. In cylindrical coordinates the Laplace differential operator is given by

$$\Delta = \frac{1}{\rho}\frac{\partial}{\partial\rho}\rho\frac{\partial}{\partial\rho} + \frac{\partial^2}{\partial z^2} + \frac{1}{\rho^2}\frac{\partial^2}{\partial\phi^2} \tag{18.6}$$

where ρ is the radius perpendicular to z. Because only the wavefunctions with the angular-momentum quantum number $m = 0$ are finite in the origin and thus contribute to the absorption spectrum, we drop the dependence on the angle ϕ,

$$\left[\frac{1}{\rho}\frac{\partial}{\partial\rho}\rho\frac{\partial}{\partial\rho} + \frac{\partial^2}{\partial z^2} + \mathcal{F}z + \frac{2}{\sqrt{\rho^2+z^2}} - \epsilon\right]\psi_{sc} = 0 \ . \tag{18.7}$$

The force linked with the potential

$$V(\rho, z) = -\mathcal{F}z - 2/\sqrt{\rho^2+z^2} \tag{18.8}$$

is

$$\mathbf{K} = -\nabla V = \mathscr{F}\mathbf{e}_z - \frac{2z\mathbf{e}_z + 2\rho\mathbf{e}_\rho}{r^3}, \tag{18.9}$$

where $\mathbf{e}_z$ and $\mathbf{e}_\rho$ are the unit vectors in z and ρ direction, respectively. The ratio of the two force components

$$\frac{K_\rho}{K_z} = -\frac{2\rho}{\mathscr{F}r^3 - 2z} \tag{18.10}$$

is always small for $r^2 >> \mathscr{F}^{-1}$. At the maximum of the potential barrier

$$z_3^2 = \frac{2}{\mathscr{F}},$$

so that one can approximately neglect K_ρ in the region $z > z_3$. It is therefore a good approximation in the whole quasiclassical region to use only a one-dimensional potential

$$V(z) = -\mathscr{F}z - \frac{2}{z} \tag{18.11}$$

instead of the full potential, Eq. (18.8). In this case Eq. (18.7) simplifies to

$$\left[\frac{1}{\rho}\frac{\partial}{\partial\rho}\rho\frac{\partial}{\partial\rho} + \frac{\partial^2}{\partial z^2} - V(z) - \epsilon\right]\psi_{sc} = 0\,. \tag{18.12}$$

Using the ansatz

$$\psi_{sc}(\rho, z) = \chi(\rho)\Psi(z) \tag{18.13}$$

we can separate Eq. (18.12) into

$$\left[\frac{1}{\rho}\frac{\partial}{\partial\rho}\rho\frac{\partial}{\partial\rho} + p_\rho^2\right]\chi = 0 \tag{18.14}$$

and

$$\left[\frac{\partial^2}{\partial z^2} - V(z) - \epsilon - p_\rho^2\right]\Psi = 0\,, \tag{18.15}$$

where p_ρ is the quasimomentum perpendicular to the z-axis.

Eq. (18.14) is a version of Bessel's differential equation and the solutions are the cylindrical Bessel functions

$$\chi(\rho) = J_0(p_\rho \rho) \ . \tag{18.16}$$

To solve Eq. (18.15), we make the quasiclassical approximation. For this purpose we introduce again formally the $\hbar$-dependence of the kinetic energy operator, use $\hbar$ as a formal expansion parameter, and put it equal to unity at the end. We write Eq. (18.15) as

$$\left[\hbar^2 \frac{\partial^2}{\partial z^2} + p^2(z)\right]\Psi = 0, \tag{18.17}$$

where we introduce the quasimomentum $p(z)$ through the relation

$$p^2(z) = - \epsilon - V(z) - p_\rho^2 \ . \tag{18.18}$$

Inserting the the ansatz

$$\Psi(z) = e^{\frac{i}{\hbar}\sigma(z)} \tag{18.19}$$

into Eq. (18.17) results in

$$i\hbar\sigma'' - (\sigma')^2 + p^2 = 0. \tag{18.20}$$

Now we expand the phase function $\sigma(z)$ formally in powers of $\hbar/i$

$$\sigma(z) = \sigma_0(z) + \hbar/i \ \sigma_1(z) + (\hbar/i)^2 \sigma_2(z) + \cdots \tag{18.21}$$

and compare the various orders of $\hbar$. In the order $O(\hbar^0)$ we obtain

$$\sigma_0'(z) = \pm \ p(z) \ , \tag{18.22}$$

with the solution

$$\sigma_0(z) = \pm \int_{z_2}^{z} d\zeta \ p(\zeta) \ . \tag{18.23}$$

The first-order equation is

$$i\sigma_0'' + 2i\sigma_0'\sigma_1' = 0 \,, \tag{18.24}$$

so that

$$\sigma_1(z) = -\ln\sqrt{p(z)} + \ln C \,, \tag{18.25}$$

where ln C is a normalization constant. Summarizing the results, we obtain the semiclassical wavefunction up to order $\hbar^2$ or higher as

$$\Psi_{sc}(z) = \frac{C}{\sqrt{|p(z)|}} \exp\left[\pm i \int_{z_2}^{z} d\zeta \; p(\zeta)\right] , \tag{18.26}$$

where we put the formal expansion parameter $\hbar \to 1$.

Because a classical particle cannot penetrate into regions in which the potential energy exceeds the total energy, we define the classical turning points by $p^2(z_{1,2}) = 0$ and find

$$z_{2,1} = \frac{1}{2\mathscr{F}}\left\{(\epsilon+p_\rho^2) \pm \sqrt{(\epsilon+p_\rho^2)^2 - 8\mathscr{F}}\right\} . \tag{18.27}$$

The quasimomentum $p(z)$ is real in the region, $z > z_2$ and the wavefunction (18.26) describes oscillatory solutions.

In the region of the potential barrier, $z_1 < z < z_2$, $\epsilon > 1$ and $-V(z) < 1$, so that $p^2(z) < 0$ and $p(z)$ is purely imaginary. Eq. (18.26) describes an exponentially increasing or decaying solution in the classically forbidden region. The exponentially increasing solution is clearly unphysical and has to be discarded, so that we have in the classically forbidden region

$$\Psi_{sc}(z) = \frac{C}{\sqrt{|p(z)|}} \exp\left[-\int_{z_2}^{z} d\zeta \; |p(\zeta)|\right] . \tag{18.28}$$

Now we can approximate

$$|p(z)| = \sqrt{\epsilon + V(z) + p_\rho^2} \cong p_0(z) + \frac{p_\rho^2}{2p_0(z)} \tag{18.29}$$

with $p_0(z) = [\epsilon+V(z)]^{1/2}$ and the integral in Eq. (18.28) can be simplified by considering that z deviates only slightly from r, i.e.,

$$z = r\cos(\theta) \cong r\left[1-\frac{1}{2}\theta^2\right], \tag{18.30}$$

so that

$$\Psi_{sc}(z) \cong \frac{C}{\sqrt{|p(z)|}}\exp\left[\int_{z_2}^{r} d\zeta\left[p_0 + \frac{p_\rho^2}{2p_0}\right] + \int_{r}^{z} d\zeta\; p_0\right]$$

$$\cong \frac{C}{\sqrt{|p(z)|}}\exp\left[-\int_{r}^{z_2} d\zeta\left[p_0 + \frac{p_\rho^2}{2p_0}\right]\right]\exp\left[-\frac{r\theta^2}{2}\sqrt{\epsilon}\right]. \tag{18.31}$$

Here z_2 can be evaluated for $p_\rho \cong 0$. Furthermore, in the vicinity of the z-axis the argument of the Bessel function can be put equal to zero $J_0(p_\rho\rho)\cong J_0(0)=1$, so that the semiclassical form of the wavefunction (18.13) becomes $\psi_{sc} = \Psi_{sc}$.

Now we turn to the solution of the exciton problem for the core region, $z<z_3$, in which we may neglect approximately the field. Eq. (10.75) shows that we may write the exciton wavefunctions with $\ell=0$ and $m=0$, which are finite at the origin, as

$$\Psi_x(\mathbf{r}) = \Psi(0)e^{-r\sqrt{\epsilon}}F\left[1-\frac{1}{\sqrt{\epsilon}};2;2\sqrt{\epsilon}r\right], \tag{18.32}$$

where we used the relations

$$L_{n+\ell}^{2\ell+1}\left[\frac{2r}{na_0}\right] = \frac{(n+\ell)!(n+\ell)!}{(n-\ell-1)!(2\ell+1)!}F\left[-n+\ell+1;2\ell+2;\frac{2r}{na_0}\right] \tag{18.33}$$

and

$$n = \frac{E_0}{E_n} = \frac{1}{\sqrt{\epsilon}}. \tag{18.34}$$

For $r \gg \epsilon^{-1/2}$, one may use the asymptotic form of the confluent hypergeometric function, second term in Eq. (10.82),

$$F(a;b;z) \rightarrow \frac{e^z \Gamma(b) z^{a-b}}{\Gamma(a)} \tag{18.35}$$

which yields

$$\Psi_x(r) = \frac{\Psi(0)\ e^{-r\sqrt{\epsilon}}}{\Gamma\left[1-\frac{1}{\sqrt{\epsilon}}\right] (2\sqrt{\epsilon}r)^{1+\frac{1}{\epsilon^{1/2}}}} . \tag{18.36}$$

This function has to be matched with the spherical part of the semiclassical wavefunction (18.31) which is obtained by averaging the angle-dependent part of (18.31) over the angles

$$\overline{\Psi}_{sc}(r) = \int \frac{d\Omega}{4\pi}\ \Psi(\mathbf{r}) = \int_0^{2\pi} \frac{d\phi}{2\pi} \int_0^{\pi} \frac{d\theta}{2} \sin\theta\ f(r)\ \exp\left[-\frac{1}{2} r\sqrt{\epsilon}\theta^2\right]$$

$$\cong \frac{1}{2} \int_0^{\pi} d\theta\ \theta\ f(r)\ \exp\left[-\frac{1}{2} r\sqrt{\epsilon}\theta^2\right]$$

$$\cong \frac{1}{4} \int_0^{\infty} dx\ f(r)\ \exp\left[-\frac{1}{2} r\sqrt{\epsilon}x\right] = f(r)\ \frac{1}{2r\sqrt{\epsilon}},$$

where $f(r)$ stands for the angel-independent parts. The total spherical part of the semiclassical wavefunction can therefore be written as

$$\overline{\Psi}_{sc}(r) \cong \frac{C}{\epsilon^{1/4}(2\sqrt{\epsilon}r)^{1+\frac{1}{\sqrt{\epsilon}}}} \exp\left[-\int_{z_1}^{z_2} d\zeta \left[p_0 + \frac{p_\rho^2}{2p_0}\right] - \frac{\ln\sqrt{\epsilon}}{\sqrt{\epsilon}} - \frac{1}{\sqrt{\epsilon}} - \sqrt{\epsilon}r\right]. \tag{18.37}$$

Here, we have introduced the tunnel integral ranging from z_1 to z_2. The integral from z_1 to r has been evaluated approximately. The semiclassical wavefunction $\overline{\Psi}_{sc}$, Eq. (18.37), and the asymptotic exciton wavefunction Ψ_x, Eq. (18.36), have approximately the same r-dependence. We used $\epsilon^{-1/2}\ \ln 2r \cong \epsilon^{-1/2}$, which holds approximately for the matching region. A comparison of the coefficients yields

$$\psi(0) = \frac{\Gamma\left[1-\frac{1}{\sqrt{\epsilon}}\right] C}{\epsilon^{1/4}} \exp\left[-T_t - \frac{p_\rho^2}{2}\alpha_t\right] , \tag{18.38}$$

where

$$T_t = \int_{z_1}^{z_2} d\zeta \; p_0(\zeta) + \frac{\ln\sqrt{\epsilon}}{\sqrt{\epsilon}} + \frac{1}{\sqrt{\epsilon}} \tag{18.39}$$

and

$$\alpha_t = \int_{z_1}^{z_2} d\zeta \; \frac{1}{p_0(\zeta)} . \tag{18.40}$$

The α_t correction term becomes unimportant, because the summation over all p_ρ values brings the correction down from the exponent, so that it enters into the result as an unimportant prefactor. The evaluation of the tunnel integral yields

$$T_t = \frac{2\epsilon^{3/2}}{3\mathcal{F}} - \frac{1}{\sqrt{\epsilon}} \ln\left[\frac{8\epsilon^{3/2}}{\mathcal{F}}\right] , \tag{18.41}$$

see problems 18.1-18.4. The first term in Eq. (18.41) gives rise to the Franz-Keldysh result, while the second term describes the quite significant modification due to the Coulomb potential. The only unknown coefficient in the result is the normalization constant C. The existence of the exciton has little influence on the normalization which is determined by the asymptotic form of the wavefunction, see Chap. 10. Thus the normalization is the same as for free carriers, which has been derived in the previous chapter.

The resulting absorption spectrum is

$$\alpha(\omega) = \alpha_{FK}(\omega) \left\{ \Gamma\left[1 - \frac{1}{\sqrt{\epsilon}}\right] \exp\left[\frac{1}{\sqrt{\epsilon}} \ln\left[\frac{8\epsilon^{3/2}}{\mathscr{F}}\right]\right]\right\}^2$$

exciton electroabsorption tail-spectrum
for bulk semiconductors (18.42)

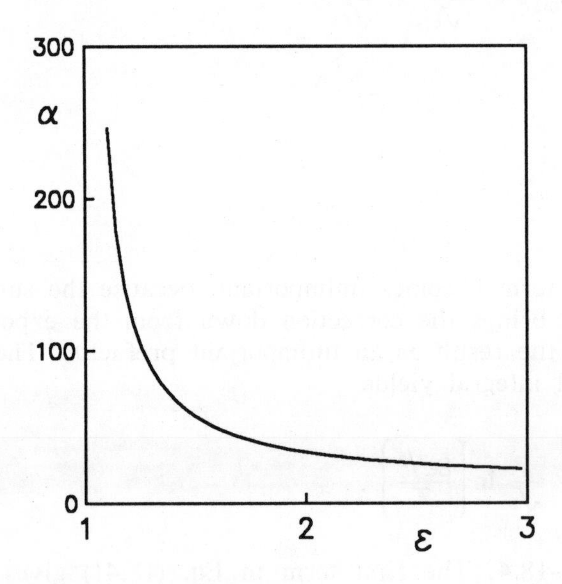

Fig. 18.2: Exciton enhancement of the electroabsorption according to Eq. (18.42). α is in units of α_{FK} *and* $\epsilon = (E_g - \hbar\omega)/E_0$.

where $\epsilon = (E_g - \hbar\omega)/E_0$ and $\alpha_{FK}(\omega)$ is the Franz-Keldysh absorption coefficient given in Eq. (17.33). An example of the absorption spectrum according to Eq. (18.42) is shown in Fig. 18.2. Depending on the detuning ϵ and the field strength $\mathscr{F}$, the exciton electroabsorption coefficient can be up to 10^3 times larger than the Franz-Keldysh absorption coefficient. The absorption approaches asymptotically the Franz-Keldysh spectrum only for very large detunings.

18-2. Quantum Wells

The spatial confinement in a quantum well prevents field ionization of the exciton up to very large field strengths. As a consequence one can observe very large Stark shifts of, e.g., the lowest exciton resonance in a field perpendicular to the layer of the quantum well (Miller *et al.*, 1985).

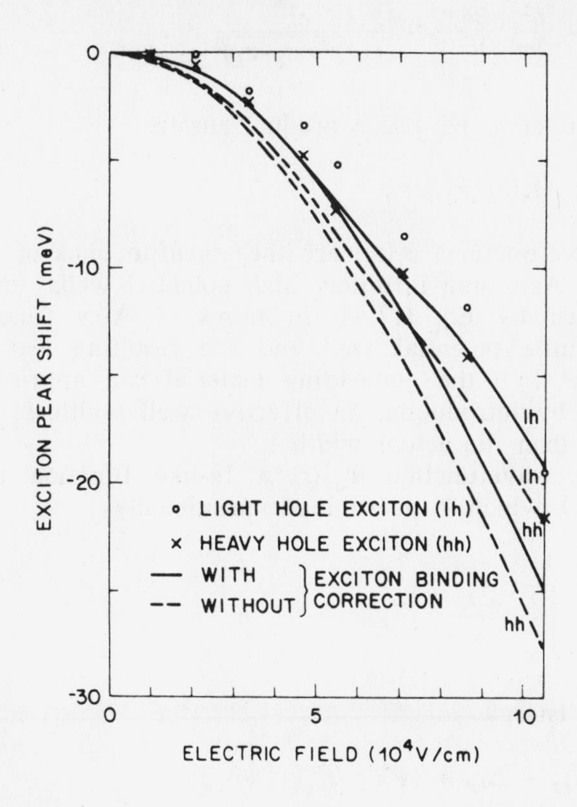

Fig. 18.3: Shift of the exciton peak position in a 95-Å multiple-quantum well as function of electric field. (From Miller *et al.* (1985).)

In order to treat the problem, we decompose it into that of the one-dimensional motion of the non-interacting electron and hole in the quantum-well potential $V_i(z_i)$ and the field, and into that of the relative electron-hole motion in the layer under the influence of the Coulomb interaction. We write the total Hamiltonian as

$$\mathcal{H} = \mathcal{H}_{ez} + \mathcal{H}_{hz} + \mathcal{H}_{eh} , \tag{18.43}$$

where

$$\mathcal{H}_{iz} = -\frac{\hbar^2}{2m_i}\frac{\partial^2}{\partial z_i^2} + V_i(z_i) \pm eEz_i \ , \ i = e, h \tag{18.44}$$

and

$$\mathcal{H}_{eh} \doteq -\frac{\hbar^2}{2m}\frac{\partial^2}{\partial r^2} + -\frac{e^2}{\epsilon_0\sqrt{r^2+(z_e-z_h)^2}} . \tag{18.45}$$

For the wavefunction we use a product ansatz

$$\psi = \psi_e(z_e)\psi_h(z_h)\Psi_{eh}(r) , \tag{18.46}$$

where the wavefunctions $\psi_i(z_i)$ are the eigenfunctions of the Hamiltonian (18.44). Assuming infinitely high potential wells, the functions $\psi_i(z_i)$ are given by Eq. (17.44) in terms of Airy functions. The effect of a finite potential well and the resulting leaking of the wave functions into the embedding material can approximately be accounted for by introducing an effective well width L_{eff} which is slightly larger than the actual width L.

For the wavefunction $\Psi_{eh}(r)$ a 1s-like function is assumed with a radius λ which is determined variationally

$$\Psi_{eh}(r) = \sqrt{\frac{2}{\pi}}\frac{e^{-\frac{r}{\lambda}}}{\lambda} . \tag{18.47}$$

The total pair energy

$$E_{eh} = E_{ez} + E_{hz} + \langle\Psi^* \mid H_{eh} \mid\Psi\rangle \tag{18.48}$$

is minimized with respect to λ for a given field F. The resulting energy shifts are shown in Fig. 18.3, together with the experimentally observed Stark shifts for the heavy and light hole (*hh* and *lh*) exciton of a GaAs quantum well with a width of 95 Å . One sees that Stark shifts up to 20 meV are obtained with an electric field of about 10^5 V/cm. Only above this large value of the electric field, field-induced tunneling sets in and broadens the exciton resonance. The quantum confined Stark shift of ≅ 20 meV is more than twice the exciton binding energy of ≅ 9meV in this quantum well.

REFERENCES

J.D. Dow and D. Redfield, Phys. Rev. **B1**, 3358 (1970)

I.A. Merkulov and V. I. Perel, Phys. Lett. **45A**, 83 (1973)

I.A. Merkulov, Sov. Phys. JETP **39**, 1140 (1974)

D. A. B. Miller, D.S. Chemla, T.C. Damen, A.C. Gossard, W. Wiegmann, T.H. Wood, and C.A. Burrus, Phys. Rev. **B32**, 1043 (1985)

PROBLEMS

Problem 18.1: Show that the tunnel integral, first term of Eq. (18.39), can be transformed into the expression

$$T_1 = \mathcal{F}^{1/2} \int_{z_1}^{z_2} \frac{dz}{\sqrt{z}} \sqrt{(z_2-z)(z-z_1)} \tag{18.49}$$

and further into the form

$$T_1 = \frac{\epsilon^{3/2}}{2^{3/2}\mathcal{F}} s^2 \int_{-1}^{+1} dt \sqrt{\frac{1-t^2}{1+st}} = \frac{\epsilon^{3/2}}{2^{3/2}\mathcal{F}} s^2 \, I(s) \, , \tag{18.50}$$

where $s = \sqrt{1-\frac{8\mathcal{F}}{\epsilon^2}} = \sqrt{1-y^2}$.

Problem 18.2: Show that the evaluation of the integral $I(s)$ defined in Eq. (18.50), for $s = 1$ gives the Franz-Keldysh result $I(1) = \frac{4}{3}\sqrt{2}$, but that $\left.\frac{dI(s)}{ds}\right|_{s=1} = \infty.$

Problem 18.3: $I(s)$ can be evaluated by noting that the most divergent part stems from the region $t \cong -1$. Calculate

$$I_1(s) = \sqrt{2} \int_{-1}^{+1} dt \sqrt{\frac{1+t}{1+st}}, \tag{18.51}$$

where the factor $\sqrt{1-t}$ has been approximated by $\sqrt{2}$. Show that

$$I_1(s) = \left(\frac{2}{s}\right)^{3/2} \left[-\frac{1-s}{4} \ln\left(\frac{1+\sqrt{2s/(1+s)}}{1+\sqrt{2s/(1+s)}} \right) + \frac{\sqrt{2s(1+s)}}{2} \right]. \tag{18.52}$$

Verify that one gets for small y^2

$$I_1(s) \cong 2\sqrt{2} - \frac{\sqrt{2}}{4} y^2 \ln\left(\frac{16}{y^2}\right) + \frac{3}{4}\sqrt{2}\ y^2 . \tag{18.53}$$

Problem 18.4: With the identity

$$I(s) = (I(s) - I_1(s)) + I_1(s) = I_2(s) + I_1(s) \tag{18.54}$$

one can determine I_2 approximately as

$$I_2(s) \cong I_2(1) + (s-1) \left.\frac{dI_2}{ds}\right|_{s=1} . \tag{18.55}$$

Show that in this approximation $I_2(s)$ is

$$I_2(s) = -\frac{2}{3}\sqrt{2} + \frac{\sqrt{2}}{2}\left(\frac{2}{3} - \ln 2\right) y^2 , \tag{18.56}$$

so that the total tunnel integral is given by

$$T_1 = \frac{1}{\sqrt{\epsilon}} \left[\frac{16}{3} \frac{1}{y^2} - \ln\left(\frac{64}{y^2}\right) - 1 \right], \tag{18.57}$$

from which one obtains Eq. (18.41).

Chapter 19
SEMICONDUCTOR QUANTUM WIRES

Motivated by the success of semiconductor quantum-well structures in permitting the study of quasi-two-dimensional phenomena, there is growing interest in structures with more pronounced quantum confinement effects. Examples are the quantum wires, where the electron and holes are free to move in one space dimension, and the quantum dots, where the carriers are confined in all space dimensions. Quantum dots will be discussed in the following Chap. 20. In this chapter we analyze some of the basic physical properties of electron-hole pairs in quantum wires. Attempts to fabricate quantum wires have been made by etching small stripes out of quantum-well material, or by indirect methods to effectively confine the carriers to quasi-one dimension.

The free-carrier absorption spectrum of a one-dimensional system has been computed in Chap. 4 and is shown in Fig. 4.1. In order to improve these results, we need to include the electron-hole Coulomb interaction. As it turns out, however, this is a somewhat tricky problem, since the Coulomb potential in one dimension has several pathological features. For example, the ground-state energy is infinite in one dimension (Loudon, 1959).

Real quantum wires are never strictly one-dimensional since they always have a finite extension in the other two space dimensions. As we will show in this section, this finite extension of the quasi-one dimensional system makes the ground-state energy finite.

19-1. One-Dimensional Electron-Hole Pair

The ideal one-dimensional hydrogen atom has been studied by Loudon (1959). Since these results are also appropriate for the electron-hole pair in an ideal one-dimensional semiconductor, we summarize the main points of Loudon's results in this section. Denoting the coordinate in the unconfined direction by z, we can write the electron-hole Schrödinger equation as

$$-\left[\frac{\hbar^2}{2m_r}\frac{d^2\psi}{dz^2}+\frac{e^2}{\epsilon_0|z|}\right]\psi(z) = E\ \psi(z)\ . \tag{19.1}$$

We are interested only in the bound-state solutions of Eq. (19.1). As in Chap. 10, we introduce the dimensionless quantities

$$\rho = z\alpha\ ,\quad \lambda = \frac{e^2 2m_r}{\epsilon_0\hbar^2\alpha}\ ,\quad E = -\frac{m_r e^4}{2\hbar^2\epsilon_0}\frac{1}{\lambda^2} \tag{19.2}$$

with α and λ given by Eqs. (10.49) and (10.51), respectively. Thus, Eq. (19.1) becomes

$$-\left[\frac{d^2\psi}{d\rho^2}+\frac{\lambda}{|\rho|}\right]\psi(\rho) = -\frac{1}{4}\ \psi(\rho)\ , \tag{19.3}$$

compare Eq. (10.50). Eq. (19.3) has the form of Whittaker's equation for the confluent hypergeometric functions (Abramowitz and Stegun, 1970). The solution of this equation for the regions $\rho > 0$ and $\rho < 0$ is relatively easy, but because of the pole in the potential at $\rho = 0$, it is not immediately obvious how the solutions in the two regions should be joined together at the origin.

To resolve this problem, one may replace the actual potential by the regularized potential

$$V(\rho) = \frac{1}{\rho_c+|\rho|} \tag{19.4}$$

and take the limit $\rho_c \to 0$ in the final results. The wave equation (19.3) now has the form

$$-\left[\frac{d^2\psi}{d\rho^2}+\frac{\lambda}{\rho_c+|\rho|}\right]\psi(\rho) = -\frac{1}{4}\ \psi(\rho)\ . \tag{19.5}$$

In his 1959 paper Loudon explicitly solved Eq. (19.5) and showed that the ground state has an even nodeless wavefunction. The wavefunctions for the excited states are alternately odd and even.

The mathematical solution of Eq. (19.5) will be presented in the following Sec. 19-2. In this section we only want to discuss the qualitative behavior of the eigenfunctions and eigenvalues and their behavior for $\rho_c \rightarrow 0$. When the cutoff distance ρ_c is reduced to zero, there is no problem with the solutions involving odd wavefunctions, since $\psi(\rho)$ in this case is zero at the origin and $\psi(\rho)/(\rho_c+|\rho|)$ is well behaved when we decrease ρ_c. However, for the states with an even wavefunction, $\psi(\rho)$ is finite at the origin. In order to satisfy Eq. (19.5), $d^2\psi/d\rho^2$ must become very large for $\rho \rightarrow 0$. In the limit $\rho_c = 0$, $d^2\psi/d\rho^2$ diverges at $\rho=0$, indicating that the even wavefunctions have infinite curvature and, hence, discontinuous slopes in the origin. It turns out that the functional form of an even state in the regions $\rho > 0$ and $\rho < 0$ becomes the same as that of the odd state which is energetically next below it in the modified potential (19.4). However, for the even states the solutions in the two regions are joined together at the origin to form a function of even parity, which is zero in the origin and has a discontinuous slope there. In the limit $\rho_c \rightarrow 0$, the odd and even states become degenerate in pairs.

The only exception to this behavior occurs for the ground state. The ground-state eigenfunction has zero slope at $\rho=0$. Therefore, as $\rho_c \rightarrow 0$, the energy eigenvalue E_0 must become large and negative to compensate the diverging potential in Eq. (19.5). The wavefunction becomes concentrated more and more in the region of the origin and in the limit $\rho_c = 0$, the ground state has infinite binding energy and its probability density is a delta function $\delta(\rho)$.

19-2. Quasi-One-Dimensional Electron-Hole Pair

The results of Sec. 19-1. for the purely one-dimensional electron-hole pair are clearly unphysical. To resolve the problem we must consider a more sophisticated model of an exciton in a quantum wire (Banyai *et al.* 1987). For this purpose, we model the quantum wire as a cylindrical wire with a finite radius R. The electrons and holes are then confined in the cylindrical potential, and the wavefunctions have to vanish at the boundaries.

The Coulomb potential is thus not simply $e^2/\epsilon_0|z_e-z_n|$ but in fact

$$V(\mathbf{r}_e, \mathbf{r}_h) = \frac{e^2}{\epsilon_0|\mathbf{r}_e - \mathbf{r}_h|}. \tag{19.6}$$

Instead of the one-dimensional problem, Eq. (19.1), we then have to solve the three-dimensional Schrödinger equation

$$\left[\frac{\hbar^2}{2m_e}\nabla_e^2 + \frac{\hbar^2}{2m_h}\nabla_h^2 + V_e(\mathbf{r}_e) + V_h(\mathbf{r}_h) + V(\mathbf{r}_e,\mathbf{r}_h)\right] \Psi(\mathbf{r}_e,\mathbf{r}_h)$$

$$= E\ \Psi(\mathbf{r}_e,\ \mathbf{r}_h)\ , \tag{19.7}$$

where $V_e(\mathbf{r}_e)$ and $V_h(\mathbf{r}_h)$ are the confining potentials due to the wire. To solve Eq. (19.7), we make the ansatz that the wavefunction $\Psi(\mathbf{r}_e,\mathbf{r}_h)$ is separable and has cylindrical symmetry about the z-axis. Changing to cylindrical coordinates $\{\mathbf{r}_e\} \rightarrow \{\rho_e,\ z_e\}$, we write

$$\Psi(\mathbf{r}_e,\mathbf{r}_h) = \phi_e(\rho_e)\ \phi_h(\rho_h)\ \psi(z_e,z_h)\quad . \tag{19.8}$$

The $\phi_{e,h}$ are the single-particle ground-state wavefunctions corresponding to solutions of

$$\left[\frac{\hbar^2}{2m_e}\ \frac{1}{\rho_e}\ \frac{\partial}{\partial\rho_e}\ \rho_e\ \frac{\partial}{\partial\rho_e} + V_e(\rho_e)\right] \phi_e(\rho_e) = E_e\ \phi_e\ (\rho_e)\ , \tag{19.9}$$

and similarly for $\phi_h(\rho_h)$ with $e \rightarrow h$. For the infinite barrier model we can write these wavefunctions analytically

$$\phi_e(\rho_e) = \frac{J_0(\alpha_0\rho_e/R)}{\sqrt{\pi}\ R\ J_1(\alpha_0)}\ , \tag{19.10}$$

where $J_n(r)$ is the Bessel function of order n. The corresponding confinement energy is

$$E_e = \frac{\alpha_0^2\hbar^2}{2m_e R^2}\ , \tag{19.11}$$

where $\alpha_0 = 2.405$ is the first zero of $J_0(x) = 0$. The denominator in Eq. (19.10) comes simply from the normalization of the wavefunction (see problem 19.2).

Returning to our original problem of Eq. (19.7) we may now use Eq. (19.8) to simplify the solution. Substituting Eq. (19.8) into Eq. (19.7), multiplying by ϕ_e, ϕ_h and integrating over the transverse plane for both electrons and holes, we obtain

$$\left[\frac{\hbar^2}{2m_e} \frac{\partial^2}{\partial z_e^2} + \frac{\hbar^2}{2m_h} \frac{\partial^2}{\partial z_h^2} + V^{eff}(z_e, z_n) \right] \psi(z_e, z_h)$$

$$= (E - E_e - E_h)\, \psi(z_e, z_h) \,, \tag{19.12}$$

where

$$V^{eff}(z_e, z_n) = - \frac{e^2}{\epsilon_0 \pi^2 R^4 J_1^4(\alpha_0)} \int_0^R d\rho_e \rho_e \int_0^R d\rho_h \rho_h \int_0^{2\pi} d\theta_e \int_0^{2\pi} d\theta_h$$

$$\times \frac{J_0^2(\alpha_0 \rho_e / R)\, J_0^2(\alpha_0 \rho_h / R)}{[(z_e - z_h)^2 + (\rho_e \cos\theta_e - \rho_h \cos\theta_h)^2 + (\rho_e \sin\theta_e - \rho_n \sin\theta_n)^2\,]^{1/2}}$$

effective quasi-one-dimensional Coulomb potential (19.13)

This is the effective potential for the quasi-one-dimensional exciton problem. Notice, that V^{eff} is simply an average over the radial coordinates of the potential between the electron and hole and depends only on their separation. With this in mind we can move into center-of-mass and relative coordinates

$$Z = \frac{z_e + z_h}{2} \quad , \quad z = z_e - z_h \,. \tag{19.14}$$

As usual, the center-of-mass motion describes the overall movement of the exciton as a unit (problem 19.3). Here, we are interested in the relative motion, z, which determines the binding energy. This yields

$$\frac{\hbar^2}{2m_r} \frac{\partial^2}{\partial z^2} \psi(z) + V_{eff}(z)\, \psi(z) = E_b \psi(z) \,, \tag{19.15}$$

where now $E_b = E - E_e - E_h$ is defined as the binding energy.

Comparing Eq. (19.15) with Eq. (19.1), we see that $V^{eff}(z)$ is no longer singular at z=0 and we expect to recover a physically reasonable ladder of exciton states. We now go on to calculate the wavefunctions and exciton binding energies which one gets using such an effective potential. At this point we have the choice of solving the Schrödinger equation with $V^{eff}(z)$ numerically, or variationally, or by making some approximation to V^{eff}, which allows an analytic solution. We follow the ladder approach and approximate

$$V(z) = \frac{e^2}{\epsilon_0(|z|+\gamma R)} \quad . \tag{19.16}$$

The fitting parameter γ may be varied to give a good approximation to $V^{eff}(z)$. Using the potential (19.16) and introducing the dimensionless variables defined in Eq. (19.2), Eq. (19.15) becomes identical to Eq. (19.5). Here we denote

$$\rho_c = R\gamma\alpha \ , \tag{19.17}$$

showing that the finite radius of the quantum wire introduces a natural regularization (cut-off) of the Coulomb potential in the origin. As mentioned above, the eigenfunctions are given in terms of

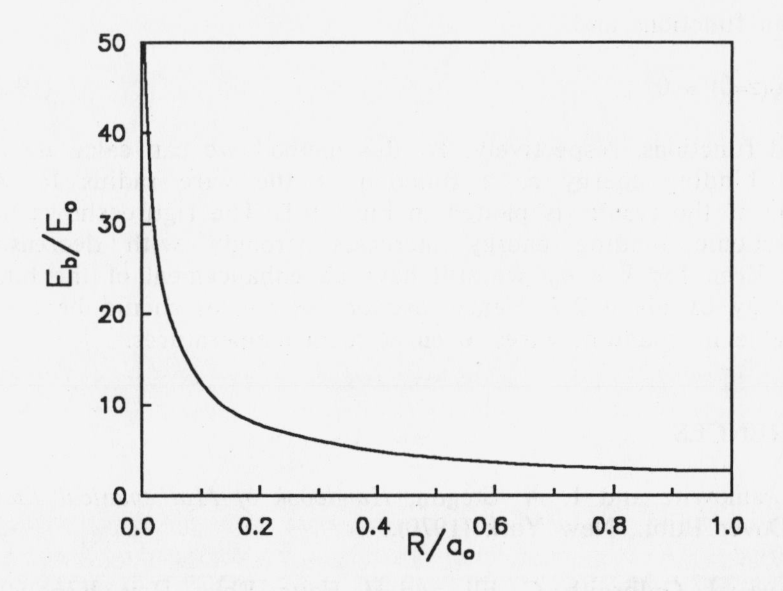

Fig. 19.1: Excitonic binding energy E_b/E_0 as function of quantum-wire radius R/a_0. Here E_0 and a_0 denote the exciton Rydberg energy and Bohr radius of the corresponding bulk material, respectively. (After Banyai *et al.*, 1987).

the Whittaker functions W as

$$\psi_\lambda(|z|) = W_{\lambda,1/2}\left[\frac{2(|z|+\alpha R)}{\lambda a_0}\right] \tag{19.18}$$

where a_0 is the bulk-exciton Bohr radius, Eq. (10.71). The Whittaker functions can be defined through

$$W_{\lambda,1/2}(x) = \frac{e^{-x/2}}{\Gamma(1-\lambda)} \int_0^\infty dy\, e^{-y} \left[\frac{x}{y} + 1\right]^\lambda . \tag{19.19}$$

The energy eigenvalues as a function of λ can be obtained from the solution of the transcendental equation

$$\left.\frac{d\psi(z)}{dz}\right|_{z=0} = 0 \tag{19.20}$$

for even functions and

$$\psi(z=0) = 0 \tag{19.21}$$

for odd functions, respectively. By this method we can calculate the exciton binding energy as a function of the wire radius R. An example of the results is plotted in Fig. 19.1. The figure shows that the excitonic binding energy increases strongly with decreasing radius. Even for $R \cong a_0$, we still have an enhancement of the binding energy by about 2.7. Hence, exciton resonances should be easily observable in quantum wires, even at room temperatures.

REFERENCES

M. Abramowitz and I. A. Stegun, *Handbook of Mathematical Functions*, Dover Publ., New York (1970).

L. Banyai, I. Galbraith, C. Ell, and H. Haug, Phys. Rev. **B36**, 6099 (1987).

R. Loudon, Am. J. Phys. **27**, 649 (1959).

PROBLEMS

Problem 19.1: Follow Loudon (1959) to solve Eq. (19.8) explicitly. Show that the ground-state energy diverges for $\rho_c \to 0$.

Problem 19.2: Solve the single-particle wave equation (19.9) and discuss the first few eigenfunctions and energy eigenvalues.

Problem 19.3: Transform the effective electron-hole Schrödinger equation (19.12) into relative and center-of-mass coordinates. Discuss the center-of-mass motion.

Chapter 20
SEMICONDUCTOR QUANTUM DOTS

The ultimate quantum-confinement effects occur in very small semiconductor micro-crystallites (quantum dots), which confine the laser-excited electron-hole pairs in all three space dimensions. Presently available examples of such systems are colloids or semiconductor microcrystallite-doped glasses, as well as microstructures obtained by sophisticated etching procedures. Special glasses doped with CdS, CdSe, CuCl, or CuBr crystallites can be fabricated, which clearly exhibit quantum confinement. The microcrystallites in these glasses form out of the supersaturated solid solution of the basic constituents originally brought into the glass melt. The crystallites are more or less randomly arranged in the glass matrix and they exhibit a certain size distribution. Average crystallite sizes from around 10Å up to several 100Å have been produced.

20-1. Effective Mass Approximation

For most of the discussion in this chapter we consider spherical semiconductor microcrystallites with a radius R and background dielectric constant ϵ_2, which are embedded in another material with background dielectric constant ϵ_1. It is reasonably straightforward to modify the results for other (simple) geometries, such as microcubes or boxes. We call a semiconductor microcrystallite a quantum dot if

$$\ell_c << R \cong a_0 \ , \tag{20.1}$$

where ℓ_c is the characteristic length of the semiconductor lattice (unit cell). Hence, the quantum dot has a macroscopic size in comparison to the unit cell, but it is small on all macroscopic scales. One often calls quantum dots, quantum wires, and also quantum wells *mesoscopic* structures.

As a consequence of the inequality (20.1), it is reasonable for mesoscopic structures to make the usual bulk-semiconductor ansatz for the single-particle wavefunctions,

$$\psi(\mathbf{r}) = \phi(\mathbf{r})\, u(\mathbf{r}), \tag{20.2}$$

where $u(\mathbf{r})$ is the Bloch function, Eq. (3.27), and $\phi(\mathbf{r})$ is the envelope function, which varies on the scale of several unit cells. As discussed in Chap. 3, the Bloch function has the periodicity of the lattice

$$u(\mathbf{r}+\mathbf{n}) = u(\mathbf{r}). \tag{20.3}$$

However, in contrast to bulk semiconductors, $\psi(\mathbf{r})$ has to satisfy the boundary conditions of the quantum dot. For simplicity, we assume ideal quantum confinement, i.e.,

$$\psi(r \geq R) = 0 \ . \tag{20.4}$$

In the spirit of the ansatz (20.2), it is reasonable to assume that the energy eigenvalues of the electron in the periodic lattice, i.e., the energy bands, are not appreciably modified through the quantum confinement. Therefore, we use the effective mass approximation to describe the free motion of electrons and holes.

The Hamiltonian for one electron-hole pair is

$$\mathcal{H} = \mathcal{H}_e + \mathcal{H}_h + V_{ee} + V_{hh} + V_{eh} \tag{20.5}$$

where the kinetic terms are

$$\mathcal{H}_e = -\frac{\hbar^2}{2m_e} \int d\mathbf{r}\ \hat{\psi}_e^\dagger(\mathbf{r})\ \nabla^2 \hat{\psi}_e(\mathbf{r}) + E_g \int d\mathbf{r}\ \hat{\psi}_e^\dagger(\mathbf{r})\ \hat{\psi}_e(\mathbf{r})\ , \tag{20.6}$$

$$\mathcal{H}_h = -\frac{\hbar^2}{2m_h} \int d\mathbf{r}\ \hat{\psi}_h^\dagger(\mathbf{r})\ \nabla^2 \hat{\psi}_h^\dagger(\mathbf{r}), \tag{20.7}$$

and the Coulomb interaction is described by

$$V_{ee} = \frac{1}{2} \iint d\mathbf{r} d\mathbf{r}' \, \hat{\psi}_e^\dagger(\mathbf{r}) \, \hat{\psi}_e^\dagger(\mathbf{r}') \, V(\mathbf{r},\mathbf{r}') \, \hat{\psi}_e(\mathbf{r}') \, \hat{\psi}_e(\mathbf{r}) \; , \tag{20.8}$$

$$V_{eh} = - \iint d\mathbf{r} \; d\mathbf{r}' \; \hat{\psi}_e^\dagger(\mathbf{r}) \hat{\psi}_h^\dagger(\mathbf{r}') \; V(\mathbf{r},\mathbf{r}') \; \hat{\psi}_h(\mathbf{r}') \; \hat{\psi}_e(\mathbf{r}) \; , \tag{20.9}$$

and

$$V_{hh} = V_{ee} \; (e \rightarrow h) \; , \tag{20.10}$$

where $V(r)$ is the effective Coulomb interaction potential inside the microsphere.

The Coulomb interaction between two point charges in a semiconductor microsphere, which is embedded in a material with different background dielectric constant, is

$$V(\mathbf{r}_1,\mathbf{r}_2)\Big|_R = V(\mathbf{r}_1,\mathbf{r}_2)\Big|_{R=\infty} + \delta V(\mathbf{r}_1,\mathbf{r}_2), \tag{20.11}$$

where $V(\mathbf{r}_1,\mathbf{r}_2)\Big|_{R=\infty}$ is the usual bulk Coulomb interaction, and the additional term is caused by the induced surface charge of the sphere,

$$\delta V(\mathbf{r}_1,\mathbf{r}_2) = Q_1(\mathbf{r}_1) + Q_1(\mathbf{r}_2) \mp Q_2(\mathbf{r}_1,\mathbf{r}_2) \; , \tag{20.12}$$

with - (+) for charges with opposite (equal) sign, respectively (Brus, 1984). The different contributions in Eq. (20.12) are

$$Q_1(\mathbf{r}) = \sum_{n=0}^{\infty} Q_{1,n}(r) \tag{20.13}$$

with

$$Q_{1,n}(r) = \frac{e^2}{2R} \, \alpha_n \, (r/R)^{2n} \tag{20.14}$$

and

$$Q_2(\mathbf{r}_1,\mathbf{r}_2) = \sum_{n=0}^{\infty} Q_{2,n}(\mathbf{r}_1,\mathbf{r}_2), \tag{20.15}$$

with

$$Q_{2,n}(\mathbf{r}_1,\mathbf{r}_2) = \alpha_n \frac{e^2}{R} \left[\frac{r_1\ r_2}{R^2}\right]^n P_n(\cos(\theta)), \tag{20.16}$$

where θ is the angle between $\mathbf{r}_1$ and $\mathbf{r}_2$, P_n is the n-th order Legendre polynomial and

$$\alpha_n = \frac{(\epsilon_2/\epsilon_1 - 1)(n+1)}{\epsilon_2(n\,\epsilon_2/\epsilon_1 + n + 1)} \,. \tag{20.17}$$

Obviously, the surface polarization term δV vanishes for $\epsilon_1 = \epsilon_2$.

20-2. Single Particle Properties

The eigenstates and energy eigenvalues for a single electron in the quantum dot are determined by the Schrödinger equation

$$\mathcal{H}\ |\psi_e\rangle = E_e\ |\psi_e\rangle\ . \tag{20.18}$$

The eigenstate is of the form

$$|\psi_e\rangle = \int d\mathbf{r}\ \phi_e(\mathbf{r})\ \hat{\psi}_e^\dagger(\mathbf{r})\ |0\rangle\ , \tag{20.19}$$

where $|0\rangle$ is the crystal ground-state, i.e., the state without excited electrons or holes. The coefficients $\phi(\mathbf{r})$ in Eq. (20.19) have to be determined from Eq. (20.18). Using the Hamiltonians (20.7) - (20.10) in Eq. (20.18), we find

$$V_{ee}\,|\psi_e\rangle = V_{eh}\,|\psi_e\rangle = V_{hh}\,|\psi_e\rangle = \mathcal{H}_h\,|\psi_e\rangle = 0\ . \tag{20.20}$$

However, we have

$$\mathcal{H}_e|\psi_e\rangle = -\frac{\hbar^2}{2m_e}\int d\mathbf{r}'\,[\nabla^2\hat{\psi}_e^\dagger(\mathbf{r}')]\,\hat{\psi}_e(\mathbf{r}')\int d\mathbf{r}\,\phi_e(\mathbf{r})\,\hat{\psi}_e^\dagger(\mathbf{r})\,|0\rangle$$

$$+\,E_g\int d\mathbf{r}'\int d\mathbf{r}\,\hat{\psi}_{e\mathbf{r}'}^\dagger\,\psi_e(\mathbf{r}')\,\phi_e(\mathbf{r})\,\hat{\psi}_e^\dagger(\mathbf{r})\,|0\rangle$$

$$= -\frac{\hbar^2}{2m_e}\int d\mathbf{r}'\int d\mathbf{r}\,\phi_e(\mathbf{r})\,\delta(\mathbf{r}-\mathbf{r}')\,[\nabla^2\hat{\psi}_e^\dagger(\mathbf{r}')]\,|0\rangle$$

$$+\,E_g\int d\mathbf{r}'\int d\mathbf{r}\,\phi_e(\mathbf{r})\,\hat{\psi}_e^\dagger(\mathbf{r})\,\delta(\mathbf{r}-\mathbf{r}')\,|0\rangle$$

$$= -\frac{\hbar^2}{2m_e}\int d\mathbf{r}\,[\nabla^2\phi_e(\mathbf{r})]\,\hat{\psi}_e^\dagger(\mathbf{r})\,|0\rangle + E_g\int d\mathbf{r}\,\phi_e(\mathbf{r})\,\hat{\psi}_e^\dagger(\mathbf{r})\,|0\rangle$$

$$= E_e\int d\mathbf{r}\,\phi_e(\mathbf{r})\,\hat{\psi}_e^\dagger(\mathbf{r})\,|0\rangle\,. \tag{20.21}$$

This equation is satisfied if

$$-\frac{\hbar^2}{2m_e}\nabla^2\phi_e(\mathbf{r}) = (E_e-E_g)\,\phi_e(\mathbf{r})\,, \tag{20.22}$$

which is the one-electron eigenvalue equation. Similarly, we find for the one-hole state

$$-\frac{\hbar^2}{2m_h}\nabla^2\,\phi_h(\mathbf{r}) = E_h\phi_h(\mathbf{r})\,. \tag{20.23}$$

The problem is completely defined with the boundary conditions

$$\phi_e(\mathbf{r}) = \phi_h(\mathbf{r}) = 0 \text{ for } |\mathbf{r}| \geq R\,. \tag{20.24}$$

The solution is

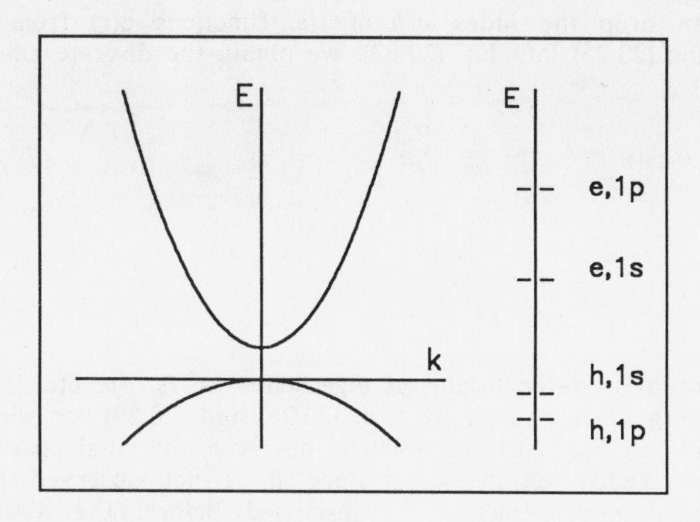

Fig. 20.1: Schematic plot of the single particle energy spectrum in bulk semiconductors (left). The single particle energies for electrons (e) and holes (h) in small quantum dots are shown in the right part of the figure.

$$\boxed{\phi_{e,n\ell m}(\mathbf{r}) = \sqrt{\frac{2}{R^3}}\,\frac{j_\ell(\alpha_{n\ell} r/R)}{j_{\ell+1}(\alpha_{n\ell})}\,Y_{\ell,m}(\Omega)}$$

single particle wavefunction in quantum dot (20.25)

where j_ℓ is the spherical Bessel function of order ℓ and $Y_{\ell,m}$ denotes the spherical harmonics, see Eq. (10.55). The boundary condition (20.24) is satisfied if

$$j_\ell(\alpha_{n\ell}) = 0 \ , \text{ for } n = 1,2,... \tag{20.26}$$

and

$$\alpha_{10} = \pi,\ \alpha_{11} = 4.4934,\ \alpha_{12} = 5.7635,\ \alpha_{20} = 6.2832,$$

$$\alpha_{21} = 7.7253,\ \alpha_{22} = 9.0950,\ \alpha_{30} = 9.4248\ ... \tag{20.27}$$

Since the wavefunction (20.25) depends only on R and not on any physical parameters which are specific for the electron, the corresponding wavefunction for the hole must have the same form, and we can drop the index e/h of the functions $\phi(\mathbf{r})$ from now on. Inserting (20.25) into Eq. (20.23), we obtain the discrete energies

$$E_{e,n\ell m} = E_g + \frac{\hbar^2}{2m_e} \frac{\alpha_{n\ell}^2}{R^2} \tag{20.28}$$

and

$$E_{h,n\ell m} = \frac{\hbar^2}{2m_h} \frac{\alpha_{n\ell}^2}{R^2} . \tag{20.29}$$

It is usual to refer to the $n\ell$ eigenstates as $1s$, $1p$, etc. The lowest two energy levels given by Eqs. (20.28) and (20.29) are shown schematically in Fig. 20.1. In practice, however, the single particle spectrum is rather uninteresting since it is not observed in optical absorption measurements. As discussed before, the absorption is always given by the electron-hole-pair excitation spectrum.

20-3. Pair States

For the electron-hole-pair eigenstates we make the ansatz

$$|\psi_{eh}\rangle = \int\!\!\int d\mathbf{r}_e \; d\mathbf{r}_h \; \psi_{eh}(\mathbf{r}_e,\mathbf{r}_h) \; \hat{\psi}_e^\dagger(\mathbf{r}_e) \; \hat{\psi}_h^\dagger(\mathbf{r}_h) \; |0\rangle \tag{20.30}$$

and obtain the Schrödinger equation for the pair

$$\left[-\frac{\hbar^2}{2m_e}\nabla_e^2 - \frac{\hbar^2}{2m_h}\nabla_h^2 - V(\mathbf{r}_e,\mathbf{r}_h)\right] \psi_{eh}(\mathbf{r}_e,\mathbf{r}_h) = (E - E_g) \; \psi_{eh}(\mathbf{r}_e,\mathbf{r}_h). \tag{20.31}$$

Because of the boundary conditions

$$\psi_{eh}(\mathbf{r}_e,\mathbf{r}_h) = 0 \quad \text{if} \quad |\mathbf{r}_e| > R \; , \; \text{or} \; |\mathbf{r}_h| > R \; , \tag{20.32}$$

it is not useful to introduce relative and center-of-mass coordinates, in contrast to the case of bulk or quantum-well semiconductor materials (Chap. 10).

If the quantum dot radius is smaller than the bulk-exciton Bohr radius, $R < a_0$, electron and hole are closer together than they would be in the corresponding bulk material. This leads to a dramatic increase of the pair energy with decreasing quantum-dot size. As function of quantum-dot radius, the kinetic part of the energy varies like

$$\langle \mathcal{H}_e + \mathcal{H}_h \rangle \propto \frac{1}{R^2} , \tag{20.33}$$

whereas the interaction part behaves like

$$\langle V_{ij} \rangle \propto \frac{1}{R} . \tag{20.34}$$

To obtain an estimate of the pair energy for small dot radii, $R \ll a_0$, it is a reasonable first-order approximation to consider the electrons and holes essentially as non-interacting and ignore the Coulomb energy in comparison to the kinetic energy. This yields

$$E_{eh,n\ell m} = E_{e,n\ell m} + E_{h,n\ell m} , \tag{20.35}$$

i.e., an energy variation proportional to R^{-2}. Experimentally, this increasing pair energy is observed as a pronounced blue shift of the onset of absorption with decreasing dot size.

It is not analytically possible to solve the pair Schrödinger equation (20.31) including the Coulomb interaction. Therefore, one has to use numerical or approximative methods. One method consists of expanding full pair-state into the eigenstates of the system without Coulomb interactions.

$$|\psi_{eh,\ell m}\rangle = \sum_{\substack{n_1,n_2 \\ \ell_1,\ell_2}} C_{n_1,n_2,\ell_1,\ell_2} \, |n_1 n_2 \ell_1 \ell_2;\ell m\rangle . \tag{20.36}$$

For such an expansion it is important to note that the total angular momentum operator $\hat{L}$ commutes with the Hamiltonian, i.e., the angular momentum is a good quantum number and the eigenstates of the Hamiltonian are also eigenstates of $\hat{L}^2$ and $\hat{L}_z$. A well converg-

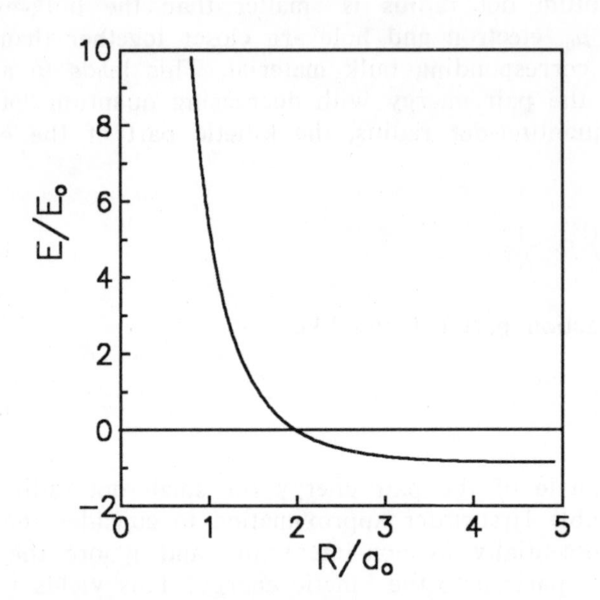

Fig. 20.2: Plot of the ground-state energy of one electron-hole pair in a quantum dot. Energy and radius are in units of the bulk-exciton Rydberg E_0 and Bohr radius a_0, respectively. The electron-hole mass ratio has been choosen as m_e/m_h=0.1.

ing expansion is obtained if one constructs the set of expansion functions already as eigenfunctions of $\hat{L}^2$ and $\hat{L}_z$. A convenient choice of the one-pair-state basis functions is

$$|n_1 n_2 \ell_1 \ell_2; \ell m\rangle = \sum_{m_1 m_2} \langle \ell_1 m_1 \ell_2 m_2 | \ell m\rangle \, | n_1 \ell_1 m_1\rangle_e \, | n_2 \ell_2 m_2\rangle_h , \qquad (20.37)$$

where $\langle \ell_1 m_1 \ell_2 m_2 | \ell m\rangle$ is the Clebsh-Gordon coefficient and ℓ and m the angular momentum quantum numbers of the pair state.

Using the expansion (20.37), we compute the expectation value of the Hamiltonian truncating the expansion at finitie values of n, ℓ and m. This transforms the Hamiltonian into a matrix. Eqs. (20.33) - (20.34) show that for the regime of sufficiently small quantum dots, the single-particle energies are much larger than the Coulomb contributions. Therefore, the off-diagonal elements in the Hamiltonian matrix are small in comparsion to the diagonal elements and the truncatation of the expansion introduces only small errors. The magnitude of these errors can be checked by using increasingly large $n_i \ell_i m_i$ values. Numerical diagonalization of the resulting matrix yields the energy eigenvalues and the expansion coefficients of the pair wavefunction (Hu *et al.*, 1990). Examples of the results are shown in Fig. 20.2 where we plot the ground-state energy E_{1s} for one electron-hole pair as function of the quantum-dot radius. This figure clearly shows the sharp energy increase for smaller dots expected from Eq. (20.35).

For convenience we also use the notation *1s*, *1p*, etc. for the situation with Coulomb interaction. This notation indicates that the leading term in the wavefunction expansion is the product of the *1s* single-particle functions,

$$\psi_{eh}(r_e, r_h) \cong \phi_{100}(r_e)\, \phi_{100}(r_h) + \text{other states} . \tag{20.38}$$

Note, however, that if one keeps only this product state and neglects the rest, one may get completely wrong answers for quantities like binding energies or transition dipoles.

In Fig. 20.3 we show an example of the radial distribution

$$Pr_{e/h} = r_{e/h}^2 \int d\Omega_{e/h}\, d\mathbf{r}_{h/e}\, r_{h/e}^2\, |\psi_{eh}(\mathbf{r}_e, \mathbf{r}_h)|^2 \tag{20.39}$$

of electron and hole in the quantum dot. Fig. 20.3 shows that, as a consequence of the electron-hole Coulomb interaction, the heavier particle, i.e., the hole, is pushed toward the center of the sphere.

In addition to the numerical solution, one can also find an approximate analytical solution for the one-electron-hole-pair state in quantum dots if the electron-hole mass ratio, m_e/m_h, is very small. In this case, the motion of the particles may be approximately decoupled, as in the hydrogen-atom problem, and

$$a_0 \cong \frac{\hbar^2}{m_e e^2} . \tag{20.40}$$

Since the electron states are well separated in energy, we may concentrate on the lowest single electron states. The single-particle hole states are closer in energy and correlations between the states are

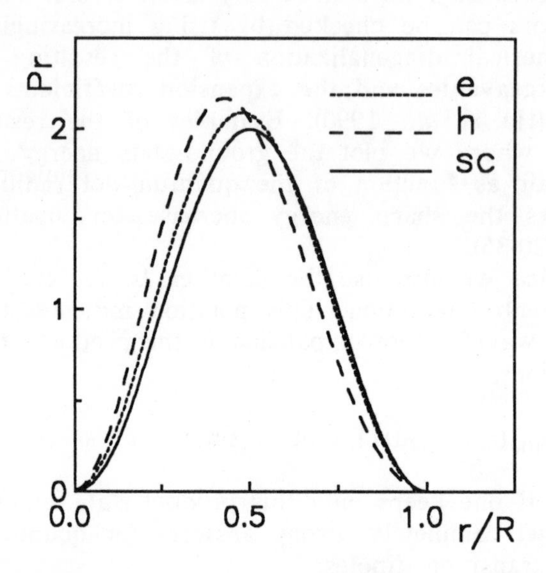

Fig. 20.3: Plot of the distribution, Eq. (20.39) of an electron (e) and a hole (h) in a quantum dot with R/a_0=1. The full line (sc) shows the distribution if one neglects the Coulomb interaction.

more important. Therefore, we use the ansatz for the pair-wave-function

$$\psi_{eh,n\ell m}(\mathbf{r}_e,\mathbf{r}_h) \cong \phi_{n\ell m}(\mathbf{r}_e)\ \psi_h(\mathbf{r}_h) \tag{20.41}$$

Inserting (20.41) into Eq. (20.31) and projecting with $\phi^*_{n\ell m}(\mathbf{r}_e)$ we obtain

$$\left[-\frac{\hbar^2}{2m_e}\nabla_h^2 - \int d\mathbf{r}_e\ |\phi_{n\ell m}(\mathbf{r}_e)|^2\ V(\mathbf{r}_e,\mathbf{r}_h)\right]\psi_h(\mathbf{r}_h)$$

$$= \left[E_{eh} - E_g - \frac{\hbar^2}{2m_e}\frac{\alpha_{n\ell}^2}{R^2}\right]\psi_h(\mathbf{r}_h)\ . \tag{20.42}$$

Eq. (20.42) describes the motion of the hole in the average potential induced by the electron. The potential is attractive and if we take $(n\ell m) = (100)$ it is spherically symmetrical. This effective potential is responsible for pushing the hole toward the center of the sphere, as shown in Fig. 20.3.

20-4. Dipole Transitions

In order to compute the optical response of semiconductor quantum dots, we need the dipole transition matrix elements between the different electron-hole-pair states. The interaction Hamiltonian is written as

$$\hat{\mathcal{H}}_{int} = \hat{P}\, \mathcal{E}(t) \ , \tag{20.43}$$

where $\hat{P}$ is the polarization operator and $\mathcal{E}(t)$ is the light field. The polarization is a single-particle operator, which can be written as

$$\hat{P} = \int d\mathbf{r} \sum_{i,j=e,h} \hat{\psi}_i^\dagger(\mathbf{r})\ e\mathbf{r}\ \hat{\psi}_j(\mathbf{r})$$

$$= \int d\mathbf{r}\ e\mathbf{r}\ [\hat{\psi}_e^\dagger(\mathbf{r})\hat{\psi}_e(\mathbf{r}) + \hat{\psi}_h(\mathbf{r})\hat{\psi}_h^\dagger(\mathbf{r}) + \hat{\psi}_e^\dagger(\mathbf{r})\hat{\psi}_h^\dagger(\mathbf{r}) + \hat{\psi}_h(\mathbf{r})\hat{\psi}_e(\mathbf{r})\ ] \tag{20.44}$$

The field operators are expanded as

$$\hat{\psi}_e(\mathbf{r}) = \sum_{n\ell m} \psi^e_{n\ell m}(\mathbf{r})\ a_{n\ell m} \tag{20.45}$$

$$\hat{\psi}_h(\mathbf{r}) = \sum_{n\ell m} \psi^h_{n\ell m}(\mathbf{r})\ b_{n\ell m}, \tag{20.46}$$

where $\psi(\mathbf{r})$ is the single-particle wavefunction (20.2) with $\phi(\mathbf{r})$ given by (20.25), and $a_{n\ell m}$ and $b_{n\ell m}$ are the annihilation operators for an electron or hole in the state $n\ell m$, respectively. Inserting the expansions (20.45) and (20.46) into (20.44) yields an explicit expression for the polarization operator. In the evaluation, we basically follow the line of argumentation explained in Sec. 10-1.A. This way, we manipulate the first term of Eq. (20.44) as

$$\int d\mathbf{r}\; e\mathbf{r}\; \hat{\psi}_e^\dagger(\mathbf{r})\, \hat{\psi}_e(\mathbf{r}) = \sum_{\substack{n\ell m \\ n'\ell' m'}} \int d\mathbf{r}\; e\mathbf{r}\; \phi^*_{n\ell m}(\mathbf{r})\phi_{n'\ell' m'}(\mathbf{r})|u_c(\mathbf{r})|^2 a^\dagger_{n\ell m} a_{n'\ell' m'}$$

$$= \sum_{\substack{n\ell m \\ n'\ell' m'}} a^\dagger_{n\ell m} a_{n'\ell' m'} \sum_{\substack{unit \\ cells}} \int_{\substack{unit \\ cell}} d\mathbf{r}\; e(\mathbf{R}+\mathbf{r})\; \phi^*_{n\ell m}(\mathbf{R}+\mathbf{r})\; \phi_{n'\ell' m'}(\mathbf{R}+\mathbf{r})\; |u_c(\mathbf{r})|^2$$

$$\cong \sum_{\substack{n\ell m \\ n'\ell' m'}} a^\dagger_{n\ell m} a_{n'\ell' m'} \sum_{\substack{unit \\ cells}} \phi^*_{n\ell m}(\mathbf{R})\; \phi_{n'\ell' m'}(\mathbf{R}) \int d\mathbf{r}\; e(\mathbf{R}+\mathbf{r})\; |u_c(\mathbf{r})|^2$$

$$= \sum_{\substack{n\ell m \\ n'\ell' m'}} a^\dagger_{n\ell m}\; a_{n'\ell' m'} \sum_{\substack{unit \\ cells}} e\mathbf{R}\; \phi^*_{n\ell m}(\mathbf{R})\; \phi_{n'\ell' m'}(\mathbf{R})\; . \tag{20.47}$$

As in Eq. (10.15), we now replace the sum over the unit cells by an integral

$$\int d\mathbf{r}\; e\mathbf{r}\; \hat{\psi}_e^\dagger(\mathbf{r})\hat{\psi}_e(\mathbf{r}) = \sum_{\substack{n\ell m \\ n'\ell' m'}} a^\dagger_{n\ell m}\; a_{n'\ell' m'} \int d\mathbf{R}\; e\mathbf{R}\; \phi^*_{n\ell m}(\mathbf{R})\phi_{n'\ell' m'}(\mathbf{R})$$

$$\equiv \sum_{\substack{n\ell m \\ n'\ell' m'}} p_{n\ell m;n'\ell' m'}\; a^\dagger_{n\ell m}\; a_{n'\ell' m'}\; . \tag{20.48}$$

Using the symmetries of the functions $\phi_{n\ell m}$, Eq. (20.25), one can verify that

$$p_{n\ell m;n\ell m} = 0$$

$$p_{n\ell m;n'\ell' m'} \neq 0 \quad \text{for } n\neq n';\; \ell-\ell'=0, \pm 1\;;\; m-m'=0, \pm 1\; . \tag{20.49}$$

The second term of Eq. (20.31) is evaluated by making the appropriate $e \rightarrow h$ replacements in Eqs. (20.47) and (20.48). The result shows that these two terms do not involve creation or destruction of electron-hole pairs. Their only effect is to change the state of either the electron or the hole, leaving the respective state of the other particle unchanged. These terms therefore describe *"intraband" transitions.*

The last two terms in Eq. (20.44) involve creation and annihilation of electron-hole pairs, i.e., *"interband" transitions.* For these terms we get

$$\int d\mathbf{r}\; e\mathbf{r}\; \hat{\psi}_e^\dagger(\mathbf{r})\hat{\psi}_h^\dagger(\mathbf{r}) = d_{cv} \sum_{\substack{n\ell m \\ n'\ell' m'}} a_{n\ell m}^\dagger b_{n'\ell' m'}^\dagger \int d\mathbf{R}\; \phi_{n\ell m}^*(\mathbf{R})\phi_{n'\ell' m'}(\mathbf{R})$$

$$= d_{cv} \sum_{n\ell m} a_{n\ell m}^\dagger\; b_{n\ell m}^\dagger \;, \tag{20.50}$$

where

$$d_{cv} = \int d\mathbf{r}\; e\mathbf{r}\; u_c^*(\mathbf{r})u_v(\mathbf{r}), \tag{20.51}$$

compare Eq. (10.8). Hence, we see that this term introduces transitions between states in different bands, creating pairs of electrons and holes with the same quantum numbers.

In Fig. 20.4 we plot schematically the energy spectrum of the energetically lowest one electron-hole-pair states with total angular momentum $\ell=0$, $\ell=1$. The arrows indicate the most important dipole allowed interband transitions and the intraband transitions involving a change of the state of the hole. Correspondingly, the intraband transitions changing the state of the electron (not shown) connect $(1s, 1s) \leftrightarrow (1p, 1s)$ and $(1p, 1p) \leftrightarrow (1s, 1p)$, respectively.

20-5. Optical Spectra

In this section we compute the optical susceptibility χ for quantum dots using density matrix theory (compare Chap. 4). The density density-matrix approach is well suited for this purpose since it allows to include also phenomenological relaxation processes, which is not possible in the wavefunction formalism. The dynamic equation for the density matrix is given by

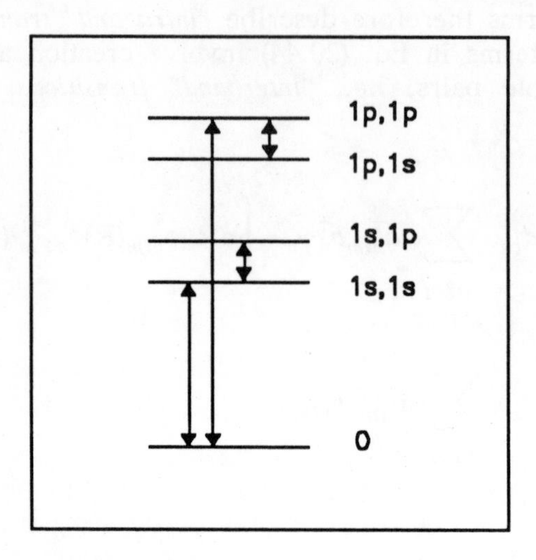

Fig. 20.4: Energy level scheme for the states with zero or one electron-hole pairs.

$$i\hbar \frac{\partial}{\partial t} \rho = [\, \mathcal{H} + \mathcal{H}_I \,, \rho] + L_R(\rho)$$
$$= L_S(\rho) + L_R(\rho) \tag{20.52}$$

where L_R models all dissipative processes, $\mathcal{H}$ is the total Hamiltonian of the electronic excitations in the quantum dot, and and $\mathcal{H}_I$ describes the dipole coupling to the light field, respectively. After the diagonalization, the quantum-dot Hamiltonian can be written in the form

$$\mathcal{H} = \sum_e \hbar\omega_e \, P_{ee} + \sum_b \hbar\omega_b \, P_{bb} \,, \tag{20.53}$$

where the indices e and b refer to the one-pair and two-pair states, and $\hbar\omega_e$ and $\hbar\omega_b$ are the numerically computed energy eigenvalues, respectively. The operators P_{ij} are projectors which in the bra-ket formalism have the form $|i\rangle\langle j|$. The interaction Hamiltonian is then (compare Eq. (4.10))

$$\mathcal{H}_I = -\sum_{e} d_{eo}\, \mathcal{E}(t)\, P_{eo} - \sum_{eb} d_{be}\, \mathcal{E}(t)\, P_{be} + \text{h.c} \, , \quad (20.54)$$

where the index o refers to the ground state without any electron-hole pairs.

In the following, we calculate the steady-state optical properties for pump-probe excitation with

$$\mathcal{E}(t) = \mathcal{E}_L(t) + \mathcal{E}_p(t) \, , \quad (20.55)$$

where L/p label pump/probe field, respectively. An analytic solution of Eq. (20.52) for more than three levels is very tedious, even for single-beam excitation. Hence, one usually uses perturbation expansions, computing the optical response for finite orders of the field $\mathcal{E}(t)$. The first-order, or linear response χ_1, is obtained if one keeps only terms proportional to $\mathcal{E}_p$, dropping all contributions containing $\mathcal{E}_L$. Similarly, for the third-order susceptibility, one neglects all terms containing third or higher powers of $\mathcal{E}_L$ and second or higher powers of $\mathcal{E}_p$.

The susceptibility χ is then obtained from the total polarization in the usual way. That part of χ which is independent of the pump field (zeroes order) yields the linear absorption and refractive index spectra. The lowest-order nonlinear effects are given by the third order susceptibility χ_3, which is proportional to the intensity of the pump field. Since the calculations of the third-order susceptibility involve very long algebraic manipulations, we restrict ourselves, for the purpose of this book, to compute the first-order susceptibility. The third-order susceptibility is evaluated by Hu *et al.* (1990).

The linear polarization in our case can be written as

$$P = \sum_{e} d_{eo}\, \rho_{oe} + \text{h.c.} \, , \quad (20.56)$$

where ρ_{oe} has to be evaluated linear in $\mathcal{E}_p$. The susceptibility is then obtained from

$$\chi_1(\omega) = P/\mathcal{E}_p \ . \tag{20.57}$$

We model the dissipative processes in our system as

$$\langle o|L_R(\rho)|e\rangle = \hbar\gamma_e \ \rho_{oe}$$

describing the decay of ρ_{oe}. For simplicity, we put all other matrix elements of $L_R \equiv 0$, since they do not enter into the final expression of the linear susceptibility. Note, that in general, however, the elements of L_R are not zero at all, but describe the decay of the populations (diagonal elements) and of the coherences (non-diagonal elements) in the system.

To compute ρ_{oe} linear in $\mathcal{E}_p$, we expand the quantities in Eq. (20.52) in orders of the field $\mathcal{E}(t)$

$$\rho = \rho^{(0)} + \rho^{(1)} + \ldots$$

$$L_S = L_S{}^{(0)} + L_S{}^{(1)} \tag{20.58}$$

where $L_S{}^{(0)}$ contains only the system Hamiltonian $\mathcal{H}$, Eq. (20.53) and $L_S{}^{(1)}$ contains the interaction part, Eq. (20.54), respectively. Inserting the expansion (20.58) into Eq. (20.52), we obtain the following hierarchy of equations

$$i\hbar \frac{\partial}{\partial t} \rho^{(0)} = L_S{}^{(0)}(\rho^{(0)}) + L_R(\rho^{(0)}), \tag{20.59}$$

$$i\hbar \frac{\partial}{\partial t} \rho^{(1)} = L_S{}^{(0)}(\rho^{(1)}) + L_S{}^{(1)}(\rho^{(0)}) + L_R(\rho^{(1)}) \tag{20.60}$$

and higher-order equations. We choose the initial conditions as

$$\rho^{(0)}(t{=}-\infty) = |0\rangle\langle 0|$$

$$\rho^{(i)}(t{=}-\infty) = 0 \ , \quad i \neq 0 \ , \tag{20.61}$$

assuming that the system was in its ground state at $t{=}-\infty$. The solution of Eq. (20.59) is

$$\rho^{(0)}(t) = |0\rangle\langle 0| \tag{20.62}$$

indicating that the Hamiltonian $\mathcal{H}$ alone does not perturb the system from its ground state. Inserting (20.62) into Eq. (20.60) and taking the $\langle o|..|e\rangle$ matrix element, we get

$$\frac{\partial}{\partial t}\rho_{oe}^{(1)} = -\,(i\omega_e + \gamma_e)\rho_{oe}^{(1)} + i d_{oe}\,\frac{\mathcal{E}_p(t)}{\hbar}\ . \tag{20.63}$$

Solving Eq. (20.63) and inserting the result into Eq. (20.56), we obtain from Eq. (20.57)

$$\chi_1(\omega) = \frac{i}{\hbar}\sum_e |d_{oe}|^2\left[\frac{1}{\gamma_e + i(\omega_e-\omega)} + \frac{1}{\gamma_e - i(\omega_e+\omega)}\right] . \tag{20.64}$$

The corresponding absorption coefficient is then

$$\boxed{\alpha_1(\omega) = \frac{4\pi\omega}{\hbar c\sqrt{\epsilon_2}}\sum_e |d_{oe}|^2\,\frac{\gamma_e}{\gamma_e^2 + (\omega_e-\omega)^2}}$$

absorption coefficient
for quantum dots (20.65)

where only the resonant part was taken into account.

Eq. (20.65) shows that the absorption spectrum of a single quantum dot consists of a series of Lorentzian peaks centered around the one-electron-hole pair energies $\hbar\omega_e$. To compare the theoretical results with experimental measurements of real quantum dot systems, however, one has to take into account that there is always a certain distribution of dot sizes $f(R)$ around a mean value $\bar{R}$. Since the single-particle energies depend strongly on R, the R-distribution introduces a pronounced inhomogeneous broadening of the observed spectra. Theoretically, this can be modelled easily by noting that α_1 in Eq. (20.65) is actually $\alpha_1|_R$, i.e., the linear absorption spectrum for a given radius R. The average absorption is then computed as

$$\alpha_1(\omega)|_{av} = \int_0^\infty dR\ f(R)\ \alpha_1(\omega)|_R\ . \tag{20.66}$$

Using for $f(R)$ simply a Gaussian distribution around $\bar{R}$ = 20 Å, we obtain for CdS quantum dots the results shown in Fig. 20.5. One clearly sees the energetically lowest one-pair resonances, which merge to a continuous structure with increasing width of the size distribution. Spectra similar to the ones with 15 - 20 % widths are observed in presently available CdS doped glasses or colloids.

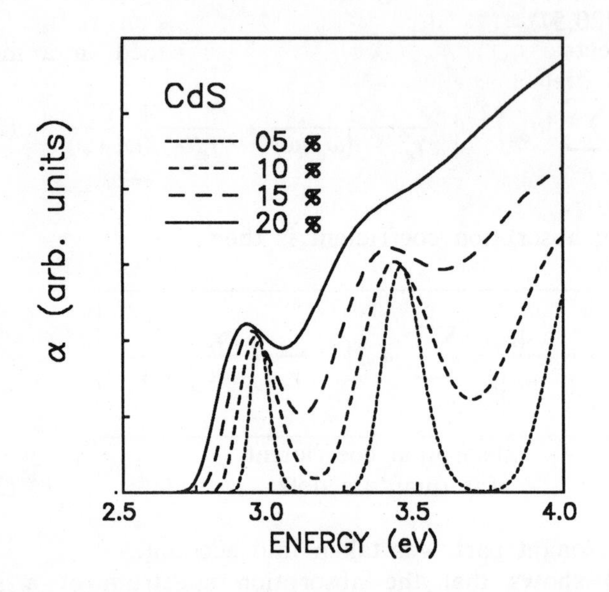

Fig. 20.5: Linear absorption for CdS quantum dots with a Gaussian size distribution around a mean radius of 20 Å. The different curves are for the widths of the Gaussian size distribution indicated in the figure.

REFERENCES

The "classical" papers introducing the theory of semiconductor quantum dots are

L.E. Brus, J. Chem. Phys. **80**, 4403 (1984)

Al. L. Efros and A.L. Efros, Sov. Phys. Semicond. **16**, 772 (1982) .

The work presented in this chapter is largely based on

Y.Z. Hu, M. Lindberg, and S.W. Koch, Phys. Rev. **B** (1990)

PROBLEMS

Problem 20.1: Compute the effective Coulomb interaction potential between two point charges in a dielectric sphere of radius R and background dielectric constant ϵ_2 which is embedded in a medium with background dielectric constant ϵ_1.

Problem 20.2: Solve Eq. (20.22) for a quantum-box. Discuss the single-particle energy eigenvalues as function of box-length.

Chapter 21
TRANSPORT THEORY - INTRABAND KINETICS

Our discussion in all previous chapters was mainly focussed on situations where the electron distribution is spatially homogeneous. In this final chapter we give an introduction into transport theory of electrons, where we have to consider electron distribution functions which vary nontrivially both in space and time.

21-1. Particle Propagator and Reduced Density Matrix

The most general single-particle distribution functions, which enter in a natural way into the Green's function description of non-equilibrium systems, are the particle propagators (see Appendix)

$$G^{<}(\mathbf{r}_1,t_1;\mathbf{r}_2,t_2) = i\ \langle\psi^{\dagger}(\mathbf{r}_1,t_1)\psi(\mathbf{r}_2,t_2)\rangle = G^{<}(1,2). \tag{21.1}$$

Note, that we suppressed the band indices. The more complete notation would be $G^{<}_{ii}(1,2)$ with $i = c, v$. The interband polarization (see Chap. 10) is nothing but the nondiagonal matrix element (in the band index) of the two-by-two particle propagator matrix. The particle propagators are two-point functions, both in space and time. If the two time arguments are equal, i.e., $t_1 = t_2 = t$, we get the reduced single-particle density matrices $\rho_1(\mathbf{r}_1,\mathbf{r}_2,t)$

$$G^{<}(\mathbf{r}_1,t;\mathbf{r}_2,t) = i\ \langle\psi^{\dagger}(\mathbf{r}_1,t)\psi(\mathbf{r}_2,t)\rangle = i\rho_1(\mathbf{r}_1,\mathbf{r}_2,t). \tag{21.2}$$

To find out how these formal distribution functions are connected with useful physical information we consider first a homogeneous noninteracting equilibrium system. We expand

$$\psi(\mathbf{r},t) = \sum_{\mathbf{k}} \frac{1}{\sqrt{L^d}}\, e^{i\mathbf{k}\cdot\mathbf{r}}\ a_{\mathbf{k}}(t),$$

and find

$$G^<(1,2) = i \sum_{\mathbf{k},\mathbf{k}'} < a^\dagger_{\mathbf{k}}(t_1) a_{\mathbf{k}'}(t_2) > \frac{1}{L^d} e^{i(\mathbf{k}\cdot\mathbf{r}_1 - \mathbf{k}'\cdot\mathbf{r}_2)}$$

$$= i \sum_{\mathbf{k}} f_k \frac{1}{L^d} e^{i(\mathbf{k}(\mathbf{r}_1-\mathbf{r}_2) - \epsilon_k(t_1-t_2))} , \tag{21.3}$$

where f_k is the Fermi equilibrium distribution. Hence, in an equilibrium system $G^<(1,2)$ depends only on the relative coordinates

$$\rho = \mathbf{r}_1 - \mathbf{r}_2 \quad \text{and} \quad \tau = t_1 - t_2 . \tag{21.4}$$

The function $G^<(1,2)$ contains rapid oscillations in ρ and τ, whose periods are given by $k = 2\pi n/L$ and ϵ_k. In an interacting many-body system the correlation between the points $\mathbf{r}_1$ and $\mathbf{r}_2$ typically decays on atomic scales, yielding a fast decay of $G^<(1,2)$ with ρ. Thus $G^<$ varies with respect to the relative coordinates on a microscopic scale. The Fourier transforms of $G^<$ with respect to ρ and τ contain the information about the spectrum, therefore ρ and τ are also called the spectral variables.

In a nonequilibrium system $G^<$ depends not only on the relative coordinates but also on the center of mass coordinates

$$\mathbf{r} = \frac{\mathbf{r}_1 + \mathbf{r}_2}{2} \quad \text{and} \quad t = \frac{t_1 + t_2}{2} . \tag{21.5}$$

For the example of local equilibrium, the dependence on $\mathbf{r}$ and t is the slow parametric variation of the macroscopic quantities, i.e., the density $n(\mathbf{r},t)$, the drift velocity $\mathbf{v}(\mathbf{r},t)$ and the temperature $T(\mathbf{r},t)$. Thus $\mathbf{r}$ and t describe the macroscopic variations and we call $\mathbf{r}$ and t the macroscopic variables.

The most general procedure to develop a transport theory starts from the Dyson equation for $G^<$ (see Appendix) and uses a Fourier transform with respect to the spectral coordinates

$$g(\mathbf{k},\omega,\mathbf{r},t) = \frac{1}{L^d} \int d^d\rho \int d\tau \; G^<(\rho,\tau;\mathbf{r},t) \; e^{-i(\mathbf{k}\cdot\rho - \omega\tau)} . \tag{21.6}$$

For this distribution function, which contains the information on the spectral properties and on the macroscopic variations, one can derive a quantum mechanical generalization of the Boltzmann transport equation. For a given problem one can then use the generalized Boltzmann equation with further specific approximations to obtain physical results.

As in the study of the interband kinetics, we only use the equation of motion for the reduced density matrix rather than that of the particle propagator. Introducing the Fourier transform of the reduced density matrix $R_1(\mathbf{r}_1, \mathbf{r}_2, t)$ with respect to the relative coordinate $\boldsymbol{\rho}$, we find

$$\boxed{f(\mathbf{k}, \mathbf{r}, t) = \frac{1}{L^d} \int d^d\rho \; R_1(\boldsymbol{\rho}, \mathbf{r}, t) \; e^{-i\mathbf{k}\cdot\boldsymbol{\rho}}}$$

Wigner distribution (21.7)

and the inverse relation

$$R_1(\boldsymbol{\rho}, \mathbf{r}, t) = \sum_{\mathbf{k}} f(\mathbf{k}, \mathbf{r}, t) \; e^{i\mathbf{k}\cdot\boldsymbol{\rho}} \; . \tag{21.8}$$

In the classical limit, the quantum mechanical Wigner distribution is nothing but the Boltzmann distribution. We see that these functions are constructed attempting to determine simultaneously the distribution in space and momentum $\mathbf{p} = \hbar\mathbf{k}$, which is not generally possible because of Heisenberg's uncertainty relation

$$\Delta r \Delta p \geq \frac{\hbar}{2} \; . \tag{21.9}$$

The position of a particle in the 6 dimensional phase space element with a volume of $\hbar^3/2^3$ cannot be determined accurately. The Wigner distribution becomes a positive definite probability distribution, only if the changes of f in such a phase space volume element are very small.

In order to see the physical content of R_1 and f in more detail let us consider limiting expressions of R_1 around the diagonal $\mathbf{r}_1 = \mathbf{r}_2$,

$$n(\mathbf{r},t) = \lim_{\rho\to 0} \; R_1(\boldsymbol{\rho},\mathbf{r},t) = \sum_{\mathbf{k}} f(\mathbf{k},\mathbf{r},t) \tag{21.10}$$

The diagonal element of R_1 is the particle density. Using

$$\frac{\partial}{\partial \rho_i} = \frac{1}{2}\left[\frac{\partial}{dr_{1i}} - \frac{\partial}{\partial r_{2i}}\right] \tag{21.11}$$

we obtain the first derivative of R_1 with respect to ρ as

$$\lim_{\rho\to 0} \; \frac{\partial}{\partial \rho_i} R_1(\boldsymbol{\rho},\mathbf{r},t) = \frac{1}{2}\left\langle \frac{\partial \psi^\dagger(\mathbf{r})}{\partial r_i}\psi(\mathbf{r}) - \psi^\dagger(\mathbf{r})\frac{\partial \psi(\mathbf{r})}{\partial r_i}\right\rangle . \tag{21.12}$$

Because $\mathbf{p} = (\hbar/i)\,\nabla$ we see that the particle current density $\mathbf{j}$ is given by

$$j_i = \langle p_i \rangle = -\frac{i}{\hbar}\lim_{\rho\to 0} \; \frac{\partial}{\partial \rho_i} R_1(\boldsymbol{\rho},\mathbf{r},t) = \sum_{\mathbf{k}} \hbar k_i f(\mathbf{k},\mathbf{r},t) \tag{21.13}$$

From Eqs. (21.12) and (21.13) we see that the Wigner distribution has the properties of a single-particle distribution function, except that it may become negative if fast variations in a phase space volume element occur.

21-2. Kinetic Equations

Let us start to calculate the kinetic equation of noninteracting particles moving in a potential $V(\mathbf{r})$ with a Hamiltonian

$$\mathscr{H} = \int d^d r \; \psi^\dagger(\mathbf{r}) \left\{ -\frac{\hbar^2\Delta}{2m} + V(\mathbf{r}) \right\} \psi(\mathbf{r}) . \tag{21.14}$$

With the Heisenberg equation

$$\frac{\partial \psi}{\partial t} = \frac{i}{\hbar}\,[\mathcal{H},\psi] = \frac{i}{\hbar}\left\{\frac{\hbar^2 \Delta}{2m} - V\right\}\psi \tag{21.15}$$

we get

$$\left\{\frac{\partial}{\partial t} + \frac{i\hbar}{2m}\,(\Delta_1 - \Delta_2) + \frac{i}{\hbar}\,[V(r_2) - V(r_1)]\right\} R_1(\mathbf{r}_1, \mathbf{r}_2, t) = 0 \ , \tag{21.16}$$

or, with $\nabla_{1,2} = \nabla_r \pm \frac{1}{2}\nabla_\rho$

$$\frac{\partial R_1}{\partial t} - \frac{i\hbar}{m}\,\nabla_r \cdot \nabla_\rho R_1 + \frac{i}{\hbar}\,\boldsymbol{\rho}\cdot(\nabla_r V(r))\ R_1 = 0 \ . \tag{21.17}$$

Multiplying Eq. (21.17) with $\exp(-i\mathbf{k}\cdot\boldsymbol{\rho})$ and integrating over $\boldsymbol{\rho}$ we find

$$\boxed{\left\{\frac{\partial}{\partial t} + \frac{\hbar \mathbf{k}}{m}\cdot\nabla_r - (\nabla_r V(r))\cdot\nabla_{\hbar k}\right\} f(\mathbf{k}, \mathbf{r}, t) = 0}$$

kinetic equation without collisions (21.18)

Using Newton's law

$$\hbar\dot{\mathbf{k}} = -\,\nabla_r V(r) \quad \text{and} \quad \dot{\mathbf{r}} = \frac{\hbar \mathbf{k}}{m} \tag{21.19}$$

one can regard the total differential operator acting on the distribution as the total time derivative

$$\frac{d}{dt} = \frac{\partial}{\partial t} + \dot{\mathbf{r}}\cdot\nabla_r + \hbar\dot{\mathbf{k}}\cdot\nabla_{\hbar k} \ . \tag{21.20}$$

If the potential is due to a spatially constant electric field in x direction, $V(r) = -\,exE(t)$, the kinetic equation becomes

$$\left\{\frac{\partial}{\partial t} + \frac{\hbar \mathbf{k}}{m}\cdot\nabla_r + eE(t)\nabla_{\hbar k_x}\right\} f(\mathbf{k}, \mathbf{r}, t) = 0 \quad . \tag{21.21}$$

In an interacting system the distribution function is further changed by the action of the collisions, so that the full Boltzmann equation reads

$$\left\{\frac{\partial}{\partial t} + \frac{\hbar \mathbf{k}}{m}\cdot\nabla_r - (\nabla_r V(r))\cdot\nabla_{\hbar k}\right\} f(\mathbf{k},\mathbf{r},t) = \left.\frac{d\ f(\mathbf{k},\mathbf{r},t)}{dt}\right|_{coll}$$

Botzmann equation (21.22)

The collision rate will now be determined considering a specific scattering mechanism, namely the intraband scattering of electrons by longitudinal optical (LO) phonons.

21-3. Electron-LO Phonon Interaction

The longitudinal optical phonons are the quanta of the longitudinal polarization oscillations due to ionic displacements in polar semiconductors. An electron can be scattered within one band by emitting or absorbing one LO phonon. The Hamiltonian for the interacting electrons in one band and the LO phonons is

$$\mathcal{H} = \sum \hbar\epsilon_k\ a^\dagger_{\mathbf{k}} a_{\mathbf{k}} + \frac{1}{2}\sum V^\infty_q\ a^\dagger_{\mathbf{k+q}} a^\dagger_{\mathbf{k'-q}} a_{\mathbf{k'}} a_{\mathbf{k}}$$

$$+ \sum \hbar\omega_0 b^\dagger_{\mathbf{q}} b_{\mathbf{q}} + \sum \hbar g_q\ a^\dagger_{\mathbf{k+q}} a_{\mathbf{k}} (b_{\mathbf{q}} + b^\dagger_{\mathbf{-q}}) \ , \tag{21.23}$$

where in *3d*

$$V^\infty_q = \frac{4\pi e^2}{\epsilon_\infty L^3 q^2} = \frac{V_q}{\epsilon_\infty} \ . \tag{21.24}$$

Notice, that we use now the high-frequency limit of the dielectric constant to screen the Coulomb potential, rather than the low frequency limit ϵ_0 which we used in earlier chapters, where the LO-phonons have not been treated explicitly. The dielectric constant ϵ_0 contains the screening due to the polarization of lattice and electronic orbits. At higher frequencies the heavy ions cannot follow, so that ϵ_∞ contains only the screening due to the electronic polarization. We use in (21.23) only the ϵ_∞ screening, because the LO-phonons and their interaction with the electrons are treated explicitly.

The matrix element g_q in Eq. (21.23) for the linear interaction of the electrons with the lattice polarization has to be derived from a microscopic model. Here, we express g_q in terms of ϵ_0 and ϵ_∞ by the following consideration. The phonon mode

$$Q_q = (b_q + b_{-q}^\dagger) \tag{21.25}$$

obeys the equation of motion (see problem (21.1))

$$\frac{d^2 Q_q}{dt^2} - \omega_0^2 Q_q = - 2\omega_0 \sum_{\mathbf{k}} a_{\mathbf{k-q}}^\dagger a_{\mathbf{k}} \ . \tag{21.26}$$

Taking only the unperturbed time development of $a_{\mathbf{k-q}}^\dagger a_{\mathbf{k}} \propto \exp[i(\epsilon_{k-q} - \epsilon_k)t]$ into account, we find the approximate solution

$$Q_q \cong D^0(q, \epsilon_{k-q} - \epsilon_k)\ g_q \sum_{\mathbf{k}} a_{\mathbf{k-q}}^\dagger a_{\mathbf{k}} \ , \tag{21.27}$$

where

$$D^0(q,\omega) = \frac{2\omega_0}{\omega^2 - \omega_0^2} \tag{21.28}$$

is the free phonon propagator. Inserting the phonon mode (21.27) into the Hamiltonian (21.23), an effective Coulomb interaction results

$$V_s(q,\omega) = \frac{V_q}{\epsilon(q,\omega)} = V_q^\infty + D^0(q, \omega = \epsilon_{k-q} - \epsilon_k)\ \hbar g_q^2 \ , \tag{21.29}$$

which explicitly contains the screening by LO-phonons. For low frequencies with $\epsilon(q,\omega) \to \epsilon_0$, we find from Eq. (21.29) the relation

$$\boxed{g_q^2 = \frac{\omega_0 V_q}{2\hbar} \left(\frac{1}{\epsilon_\infty} - \frac{1}{\epsilon_0} \right)} \ .$$

Froehlich electron-LO phonon coupling (21.30)

This result, originally due to Froehlich, expresses the interaction matrix element in terms of the difference of high- and low-frequency limits of the screened Coulomb potential. The constants ϵ_∞ and ϵ_0 can easily be measured.

In passing we mention, that the electron-LO phonon interaction has also an influence on the interband kinetics. It determines the lineshape of the low-frequency tail of the band edge absorption (Urbach tail), because it gives rise to phonon sidebands, in which not only a photon but also one or several thermal phonons are absorbed.

An electron interacting with LO-phonons via the Froehlich coupling is often called a *polaron*. This quasi-particle has a mass, the polaron mass, which differs from the unrenormalized mass of a band electron. The self-energy of an electron due to the interaction with LO-phonons causes the so-called polaron shift. In the previous chapters, where phonons were not treated explicitly, the polaron corrections were implicitly included in the used parameters.

21-4. The Collision Integral

In this section we present an elementary derivation of the collision integral, which can be justified and extended by the use of nonequilibrium many-body techniques. Because the Coulomb scattering does not give rise to a resistance, we treat only the scattering by LO-phonons. We will simplify the derivation of the collision integral in the Boltzmann equation by assuming spatially homogeneous electron and phonon distributions and replace the corresponding densities at the end by the local distributions. As in Sec. 15-2. we calculate the dissipative rates by second-order perturbation theory, starting with the Heisenberg equation for $n_{\mathbf{k}} = a_{\mathbf{k}}^\dagger a_{\mathbf{k}}$ in the interaction representation

$$\frac{dn_{\mathbf{k}}}{dt} = \frac{i}{\hbar} \left[\tilde{\mathcal{H}}_I(t) \, , \, n_{\mathbf{k}} \right] \tag{21.31}$$

with the electron - LO phonon interaction Hamiltonian

$$\tilde{\mathcal{H}}_I(t) = \sum \hbar g_q \, a_{\mathbf{k+q}}^\dagger a_{\mathbf{k}} \, e^{i(\epsilon_{\mathbf{k+q}} - \epsilon_{\mathbf{k}})t} \left(b_{\mathbf{q}} \, e^{-i\omega_0 t} + b_{-\mathbf{q}}^\dagger e^{i\omega_0 t} \right) . \tag{21.32}$$

The formal solution of Eq. (21.31)

$$n_{\mathbf{k}}(t) - n_{\mathbf{k}}(0) = \frac{i}{\hbar} \int_0^t dt_1 \left[\tilde{\mathscr{H}}_I(t_1) \, , \, n_{\mathbf{k}}(t_1) \right] \tag{21.33}$$

is now iterated twice and averaged over the phonon bath, which we assume for simplicity to be in thermal equilibrium. The result is

$$\left. \frac{dn_{\mathbf{k}}}{dt} \right|_{coll} \cong \frac{n_{\mathbf{k}}(t) - n_{\mathbf{k}}(0)}{t}$$

$$= \left(\frac{i}{\hbar} \right)^2 \frac{1}{t} \int_0^t dt_2 \int_0^{t_2} dt_1 \left[\tilde{\mathscr{H}}_I(t_2), \left[\tilde{\mathscr{H}}_I(t_1), n_{\mathbf{k}}(0) \right] \right] . \tag{21.34}$$

Evaluating the commutators and taking the expectation values yields terms of the structure (see problem (21.2))

$$T = \mathrm{Re} \sum_{\mathbf{q}} P_{\mathbf{k},\mathbf{q}} \frac{1}{t} \int_0^t dt_2 \, e^{-i\Delta_{\mathbf{k},\mathbf{q}} t_2} \int_0^{t_2} dt_1 \, e^{i\Delta_{\mathbf{k},\mathbf{q}} t_1} , \tag{21.35}$$

where $P_{\mathbf{k},\mathbf{q}}$ are population factors and $\Delta_{\mathbf{k},\mathbf{q}}$ are frequency differences. We assume that the main contributions to the t_1 integral come from large t_2, so that

$$T \cong \mathrm{Re} \sum_{\mathbf{q}} P_{\mathbf{k},\mathbf{q}} \frac{1}{t} \int_0^t dt_2 \, e^{-i\Delta_{\mathbf{k},\mathbf{q}} t_2} \, \pi \, \delta(\Delta_{\mathbf{k},\mathbf{q}})$$

$$\cong \sum_{\mathbf{q}} P_{\mathbf{k},\mathbf{q}} \, \pi \, \delta(\Delta_{\mathbf{k},\mathbf{q}}) . \tag{21.36}$$

Collecting all terms and replacing $n_{\mathbf{k}}$ by $f(\mathbf{k}, \mathbf{r}, t)$ yields the collision integral

$$\left.\frac{df(\mathbf{k},\mathbf{r},t)}{dt}\right|_{coll} = -2\pi \sum_{q} g_q^2$$
$$\times \Big\{ \delta(\Delta^-_{\mathbf{k},\mathbf{q}}) \Big[f(\mathbf{k},\mathbf{r},t)(1-f(\mathbf{k}-q,\mathbf{r},t)\ (1+g(\mathbf{q},\mathbf{r},t))$$
$$- (1-f(\mathbf{k},\mathbf{r},t))f(\mathbf{k}-q,\mathbf{r},t)g(\mathbf{q},\mathbf{r},t) \Big]$$
$$+ \delta(\Delta^+_{\mathbf{k},\mathbf{q}}) \Big[f(\mathbf{k},\mathbf{r},t)(1-f(\mathbf{k}-q,\mathbf{r},t)\ g(-q,\mathbf{r},t)$$
$$- (1-f(\mathbf{k},\mathbf{r},t))f(\mathbf{k}-q,\mathbf{r},t)(1+g(-q,\mathbf{r},t)) \Big] \Big\}$$

collision integral for
electron-LO phonon scattering (21.37)

Here the frequency differences $\Delta^{\pm}_{\mathbf{k},\mathbf{q}}$ are given by

$$\Delta^{\pm}_{\mathbf{k},\mathbf{q}} = \epsilon_{\mathbf{k}} - \epsilon_{\mathbf{k}-\mathbf{q}} \pm \omega_0 \qquad (21.38)$$

In accordance with Fermi's golden rule we find that the collision rate is given by the differences between the transition rates in and out of the state **k** under absorption or emission of LO-phonons. In detail, the first (third) term in Eq. (21.37) describes a transition **k** → **k**−**q** under emission (absorption) of a phonon, the second (fourth) term describes the transition **k**−**q** → **k** under absorption (emission) of a phonon.

The function $g(\mathbf{q},\mathbf{r},t)$ is the phonon population function. In thermal equilibrium it is given by the Bose function

$$g(\mathbf{q},\mathbf{r},t) = \frac{1}{e^{\beta\hbar\omega_0}-1} = g^0(\omega_0) \ . \qquad (21.39)$$

For thermal phonons we can rewrite the collision rate with the identity $g^0(-\omega_0) = -[1+g^0(\omega_0)]$ as

$$\left.\frac{df(\mathbf{k},\mathbf{r},t)}{dt}\right|_{\text{coll}} = -2\pi \sum_{q,\lambda=\pm 1} g_q^2\, \lambda\, \delta(\Delta_{\mathbf{k},\mathbf{q}}^{\lambda})$$

$$\times \Big[f(\mathbf{k},\mathbf{r},t)\, [1-f(\mathbf{k}-\boldsymbol{q},\mathbf{r},t)]\, g^0(\lambda\omega_0)$$

$$- [1-f(\mathbf{k},\mathbf{r},t)]\, f(\mathbf{k}-\boldsymbol{q},\mathbf{r},t)\, [1+g^0(\lambda\omega_0)] \Big] \ . \qquad (21.40)$$

In thermal equilibrium, where f^0 and g^0 are thermal Fermi and Bose functions, respectively, the collision integral vanishes (see problem (21.2)). We now calculate the collision integral under the assumption that f deviates only little from the thermal distribution, i.e.,

$$f = f^0 + \delta f \ . \qquad (21.41)$$

In this case we can linearize Eq. (21.40) in δf to obtain

$$\left.\frac{d\ \delta f(\mathbf{k},\mathbf{r},t)}{dt}\right|_{\text{coll}} = -2\pi \sum_{q,\lambda=\pm 1} g_q^2\, \lambda\, \delta(\Delta_{\mathbf{k},\mathbf{q}}^{\lambda})$$

$$\times \Big[\delta f(\mathbf{k})\{ [1-f^0(\mathbf{k}-\mathbf{q})]g^0(\lambda\omega_0) + f^0(\mathbf{k}-\mathbf{q})[1+g^0(\lambda\omega_0)]\}$$

$$- \delta f(\mathbf{k}-\mathbf{q})\ \{ f^0(k)g^0(\lambda\omega_0) + [1-f^0(k)][1+g^0(\lambda\omega_0)]\ \} \Big] , \qquad (21.42)$$

or

$$\left.\frac{d\ \delta f(\mathbf{k},\mathbf{r},t)}{dt}\right|_{\text{coll}} = -2\pi \sum_{q,\lambda=\pm 1} g_q^2\, \lambda\, \delta(\Delta_{\mathbf{k},\mathbf{q}}^{\lambda})$$

$$\times \Big[\delta f(\mathbf{k})\ [\ f^0(\mathbf{k}-\mathbf{q}) + g^0(\lambda\omega_0)\ ] - \delta f(\mathbf{k}-\mathbf{q})\ [1 - f^0(\mathbf{k}) + g^0(\lambda\omega_0)] \Big] \ . \qquad (21.43)$$

The linearized Boltzmann equation in the presence of an electric field is

$$\left\{\frac{\partial}{\partial t} + \frac{\hbar\mathbf{k}}{m}\cdot\nabla_r\right\}\delta f(\mathbf{k},\mathbf{r},t) + eE(t)v_{k,x}\frac{\partial f^0(k)}{\partial\epsilon_k} = \left.\frac{d\ \delta f(\mathbf{k},\mathbf{r},t)}{dt}\right|_{coll} . \tag{21.44}$$

Here we used $\partial\epsilon_k/\partial\hbar k_i = v_{k_i}$ with v_{k_i} being the electron velocity. An elementary approximation for stationary and homogeneous situations is obtained if we assume

$$\left.\frac{d\ \delta f(\mathbf{k})}{dt}\right|_{coll} = -\frac{\delta f(\mathbf{k})}{\tau} , \tag{21.45}$$

where τ is the relaxation time for a small deviation of the distribution from the equilibrium function. $\delta f(\mathbf{k})$ is obtained from (21.44) as

$$\delta f(\mathbf{k}) = -\ eE\ v_{k,x}\tau\ \frac{\partial f^0(\mathbf{k})}{\partial\epsilon_k} . \tag{21.46}$$

With this perturbed distribution we find the current density (see problem (21.3))

$$j_x = -e^2E\tau\sum_{\mathbf{k}}\hbar k_x v_{k,x}\tau\ \frac{\partial f^0(\mathbf{k})}{\partial\epsilon_k} = \frac{e^2\tau n}{m}E , \tag{21.47}$$

where n is the electron density. This result corresponds to the simple static conductivity

$$\boxed{\sigma = \frac{e^2\tau n}{m}}$$

conductivity (21.48)

The result (21.48) can be also derived simply from the classical motion of a charged particle in a viscous medium. Inserting Eq. (21.45) into the collision integral (21.41) allows to calculate the relaxation time. Other scattering mechanisms such as electron-impurity scattering or the electron-acoustic phonon interaction and electron-electron Coulomb scattering can be treated similarly. The electron-electron Coulomb scattering, however, conserves the total momentum of the flowing electron gas and does not give rise to a finite static conductivity.

A more complete quantum mechanical derivation of a transport equation starting with the Dyson equation for the particle propagator takes into account the modifications due to many-body effects, so that, e.g., no longer the free-particle energies but the fully renormalized quasi-particle energies enter into the resulting quantum Boltzmann equation.

REFERENCES

For the treatment of the Boltzmann equation see e.g. :

J. M. Ziman, *Electrons and Phonons*, Clarendon Press, Oxford (1960)

J. Devreese *ed.*, *Linear and Nonlinear Transport in Solids*, Plenum, New York (1976)

J. Jaeckle, *Einfuehrung in die Transporttheorie*, Vieweg, Braunschweig (1978)

For the derivation and treatment of the Quantum Boltzmann equation see e.g.:

L.P. Kadanoff and G. Baym, *Quantum Statistical Mechanics*, Benjamin, New York (1962)

J. Rammer and H. Smith, Rev. Mod. Phys. **58**, 323 (1986)

G.D. Mahan, Phys. Rep. **145**, 251 (1987) .

PROBLEMS

Problem 21.1 : Derive the equation of motion (21.26) for the phonon mode $Q_\mathbf{q}$ from the Heisenberg equations for $b_\mathbf{q}$ and $b^\dagger_{-\mathbf{q}}$.

Problem 21.2 : Use (21.40) to show that the collision integral vanishes in thermal equilibrium for the electrons and phonons.

Problem 21.3 : Show that the current density

$$j_i = e \sum_{\mathbf{k}} \hbar k_i f(\mathbf{k}, r, t)$$

reduces to the expression in Eq. (21.47) for a particle energy $\epsilon_k = \hbar^2 k^2/2m$. Hint: Use a partial integration for the energy to derive this result.

Appendix
NONEQUILIBRIUM GREEN'S FUNCTIONS

In this Appendix we make connection with the many-body techniques which are appropriate to describe a nonequilibrium system on an advanced level. We present a short introduction to nonequilibrium Green's functions and show that this formalism yields quite naturally a method to determine the spectral and the kinetic properties of a many-body system, which is driven away from equilibrium by time-dependent external fields. Naturally, we can only give a brief introduction, for further extensions we refer to the existing literature.

A-1. Contour Time Ordering and Matrix Propagators

We assume a system described by the Hamiltoninan $\mathscr{H}_0$ interacting with time-dependent external fields contained in the perturbation Hamiltonian $\mathscr{H}_I(t)$, which is finite for $t > t_0$ and zero for $t < t_0$. The expectation value of any operator Q(t) is given by

$$\langle Q(t) \rangle = \mathrm{tr}\,[\ \rho(t)\ Q(t)\ ]\ . \tag{A.1}$$

In the Heisenberg picture the quantum statistical operator $\rho(t)$ is time-independent and may be taken as $\rho(t_0)$, before the perturbation was acting. It will be assumed that $\rho(t_0{=}-\infty) = \rho_0$ is a thermal equilibrium distribution. Other assumptions about initial correlations lead to complications of the theory.

The time-dependence of the operator Q(t) is determined by the Heisenberg equation. In the interaction picture one finds

$$Q(t) = S^{\dagger}(t,-\infty)\ \tilde{Q}\,(t)\ S(t,-\infty)\quad , \tag{A.2}$$

where

$$\tilde{Q}(t) = \exp\left[\frac{i}{\hbar}\mathcal{H}_0 t\right] Q \exp\left[-\frac{i}{\hbar}\mathcal{H}_0 t\right] \tag{A.3}$$

with

$$S(t,-\infty) = T \exp\left[-\frac{i}{\hbar}\int_{-\infty}^{t} dt' \, \tilde{\mathcal{H}}_I(t')\right] . \tag{A.4}$$

$S(t, t_0)$ describes the time-development of the system under the interaction $\mathcal{H}_I(t)$. T is the time-ordering operator which puts an operator with an earlier time argument to the right of an operator with a later time argument. For the second-order expansion term we get from (A.4)

$$\left[-\frac{i}{\hbar}\right]^2 \frac{1}{2}\int_{-\infty}^{t} dt_1 \int_{-\infty}^{t} dt_2 \; T\, \tilde{\mathcal{H}}_I(t_1)\, \tilde{\mathcal{H}}_I(t_2)$$

$$= \left[-\frac{i}{\hbar}\right]^2 \frac{1}{2}\left\{\int_{-\infty}^{t} dt_1 \int_{-\infty}^{t_1} dt_2 \; \tilde{\mathcal{H}}_I(t_1)\, \tilde{\mathcal{H}}_I(t_2) + \int_{-\infty}^{t} dt_1 \int_{t_1}^{t} dt_2 \; \tilde{\mathcal{H}}_I(t_2)\, \tilde{\mathcal{H}}_I(t_1)\right\}. \tag{A.5}$$

The second double integral can also be evaluated by integrating first over t_1 from $-\infty$ to t_2 and than over t_2 from $-\infty$ to t, which yields for the second term of Eq. (A.5)

$$\int_{-\infty}^{t} dt_2 \int_{-\infty}^{t_2} dt_1 \; \tilde{\mathcal{H}}_I(t_2)\, \tilde{\mathcal{H}}_I(t_1) .$$

Interchanging the notation for the two integration variables $t_1 \leftrightarrows t_2$ we see that the second term equals the first one so that we get

$$\left(-\frac{i}{\hbar}\right)^2 \frac{1}{2}\int_{-\infty}^{t} dt_1 \int_{-\infty}^{t} dt_2 \, T\, \tilde{\mathscr{H}}_I(t_1)\, \tilde{\mathscr{H}}_I(t_2)$$

$$= \left(-\frac{i}{\hbar}\right)^2 \int_{-\infty}^{t} dt_1 \int_{-\infty}^{t_1} dt_2 \, \tilde{\mathscr{H}}_I(t_1)\, \tilde{\mathscr{H}}_I(t_2) \tag{A.6}$$

The structure of this result can be generalized to nth order. The Hermitian conjugate time development operator $S^\dagger(t,-\infty)$ can be found with the help of Eq. (A.6) to be

$$S^\dagger(t,-\infty) = S(-\infty,t) = S^{-1}(t,-\infty) = \bar{T} \exp\left[\frac{i}{\hbar}\int_{-\infty}^{t} dt' \tilde{\mathscr{H}}_I(t')\right] , \tag{A.7}$$

where $\bar{T}$ is an anti-time-ordering operator, ordering operators with earlier time arguments further to the left. Now we can write the expectation value (A.3) as

$$\boxed{\langle Q(t)\rangle = \mathrm{tr}\left[\rho_0\, S(-\infty,t)\, \tilde{Q}(t)\, S(t,-\infty)\right]}$$

quantum statistical expectation value (A.8)

Eq. (A.8) contains two different time orderings, the anti-time ordering on the LHS and the time ordering on the RHS. This complication can be overcome by introducing a time contour which runs from $-\infty \to t$ and than back again from $t \to -\infty$

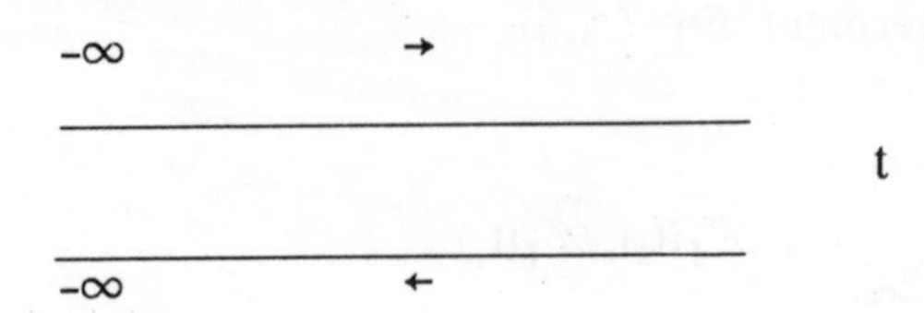

Introducing a contour time-ordering operator T_c, which orders the operators with earlier time arguments on this contour further to the right, one gets

$$\langle Q(t) \rangle = \mathrm{tr}\left[\rho_0\ T_c S_c\ \tilde{Q}(t)\right] \quad , \tag{A.9}$$

where

$$S_c = T_c\ \exp\left[-\frac{i}{\hbar}\int_c dt'\ \tilde{\mathcal{H}}_I(t')\right] \quad . \tag{A.10}$$

Here, we consider formally $\mathcal{H}_I(t,+)$ and $\mathcal{H}_I(t,-)$ as different operators (+ indicates the upper part of the contour, - the lower part). Without changing the result, we can extend the contour to + ∞

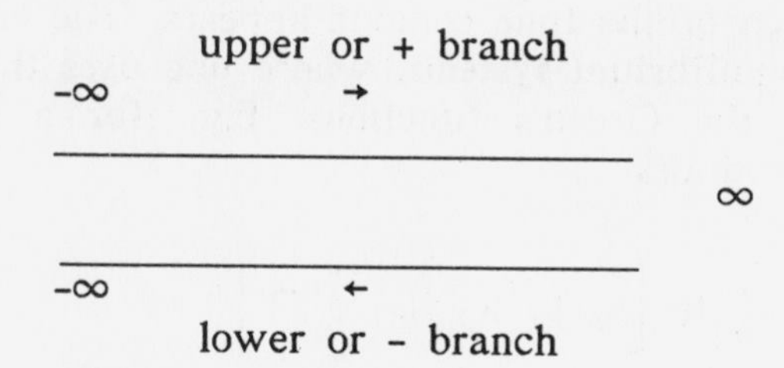

If t lies on the + branch, which is earlier on the contour, we get

$$\langle Q(t) \rangle = \mathrm{tr}\left[\rho_0\ S(-\infty,t)\ S(t,\infty)\ S(\infty,t)\ \tilde{Q}(t)\ S(t,-\infty)\right]$$

$$= \mathrm{tr}\left[\rho_0\ S(-\infty,\infty)\ S(\infty,t)\ \tilde{Q}(t)\ S(t,-\infty)\right] \quad . \tag{A.11}$$

Similarly, if t lies on the - branch, which is later on the contour, we get

$$\langle Q(t) \rangle = \mathrm{tr}\left[\rho_0\ S(-\infty,t)\ \tilde{Q}(t)\ S(t,\infty)\ S(\infty,-\infty)\right] \quad . \tag{A.12}$$

So far we did not make any specific assumptions which are special to a nonequilibrium system. The results are therefore generally valid. However, in an equilibrium system, e.g., at T=0, in which $\rho_0 = |0\rangle\langle 0|$, where $|0\rangle$ is a non-degenerate ground state, an adiabatically switched-on perturbation cannot introduce any transitions from $|0\rangle$ to $|m\rangle$

$$\langle Q(t)\rangle_0 = \sum_m \langle 0|\ S(-\infty,\infty)\ |m\rangle\langle m|\ S(\infty,t)\ \tilde{Q}(t)\ S(t,-\infty)\ |0\rangle$$

$$= \langle 0|\ S(-\infty,\infty)\ |0\rangle\langle 0|\ S(\infty,t)\ \tilde{Q}(t)\ S(t,-\infty)\ |0\rangle$$

$$= \frac{\langle 0|\ S(\infty,t)\ \tilde{Q}(t)\ S(t,-\infty)\ |0\rangle}{\langle 0|\ S(\infty,-\infty)\ |0\rangle} , \tag{A.13}$$

because $\langle 0|\ S(-\infty,+\infty)\ |0\rangle$ is merely a phase factor. Thus, the contour ordering does not appear in an equilibrium system, only one, here the + branch of the time contour appears. The contour order is typical for nonequilibrium systems, where one uses the contour ordering to define the Green's functions. E.g., for a Fermi Green's function one introduces

$$g(\underline{1},\underline{2}) = -\frac{i}{\hbar}\ \mathrm{tr}\left[\rho_0\ T_c\ S_c\ \tilde{\psi}(\underline{1})\ \tilde{\psi}^\dagger(\underline{2})\right]$$

$$= -\frac{i}{\hbar}\ \mathrm{tr}\ [\rho_0\ T_c \psi(\underline{1})\ \psi^\dagger(\underline{2})]$$

or

$$\boxed{g(\underline{1},\underline{2}) = -\frac{i}{\hbar}\ \langle T_c\ \psi(\underline{1})\ \psi^\dagger(\underline{2})\rangle}$$

nonequilibrium Fermi Green's function (A.14)

Here, $\underline{1} = r_1,t_1,\eta_1$ with $\eta_1 = \pm$ depending on whether t_1 is on the upper or lower branch of the time contour c. Next, one can also express $g(\underline{1},\underline{2})$ in terms of a two-by-two matrix

$$g^{\eta_1\eta_2}(1,2) = -\ i\ \langle T_c\ \psi(1,\eta_1)\ \psi^\dagger(2,\eta_2)\rangle\quad . \tag{A.15}$$

For $g(\underline{1},\underline{2})$ one gets a Dyson equation in the same way as for usual time-ordered Green's functions, which has the structure

$$g(\underline{1},\underline{2}) = g^0(\underline{1},\underline{2}) + \int_c d\underline{3} \;\; d\underline{4} \;\; g^0(\underline{1},\underline{3}) \; \sigma(\underline{3},\underline{4}) \; g(\underline{4},\underline{2}) \tag{A.16a}$$

$$= g^0(\underline{1},\underline{2}) + \int_c d\underline{3} \;\; d\underline{4} \;\; g(\underline{1},\underline{3}) \; \sigma(\underline{3},\underline{4}) \; g^0(\underline{4},\underline{2}), \tag{A.16b}$$

where the time integrals are taken along the full contour c. Here, σ is the irreducible self-energy. The meaning of irreducible will be explained below.

The Dyson equation can also be written as

$$\int_c d\underline{3} \; g^{-1}(\underline{1},\underline{3}) \; g(\underline{3},\underline{2}) = \delta(r_1-r_2) \; \delta^c(\underline{t}_1 - \underline{t}_2) \tag{A.17a}$$

or

$$\int_c d\underline{3} \; g(\underline{1},\underline{3}) \; g^{-1}(\underline{3},\underline{2}) = \delta(r_1-r_2) \; \delta^c(\underline{t}_1 - \underline{t}_2) \tag{A.17b}$$

with

$$g^{-1}(\underline{1},\underline{2}) = g^{0-1}(\underline{1},\underline{2}) - \sigma(\underline{1},\underline{2}). \tag{A.18}$$

Here, $\delta^c(\underline{t_1}-\underline{t_2})$ is the δ-function for the contour time $\underline{t}$.

$$\int_c d\underline{t'}_1 \; \delta^c(\underline{t'}_1-\underline{t}_2) = \int_{-\infty}^{\infty} dt_1 \; \delta(t_1-t_2) \; \delta_{\eta_2,+1} + \int_{\infty}^{-\infty} dt_1 \; \delta(t_1-t_2) \; \delta_{\eta_2,-1}$$

$$= \int_{-\infty}^{\infty} dt_1 \; \delta(t_1-t_2) \; (\; \delta_{\eta_2,+1} - \delta_{\eta_2,-1} \;) \; . \tag{A.19}$$

In matrix notation $\delta^c(\underline{t}_1,\underline{t}_2)$ can be represented as

$$\delta^c(\underline{t_1}, \underline{t_2}) = \delta(t_1 - t_2)\ \tau_3^{\eta_1\eta_2}\ , \tag{A.20}$$

where

$$\tau_3 = \begin{bmatrix} 1 & 0 \\ 0 & -1 \end{bmatrix} \tag{A.21}$$

is a Pauli matrix. In matrix notation, the LHS of Eq. (A.17) can be written as

$$\int d^3r_3 \int_{-\infty}^{\infty} dt_3\ g^{-1}(1,3)^{\eta_1,+}\, g(3,2)^{+,\eta_2}$$

$$+ \int d^3r_3 \int_{\infty}^{-\infty} dt_3\ g^{-1}(1,3)^{\eta_1,-}\ \ g(3,2)^{-,\eta_2}$$

$$= \int d^3r_3 \int_{-\infty}^{\infty} dt_3\ [g^{-1}(1,3)^{\eta_1,+}\, g(3,2)^{+,\eta_2}\ -\ g^{-1}(1,3)^{\eta_1,-}\, g(3,2)^{-,\eta_2}]$$

$$= \int d3\ g^{-1}(1,3)^{\eta_1\eta_3}\ \tau_3^{\eta_3\eta_4}\ g(3,2)^{\eta_4\eta_2}$$

$$= \delta(1,2)\ \tau_3^{\eta_1\eta_2}\ , \tag{A.22}$$

where the summation convention is used.

Following Craig (1968) we eliminate the Pauli matrices by defining

$$G(1,3)^{\eta_1,\eta_2}\ =\ g(1,2)^{\eta_1\eta_3}\ \tau_3^{\eta_3\eta_2} \tag{A.23}$$

and

$$G^{-1}(1,2)^{\eta_1\eta_2}\ =\ g^{-1}(1,2)^{\eta_1\eta_3}\ \tau_3^{\eta_3\eta_2} \tag{A.24}$$

to find

$$\int d3\ G^{-1}(1,3)^{\eta_1\eta_3}\ G(3,2)^{\eta_3\eta_2} = \delta(1,2)\ \delta^{\eta_1,\eta_2}\ . \qquad \text{(A.25)}$$

In short-hand notation we write this equation as

$$G^{-1}(1,3)^{\eta_1\eta_3}\ G(3,2)^{\eta_3\eta_2} = \delta(1,2)\ \delta^{\eta_1,\eta_2} = \delta(\underline{1,2})\ . \qquad \text{(A.26)}$$

implying the summation convention for all repeated variables. In detail, we have to integrate over the continuous variable 3,

$$\int d^3r_3 \int_{-\infty}^{+\infty} dt_3\ ,$$

note the time integral is no longer a contour integral, and to sum over the discrete index η_3, which may be + or –.

The transformed Green's function $G^{\eta_1\eta_2}$ has the following matrix elements

$$\boxed{\underline{G}(1,2) = \begin{bmatrix} g^{++}(1,2) & -\ g^{+-}(1,2) \\ g^{-+}(1,2) & -\ g^{--}(1,2) \end{bmatrix}}\ .$$

Keldysh nonequilibrium Green's function matrix (A.27)

Note the - signs in the second column! In the physical limit, where $\mathcal{H}_I(t,+) = \mathcal{H}_I(t,-)$ these elements are

$$G^{++}(1,2) = g^{++}(1,2) = -\frac{i}{\hbar}\langle T\ \psi(1)\ \psi^\dagger(2)\rangle = G^t(1,2)\ , \qquad \text{(A.28)}$$

which is the usual time-ordered Green's function

$$G^{+-}(1,2) = -g^{+-}(1,2) = -\frac{i}{\hbar}\langle\psi^\dagger(2)\ \psi(1)\rangle = -\ G^<(1,2)\ , \qquad \text{(A.29)}$$

which is the negative of the particle propagator $G^<$. For $\underline{1} = \underline{2}$ the time-ordered Green's function is proportional to the particle density, and for equal times $t_1 = t_2$ it is the reduced single-particle density matrix. Hence, this quantity describes the kinetics of the system.

$$G^{-+}(1,2) = g^{-+}(1,2) = -\frac{i}{\hbar}\langle\psi(1)\ \psi^{\dagger}(2)\rangle = G^{>}(1,2)\ , \tag{A.30}$$

is the hole propagator since for $\underline{1} = \underline{2}$ it is proportional to the density of the missing particles.

$$G^{--}(1,2) = -g^{--}(1,2) = \frac{i}{\hbar}\langle\overline{T}\ \psi(1)\ \psi^{\dagger}(2)\rangle = -\ \widetilde{G}^{t}(1,2)\ , \tag{A.31}$$

is the negative of the anti-time-ordered Green's function.

DuBois (1967) also introduced the above given matrix with the notation

$$G^{\eta_1\eta_2}(1,2) = -\frac{i}{\hbar}\eta_2\ \langle T_c\ \psi(1,\eta_1)\ \psi^{\dagger}(2,\eta_2)\rangle. \tag{A.32}$$

From these definitions it follows that

$$G^{++}(1,2) + G^{+-}(1,2) = g^{++}(1,2) - g^{+-}(1,2)$$

$$= -\frac{i}{\hbar}\theta(t_1-t_2)\ \langle\psi(1)\ \psi^{\dagger}(2)\rangle + \frac{i}{\hbar}\theta(t_2-t_1)\ \langle\psi^{\dagger}(2)\ \psi(1)\rangle$$

$$= -\frac{i}{\hbar}\theta(t_1-t_2)\ \langle\psi^{\dagger}(2)\ \psi(1)\rangle - \frac{i}{\hbar}\theta(t_2-t_1)\ \langle\psi^{\dagger}(2)\ \psi(1)\rangle$$

$$= -\frac{i}{\hbar}\theta(t_1-t_2)\ \langle[\ \psi(1),\ \psi^{\dagger}(2)\ ]_{+}\rangle$$

$$= G^{r}(1,2), \tag{A.33}$$

which is the retarded Green's function. The retarded Green's function contains the spectral properties, i.e., the information of the single-particle energy renormalizations.

We find the following relations

$$\boxed{G^{r}(1,2) = G^{++}(1,2) + G^{+-}(1,2)}$$

retarded Green's function (A.34)

or

$$G^r(1,2) = G^t(1,2) - G^<(1,2)$$

$$= G^{--}(1,2) + G^{-+}(1,2)$$

$$= - \tilde{G}^t(1,2) - G^>(1,2). \qquad (A.35)$$

On the other hand one gets

$$G^{++}(1,2) - G^{-+}(1,2) = g^{++}(1,2) - g^{-+}(1,2)$$

$$= -\frac{i}{\hbar}\theta(t_1-t_2)\ \langle\psi(1)\ \psi^\dagger(2)\rangle + \frac{i}{\hbar}\theta(t_2-t_1)\ \langle\psi^\dagger(2)\ \psi(1)\rangle$$

$$= \frac{i}{\hbar}\theta(t_1-t_2)\ \langle\psi(1)\ \psi^\dagger(2)\rangle + \frac{i}{\hbar}\theta(t_2-t_1)\ \langle\psi(1)\ \psi^\dagger(2)\rangle$$

$$= \frac{i}{\hbar}\theta(t_2-t_1)\ \langle[\ \psi(1),\ \psi^\dagger(2)\ ]_+\rangle$$

$$= G^a(1,2)\ , \qquad (A.36)$$

which is the advanced Green's function. We can write

$$G^a(1,2) = G^{++}(1,2) - G^{-+}(1,2) = G^t(1,2) - G^>(1,2)$$

$$= G^{--}(1,2) - G^{+-}(1,2) = -\tilde{G}^t(1,2) + G^<(1,2). \qquad (A.37)$$

The self-energy matrix $\Sigma^{\eta_1\eta_2}$ fulfills exactly the same relations:

$$\Sigma^r(1,2) = \Sigma^{++}(1,2) + \Sigma^{+-}(1,2) = \Sigma^t(1,2) - \Sigma^<(1,2)$$

$$= \Sigma^{--}(1,2) + \Sigma^{-+}(1,2) = -\ \bar{\Sigma}^t(1,2) + \Sigma^>(1,2) \qquad (A.38)$$

and

$$\Sigma^a(1,2) = \Sigma^{++}(1,2) + \Sigma^{-+}(1,2) = \Sigma^t(1,2) - \Sigma^>(1,2)$$

$$= \Sigma^{--}(1,2) - \Sigma^{+-}(1,2) = -\ \bar{\Sigma}^t(1,2) + \Sigma^<(1,2)\ . \qquad (A.39)$$

Adding the Dyson equations of G^{++} and G^{+-}, one gets from the Dyson equation (A.16a)

$$G^r(1,2) = G^{0r}(1,2) + \int d3 \int d4\, G^{0+\eta_1}(1,2)\Sigma^{\eta_1\eta_2}(3,4)\left[G^{\eta_2,+}(1,2) + G^{\eta_2,-}(1,2)\right]$$

$$= G^{0r}(1,2) + \int d3 \int d4\, G^{0+\eta_1}(1,2)\,\Sigma^{\eta_1\eta_2}(3,4)\,G^r(1,2)$$

$$= G^{0r}(1,2) + \int d3 \int d4\, G^{0+\eta_1}(1,2)\,\Sigma^r(3,4)\,G^r(1,2)$$

$$= G^{0r}(1,2) + \int d3 \int d4\, G^{0r}(1,2)\,\Sigma^r(3,4)\,G^r(1,2) \qquad \text{(A.40a)}$$

and from (A.16b)

$$G^r(1,2) = G^{0r}(1,2) + \int d3 \int d4\, G^r(1,2)\,\Sigma^r(3,4)\,G^{0r}(1,2), \qquad \text{(A.40b)}$$

which shows that G^r obeys a Dyson equation too, with the self-energy Σ^r.

For field operators A(r,t), which are Bosons (the vector character is not discussed here), we define in analogy the Fermi Green's function the Boson Green's function

$$d(\underline{1},\underline{2}) = -\frac{i}{\hbar}\,\mathrm{tr}\left[\rho_0 T_c S_c \tilde{A}(\underline{1})\tilde{A}(\underline{2})\right]$$

$$= -\frac{i}{\hbar}\,\mathrm{tr}\,[\rho_0 T_c A(\underline{1})A(\underline{2})] \qquad \text{(A.41)}$$

or

$$\boxed{d(\underline{1},\underline{2}) = -\frac{i}{\hbar}\langle T_c A(\underline{1})A(\underline{2})\rangle}$$

Boson nonequilibrium Green's function (A.42)

The Boson Green's function obeys a Dyson equation of the form

$$d(\underline{1},\underline{2}) = d^0(\underline{1},\underline{2}) + \int d\underline{3} \int d\underline{4}\; d^0(\underline{1},\underline{3})\; \pi(\underline{3},\underline{4})\; d(\underline{4},\underline{2}) \; , \qquad \text{(A.43)}$$

where $\pi(\underline{3},\underline{4})$ is the field self-energy, which is also called the polarization function.

Again we introduce transformed matrices

$$D^{\eta_1\eta_2}(1,2) = d^{\eta_1\eta_3}(1,2)\; \tau_3^{\eta_3\eta_2}$$

$$\Pi^{\eta_1\eta_2}(1,2) = \pi^{\eta_1\eta_3}(1,2)\; \tau_3^{\eta_3\eta_2} \; , \qquad \text{(A.44)}$$

which obey the equation

$$D^{\eta_1\eta_2}(1,2) = D^0(1,2)^{\eta_1\eta_2} + \int d3 \int d4\; D^0(1,3)^{\eta_1\eta_3}\; \Pi^{\eta_3\eta_4}(3,4)\; D^{\eta_4\eta_2}(4,2). \qquad \text{(A.45)}$$

A-2. Feynman Rules

According to Feynman the various terms of the self-energy which one obtains by a perturbational expansion can be represented by simple diagrams. As illustrated below, the Feynman diagrams are a very efficient book-keeping technique. By comparing the lowest order perturbation results with the simplest self-energy diagrams (see e.g. DuBois (1967)) one gets the following Feyman rules for the transformed matrices $\Sigma^{\eta_1\eta_2}(1,2)$ and $\Pi^{\eta_1\eta_2}(1,2)$:

1) Take into account all possible irreducible self-energy diagrams. Reducible Fermion self-energy diagrams, e.g., would fall into two parts if a Fermi-particle line will be cut, these diagrams are already summed up by the Dyson equation and cannot be considered as self-energy diagrams.

2) Multiply at each vertex with the appropriate interaction factor.

3) For each particle line connecting the vertices $\underline{3}$ and $\underline{4}$ insert a factor $G^{\eta_3\eta_4}(3,4)$ and for each field line $-iD^{\eta_3\eta_5}(3,4)\tau_3^{\eta_5\eta_4} = -iD(\underline{3},\underline{4})\eta_4$.

4) Sum over all internal variables including the time-ordering indices.

6) Interpret $G(r_3t_3;r_4t_3)$ as $G(r_3t_3,r_4t_3+\epsilon)$.

7) Introduce an overall factor of $(-1)^\ell$,where ℓ is the number of closed Fermion loops and a factor of $i\tau_3^{\eta_1\eta_3}$ (or η_1) for the Boson self-energy $\Pi^{\eta_1\eta_2}(1,2)$.

Feynman rules (A.46)

These nonequilibrium Feynman rules are a straightforward generalization of the conventional rules which hold for time-order equilibrium Green's functions. These nonequilibrium rules can e.g. be used to evaluate diagrammatically the retarded self-energy which is needed to calculate the retarded Green's function and thus the spectral properties of the system.

As an example we will evaluate the retarded electron self-energy due to the intraband Coulomb scattering in the random phase approximation which we used in Chaps. 9 and 12. It is obtained from by the following Feynman diagram

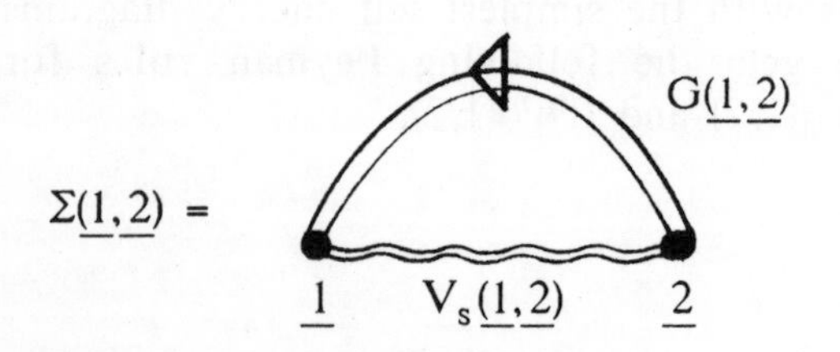

$V_s(\underline{1},\underline{2})$ is the screened Coulomb potential, which is a Boson Green's function. Following the rules 1) to 7) we get for this diagram

$$\Sigma^{\eta_1\eta_2}(1,2) = i\eta_2 G^{\eta_1\eta_2}(1,2)\, V_s{}^{\eta_1\eta_2}(1,2) \tag{A.47}$$

The retarded electron self-energy is according to Eq. (A.35)

$$\begin{aligned}\Sigma^r(1,2) &= \Sigma^{++}(1,2) + \Sigma^{+-}(1,2)\\ &= iG^{++}(1,2)V_s^{++}(1,2) - iG^{+-}(1,2)V_s^{+-}(1,2).\end{aligned} \tag{A.48}$$

In this book we discuss only a statically screened Coulomb potential which is instantaneous in time and has only diagonal elements in the Keldysh indexes. Thus, the self-energy (A.48) reduces to

$$\begin{aligned}\Sigma^r(1,2) &= iG^{++}(1,2)\, V_s^{++}(1,2)\\ &= iG^{++}(1,2)\, V(1,2) + iG^{++}(1,2)\, [V_s(1,2) - V(1,2)]\\ &= \Sigma_x^r(1,2) \qquad\qquad + \Sigma_c^r(1,2)\,,\end{aligned} \tag{A.49}$$

where we have added and subtracted the Coulomb potential

$$V(1,2) = \delta(t_1-t_2)\, V(\mathbf{r}_1-\mathbf{r}_2). \tag{A.50}$$

The two parts of the retarded self-energy are the exchange energy Σ_x^r and the correlation energy Σ_c^r. The exchange energy has to be calculated by including an infinitesimal retardation in the Coulomb potential

$$\begin{aligned}\Sigma_x^r(1,2) &= i\,[\,G^r(1,2) - G^{+-}(1,2)\,]\,V(1,2)\\ &= -\,iG^{+-}(1,2)\,V(1,2),\end{aligned} \tag{A.51}$$

so that the product of the retarded Green's function ($\propto \theta(t_1-t_2)$) and the infinitesimally retarded Coulomb potential ($\propto \delta(t_1-t_2+\epsilon)$) does not contribute. The correlation energy is

$$\Sigma_c^r(1,2) = i\,[G^r(1,2) - G^{+-}(1,2)]\,[V_s(1,2) - V(1,2)]. \tag{A.52}$$

The term proportional to the particle propagator G^{+-} can be combined with the exchange energy (A.51) to give the screened exchange energy Σ^r_{sx}

$$\Sigma^r_{sx}(1,2) = -iG^{+-}(1,2)V_s(1,2) . \tag{A.53}$$

The remaining part of the correlation energy is called the Coulomb-hole self-energy $\Sigma_{Ch}{}^r$

$$\Sigma^r_{Ch}(1,2) = iG^r(1,2) [V_s(1,2) - V(1,2)]$$

$$= iG^r(1,2)\, \delta(t_1-t_2) [V_s(\mathbf{r}_1-\mathbf{r}_2) - V(\mathbf{r}_1-\mathbf{r}_2)] , \tag{A.54}$$

where $V_s(\mathbf{r}_1-\mathbf{r}_2)$ is the statically screened Coulomb potential. With the definition of the retarded Green's function (A.33) we find

$$\Sigma^r_{Ch}(1,2) = \frac{1}{2\hbar} \langle[\psi(\mathbf{r}_1,t_1),\psi^\dagger(\mathbf{r}_2,t_2)]_+\rangle\delta(t_1-t_2) [V_s(\mathbf{r}_1-\mathbf{r}_2) - V(\mathbf{r}_1-\mathbf{r}_2)]$$

$$= \frac{1}{2\hbar} \delta(\mathbf{r}_1-\mathbf{r}_2)\delta(t_1-t_2) [V_s(\mathbf{r}_1-\mathbf{r}_2) - V(\mathbf{r}_1-\mathbf{r}_2)]$$

$$= \frac{1}{2\hbar} \delta(1,2) \lim_{r\to 0} [V_s(r) - V(r)] , \tag{A.55}$$

If we take the Fourier transform with respect to the relative coordinates we get finally for the Coulomb-hole energy

$$\hbar\Sigma^r_{Ch}(\mathbf{k},\omega) = \frac{1}{2} \lim_{r\to 0} [V_s(r) - V(r)] , \tag{A.56}$$

i.e., we obtain a constant energy reduction due to the Coulomb hole in a plasma, see Eq. (9.29).

To conclude the comparison with chapters 9 and 12, we evaluate the screened exchange self-energy, Eq. (A.53), for a quasi-equilibrium situation. For a homogeneous system we Fourier-transform to k-space and find in the quasi-particle approximation

$$\psi(\mathbf{r},t) = \sum_{\mathbf{k}} a_{\mathbf{k}}\, e^{i(\mathbf{k}\cdot\mathbf{r}-e_k t)} \tag{A.57}$$

the particle propagator

$$G^{+-}(1,2) = -\frac{i}{\hbar} \sum_{\mathbf{k}} \langle a^{\dagger}_{\mathbf{k}} a_{\mathbf{k}} \rangle \, e^{i[\mathbf{k}\cdot(\mathbf{r}_2-\mathbf{r}_1) \, - \, e_{\mathbf{k}}(t_2-t_1)]}$$

$$= -\frac{i}{\hbar} \sum_{\mathbf{k}} f_{\mathbf{k}} \, e^{i[\mathbf{k}\cdot(\mathbf{r}_2-\mathbf{r}_1) \, - \, e_{\mathbf{k}}(t_2-t_1)]} \,, \qquad (A.58)$$

where $f_{\mathbf{k}}$ is the Fermi function. With an instantaneous, screened Coulomb potential we obtain for the self-energy, Fourier transformed with respect to the relative coordinates,

$$\hbar\Sigma^{r}_{sx}(\mathbf{k},\omega) = - \int d^3r \sum_{\mathbf{k}'} f_{\mathbf{k}'} \, e^{i\mathbf{k}'\cdot\mathbf{r}} \, e^{-i\mathbf{k}\cdot\mathbf{r}} \, V_s(r)$$

$$= - \sum_{\mathbf{k}'} f_{\mathbf{k}'} \, V_{s,|\mathbf{k}-\mathbf{k}'|} \,, \qquad (A.59)$$

which is just the result of Eq. (9.25).

REFERENCES

For a general introduction to nonequilibrium Green's functions see e.g. :

R. A. Craig, J. Math. Phys. 3, 605 (1968)

D.F. DuBois in: Lectures in Theoret. Phys., Vol. 9c, eds. W.E. Brittin and A.D. Barut, Gordon-Breach, New York 1967, p. 469

Landau-Lifshitz, Theoret. Physics, Vol. X, Chap. 10, Pergamon Press, New York.

For applications to nonequilibrium superconductors see e.g. :

A. Schmid in: "Nonequilibrium Superconductivity, Phonons and Kapitza Boundaries", Nato Adv. Study Inst. Series B, Vol. 65, ed. K.E. Gray, Plenum Press, New York 1981, p. 423.

For applications to Quantum Fluids see, e.g. :

P. Woelfle, Prog. Low Temp. Phys. 7, 191 (1978);

A.E. Ruckenstein and L.P. Levi, Phys. Rev. B 39, 183 (1989) .

For applications to transport in metals see, e.g. :

J. Rammer and H. Smith, Rev. Mod. Phys. 58, 323 (1986);

G.D. Mahan, Phys. Rep. 145, 251 (1987).

For applications to laser exited semiconductors see, e.g. :

K. Henneberger, Physica A 150, 419 (1988)

H. Haug in: "Optical Nonlinearities and Instabilities in Semiconductors", ed. H. Haug, Acad. Press, New York 1988, p. 53

W. Schaefer, ibid., p. 133;

W. Schaefer and J. Treusch, Z. Physik B 63, 407 (1986)

K. Henneberger and H. Haug, Phys. Rev. B 38, 9759 (1988).

For applications to gas lasers see, e.g. :

V. Korenman, Annals of Physics 39, 72 (1966)

For applications to Langevin equations see, e.g. :

A. Schmid, J. Low Temp. Phys. 49, 609 (1982)

INDEX